Roman Gr. Maev
Acoustic Microscopy

Related Titles

Sarid, D.

Exploring Scanning Probe Microscopy with MATHEMATICA

2007. Hardcover
ISBN: 978-3-527-40617-3

Schalley, C.A. (Ed.)

Analytical Methods in Supramolecular Chemistry

2007. Hardcover
ISBN: 978-3-527-31505-5

Heath, J.P.

Dictionary of Microscopy

2005. Softcover
ISBN: 978-0-470-01199-7

Tilley, R.J.D.

Crystals and Crystal Structures

2006. Hardcover
ISBN: 978-0-470-01820-0

Tilley, R.J.D.

Understanding Solids

The Science of Materials

2004. Hardcover
ISBN: 978-0-470-85275-0

Kaufmann, E.N. (Ed.)

Characterization of Materials

2 Volume Set

2003. Hardcover
ISBN: 978-0-471-26882-6

Hornak, J.P.

Encyclopedia of Imaging Science and Technology

2 Volume Set

2002. Hardcover
ISBN: 978-0-471-33276-3

Roman Gr. Maev

Acoustic Microscopy

Fundamentals and Applications

WILEY-VCH Verlag GmbH & Co. KGaA

The Author

Prof. Roman Gr. Maev
NSERC
Industrial Research Chair
University of Windsor
Windsor, Canada

All books published by **Wiley-VCH** are carefully produced. Nevertheless, authors, editors, and publisher do not warrant the information contained in these books, including this book, to be free of errors. Readers are advised to keep in mind that statements, data, illustrations, procedural details or other items may inadvertently be inaccurate.

Library of Congress Card No.:
applied for

British Library Cataloguing-in-Publication Data
A catalogue record for this book is available from the British Library.

Bibliographic information published by the Deutsche Nationalbibliothek
The Deutsche Nationalbibliothek lists this publication in the Deutsche Nationalbibliografie; detailed bibliographic data are available in the Internet at <http://dnb.d-nb.de>.

© 2008 WILEY-VCH Verlag GmbH & Co. KGaA, Weinheim

All rights reserved (including those of translation into other languages). No part of this book may be reproduced in any form – by photoprinting, microfilm, or any other means – nor transmitted or translated into a machine language without written permission from the publishers. Registered names, trademarks, etc. used in this book, even when not specifically marked as such, are not to be considered unprotected by law.

Typesetting   VTEX, Litauen

Printing   betz-druck GmbH, Darmstadt

Binding   Litges & Dopf Buchbinderei GmbH, Darmstadt

Printed in the Federal Republic of Germany
Printed on acid-free paper

ISBN   978-3-527-40744-6

Dedicated to the memory of my parents
Grigorii Maev and Miriam Maeva

Contents

Foreword (C. F. Quate) *XI*
Preface (Yu. V. Gulyaev) *XV*
Introductory Comments *1*
Introduction *5*

1   **Scanning Acoustic Microscopy. Physical Principles and Methods. Current Development** *9*
1.1   Basics of Acoustic Wave Propagation in Condensed Media *9*
1.2   Physical Principles of Scanning Acoustic Microscopy *13*
1.3   Acoustic Imaging Principles and Quantitative Methods of Acoustic Microscopy *15*
1.4   Methodological Limitations of Acoustic Microscopy *18*

2   **Acoustic Field Structure in a Lens System of a Scanning Acoustic Microscope** *21*
2.1   Calculation of the Focal Area Structure with Due Regard for Aberrations and Absorption in a Medium *21*
2.2   The Field of a Spherical Focusing Transducer with an Arbitrary Aperture Angle *24*
2.3   Analysis of Acoustic Field Spatial Structure with a Spherical Acoustic Transducer *29*
2.4   Experimental Study of the Focal Area Structure of a Transmission Acoustic Microscope *37*
2.5   Formation of a Focused Beam of Bulk Acoustic Waves by a Planar System of Transducers *39*
2.6   About the Possibility of Using Scholte–Stoneley Waves for Surface Waves' Acoustic Microscopy *46*

3   **Output Signal Formation in a Transmission Raster Acoustic Microscope** *53*
3.1   Outline of the Problem *53*

Acoustic Microscopy. Roman Gr. Maev
Copyright © 2008 WILEY-VCH Verlag GmbH & Co. KGaA, Weinheim
ISBN: 978-3-527-40744-6

| 3.2 | Transmission Acoustic Microscope: Formation of the Output Signal as a Function of Local Properties of Flat Objects. General Concepts   54
| 3.3 | General Representation of the Output Signal of the Transmission Acoustic Microscope   56
| 3.4 | Formation of the $A(z)$ Dependence for Objects with a Small Shear Modulus   58

4 Quantitative Acoustic Microscopy Based on Lateral Mechanical Scanning   65

| 4.1 | Methods of Quantitative Ultrasonic Microscopy with Mechanical Scanning: Review   65
| 4.2 | Ray Models of $V(z)$ and $V(x)$ QSAM Systems   66
| 4.3 | Wave Theory of $V(z)$ and $V(x)$ QSAM Systems   68
| 4.4 | Angular Resolution of QSAM Systems   71
| 4.5 | Application of the $V(x)$ QSAM System to LSAW Measurement   73
| 4.6 | Temperature Stability of the $V(x)$ QSAM System   78

5 Acoustic Microscopy and Nonlinear Acoustic Effects   81

| 5.1 | Nonlinear Acoustic Applications for Characterization of Material Microstructure   81
| 5.1.1 | Schematic of Experiment   81
| 5.1.2 | Visualization by Nonlinear Acoustic Methods   86
| 5.1.3 | Parametric Representation of Acoustic Nonlinearity   89
| 5.2 | Peculiarities of Nonlinear Acoustic Effects in the Focal Area of an Acoustic Microscope   92
| 5.3 | Temperature Effects in the Focal Area of an Acoustic Microscope   94
| 5.4 | Effects of Radiation Pressure on Samples Examined with an Acoustic Microscope   101
| 5.5 | The Theory of Modulated Focused Ultrasound Interaction with Microscopic Entities   108
| 5.5.1 | Shell Model of a Cell   109
| 5.5.2 | Interaction of a Cell with a High-Frequency Field within the Framework of the Shell Model. Equation for the Radiation Force   111
| 5.5.3 | Oscillations of a Microparticle under the Action of a Nonlinear Force   112

6 Investigation of the Local Properties and Microstructure of Model Systems and Composites by the Acoustic Microscopy Methods   119

| 6.1 | Study of the Viscoelastic Properties of Model Collagen Systems by the Acousto-Microscopic Methods. Experimental Setup   119

6.2	Microstructure Investigations of Multilayer Photographic Film Structures Using Scanning Acoustic Microscopy Methods	124
6.3	Investigation of the Microstructure Peculiarities of High-temperature Superconducting Materials by Scanning Acoustic Microscopy Methods	127
6.4	Application of Acoustic Microscopy to the Study of Multilayer Reinforced Fiber–Glass Graphite Composites	137

7	**Scanning Acoustic Microscopy of Polymer Composite Materials**	**141**
7.1	Acoustic Methods for the Investigation of Polymers	142
7.2	Methods for Studying and Visualizing the Dispersed Phase in Polymer Blends	144
7.3	Objects of Investigation	146
7.4	Basic Requirements Imposed on Polymer Mixtures and Methods for their Study by Acoustic Microscopy	147
7.5	Investigation into the Mechanisms of Acoustic Contrast in Polymers	147
7.6	Acoustic Imaging of the Spatial Phase Distribution in Polymer Mixtures	158
7.7	Investigation of the Structure and Homogeneity of the Mixture Components Distribution within each other. Measure of Homogeneity	159
7.8	Numerical Processing of Acoustic Images of Granulated Structures	163
7.9	Exploring the Microstructure of Polymer Blends in an Acoustic Microscope and Comparison with other Techniques	165
7.9.1	Studies of the Microstructure of Individual Particles in a Blend	165
7.9.2	Studies of Film Structure and the Homogeneity of Phase Distribution in Polymer Blend Films	167
7.9.3	Assessment of the Component Distribution in Polymer Blends at Various Sizes of the Mixture Particle Fractions	168
7.9.4	Investigation of the Distribution Homogeneity and the Physical and Mechanical Polymer Blend Properties	171
7.9.5	Examination of the Polymer Film Structure via Surface Defects	174
7.10	Application of Acoustic Microscopy Techniques for Investigation of the Multi-layered Polymer System Structure	175
7.11	Using the Short-pulse Ultrasound Scanning Technique to Measure the Thickness of Individual Components of Multi-layer Polymer Systems	178

8	**Investigation of the Microstructure and Physical–Mechanical Properties of Biological Tissues**	**187**
8.1	Application of Acoustic Microscopy Methods in Studies of Biological Objects	187

8.2	Selection of Immersion Media for Acoustic Microscopy Studies of Biological Objects *191*
8.3	Imaging and Quantitative Data Acquisition of Biological Cells and Soft Tissues with Scanning Acoustic Microscopy *194*
8.3.1	Introduction *194*
8.3.2	Brief Description of the System *195*
8.3.3	Contrast Factor for Acoustic Imaging of Biological Cells and Tissues *197*
8.3.4	Thermal Insult *200*
8.3.5	Shock Wave Insult *201*
8.3.6	Velocity Measurement for Biological Tissue *205*
8.3.7	Concluding Remarks *210*
8.4	Methods for Tissue Preparation and Investigation *211*
8.5	Acoustic Properties of Biological Tissues and their Effect on the Image Contrast *212*
8.6	Investigation of Soft Tissue Sections *213*
8.6.1	Skin *213*
8.6.2	Eye Sclera *215*
8.6.3	Liver *217*
8.6.4	Cardiac Muscle *218*
8.7	Investigation of Hard Mineralized Tissues *219*
8.7.1	Bone Tissue and the Bone–Implant System *219*
8.7.2	Dental Tissue *222*
8.8	Acoustic Properties of Collagen *232*
8.8.1	The Effect of Collagen Anisotropy on Propagation of an Ultrasound Wave *232*
8.8.2	Experimental Investigation into Acoustic Properties of an Isolated Collagen Thread *238*

References *241*
Additional Reading *260*

Index *271*

Foreword

Acoustics is now a part of our technology base and that makes it imperative to understand acoustics if we are to appreciate the marvels of technology. Acoustics enters the technology base in multiple avenues and in this introduction, we will point to those areas with the greatest impact.

Much of science and technology is centered on extending our natural vision. The universal desire to understand the world around us has led to the development of multiple instruments for extending our knowledge of the real world. Optical instruments dominate an enormous space from the very small to the very large, from microscopes to telescopes. For those areas where optics cannot provide information, other technologies, such as acoustics, have been introduced.

Imaging technology is largely based on manipulating optical waves, but optics does not provide all of the information we need and we must turn to other technologies. Acoustic imaging is now an integral part of our continuing effort to extend our ability to 'see.' We can easily convert electromagnetic energy to acoustic energy and use this energy in imaging devices. Acoustic devices are extensively used for imaging inside the human body. Although ultrasonic images do not record the fine detail found in magnetic resonance images (MRI), the system provides us with primary information at one-tenth the cost of MRI. This form of imaging is particularly useful for delineating the interface between solids and spaces in muscles and soft tissues. Ultrasound renders live images where the operator can dynamically select the most useful sections for documenting the changes in structure without long-term side effects in the patient.

Other than imaging, acoustics has played a role in the removal of diseased tissue and other components within the human body. This process requires bursts of ultrasound three orders of magnitude more intense than that used for imaging. When the intensity is concentrated within the focal region, the temperature in that region is raised to the level where the constituents cells are killed. This procedure is emerging as a noninvasive option for treating cancer and ablating and breaking gall stones.

The systems used for transforming electromagnetic energy into sound are changing with the new technologies now being developed. Present systems use piezoceramics for converting electromagnetic change into sound energy. However, new systems based on vibrating membranes are being introduced and these may

Acoustic Microscopy. Roman Gr. Maev
Copyright © 2008 WILEY-VCH Verlag GmbH & Co. KGaA, Weinheim
ISBN: 978-3-527-40744-6

offer strong competition to piezoceramics. Micromachined ultrasonic transducers use MEMS as the enabling technology. Micromachined ultrasonic transducers are fabricated across a silicon wafer with the inherent advantages of wafer scale fabrication. Arrays of multiple transducers are easily fabricated in this manner. With the present two-dimensional systems, it is necessary to resort to multiple scans where each image is focused at a slightly different depth and this can be a lengthy and tedious process. Micromachined ultrasonic transducers arrays will record three-dimensional images directly without the need for multiple scans.

In another area, medical investigators, such as L. Mazotti in Italy, have found that ultrasonics can be used to distinguish between cancerous and benign tissue. This is accomplished by carrying out a careful spectral analysis of the reflected waveforms in the laboratory. If this technique can be perfected, it will have an impact on the present procedures used in biopsies.

We now return to imaging systems. Ultrasonic imaging as discussed above is carried out in the frequency range of 5–10 kHz with a resolution measured in millimeters. In order to improve this and examine smaller features, it is necessary to move to higher frequencies (∼300 MHz) and operate in a mode labeled Scanning Acoustic Microscope (SAM). In this mode, sound waves are focused to a diffraction limited spot with concave spherical lens immersed in water. The image is formed by scanning the entire unit over the surface of the sample in a raster pattern to form the image. These systems are used to examine internal microstructure of nontransparent solids and to monitor internal stress. In addition to measuring the elastic properties, they are used to examine adhesion in multilayered structures. In semiconductor devices, they pinpoint cracking and nonbonding in overlayers. They also reveal nonwetting solder balls, delaminations in the die attachment and internal discontinuities in flip chip packaging. A resolution of 0.5 µm is often achieved in these applications.

These microscopes are also valuable for inspecting chip scale semiconductor devices. They have the ability to identify minute changes in density, small air gaps, as well as die cracking, underfill voids and interlaminate delaminations. In some of applications, the acoustic microscope can detect defects that do not appear in X-ray studies. Inspections with the SAM are usually limited by slow scanning speeds and low throughput, but advances such as the multieyed and further even matrix array microscopes developed by R. Maev and high-speed scanners developed by others largely overcome these deficiencies.

The ultrasonic imaging systems operate in a narrow frequency range (5–10 MHz) with a resolution of 0.5 µm. The scanning is done by controlling the relative phase of the multiple heads inside the scanner. When we move to a frequency of 300 MHz, the SAM operates at a higher frequency (300 MHz) with improved resolution (0.5 µm). It incorporates a highly converging beam from a single surface spherical lens immersed in fluid. The beam is focused to a single spot 0.5 µm in diameter and is mechanically scanned across the surface of the sample in a raster pattern.

The introduction to scanning microscopes occurred at Stanford in the early 1960s when transducers were designed and fabricated for operation in gigahertz range of frequencies. In the research laboratories, images have now been recorded at a frequency of several gigahertz with submicron resolution.

In St. Petersburg, Russia, S. Sokolov made an important observation in the early 1940s when he noted that the wavelength of sound at microwave frequencies was comparable to the optical wavelength. At that time suitable transducers were unavailable, so his first microscope operated at ultrasonic frequencies. In the mid-1950s, H. Bommel and K. Dransfeld were able to use deposited piezoelectric films to generate microwave sound. This was followed by the construction of SAM at Stanford in the early 1970s. These were exciting instruments – a new form for microscopes that did not compromise the resolving power. Acoustic microscopes operating in the microwave frequency range have been built with a resolving power that surpasses the optical microscope.

Ten years later, the acoustic instruments operating at microwave frequencies were eclipsed by the emergence of the probe microscopes with scanning tips sharp enough to distinguish single atoms.

More recently, we have learned that the SAM can be combined with the scanning probe microscope to form the ultrasonic atomic force microscope (UAFM). In the combined instrument, they vibrate the tip of the atomic force microscope, or sample itself, at ultrasonic frequencies. The ultrasound is modulated at a low-frequency corresponding to the mechanical resonance of the cantilever. The resolution of the UAFM is not limited by the sound wavelength, rather by the radius of the scanning tip and this improves the resolution by orders of magnitude as compared to the sound wavelength.

These microscopes have proved to be valuable tools for inspecting chip scale semiconductor devices. Defects such as delaminations of the interlaminations of thin metal films on flexible polyimide substrates and thin layers of plastics on semiconductor packaging are easily detected with UAFM. Nanocomposites consisting of brittle glass films deposited on polymer substrates are often used in packaging materials since they are transparent and compatible with microwaves. These materials serve as barriers to water. Cracking and debonding can destroy the barriers. Inadequate adhesion will lead to debonding and, in turn, this will lead to stress concentrations and further cracking. Several techniques are available for studying cracking but heretofore, the regions of debonding have been difficult to identify. The UAFM can easily detect the debonding regions. It has also been used to detect delaminations in composite structures with layers a few tens of nanometers in thickness.

This rich background of experimental work provides the incentive for understanding the material in the presentation that follows. I was asked by Dr. Maev to prepare an introductory for his new English book and was pleased to do this again. I am confident that this new monograph prepared by Dr. Maev will convey some of the very exciting advances that are being made in acoustic microscopy.

Calvin F. Quate
Leland T. Edwards Professor, Emeritus
Stanford University

Preface

It seems incredible today that the majority of the most frequently used and widely different practices of modern ultrasonoscopy were developed by a single human. Nevertheless, this is so indeed: the most celebrated Soviet master in acoustics, Prof. S. Ya. Sokolov, had not only pioneered the use of sonic and ultrasonic waves for imaging of various objects but also was the author of the surface pattern technique, the method of light diffraction by ultrasound and electroacoustic conversion (Sokolov's tube). In addition, S. Ya. Sokolov proposed even in 1934 that high-frequency acoustic radiation could be used to visualize the inner structure of matter and this concept was embodied in acoustic microscopy. It was probably him who introduced the very term **ultrasonoscopy**.

In 1942, Firestone (of tire fame) filed a patent for a pulse-echo ultrasonic device to detect flaws in materials such as rubber (or tires). However, the patent also suggested that the same device could be used to "detect flaws in the human body." During the 1950s, different research groups in Europe, the USA, and Japan all developed the first generation of various ultrasonic imaging devices for medical diagnostics.

Since about this time, the methods of sound visualization started progressing rapidly and have come into advantageous use for nondestructive ultrasonic testing of materials and products and for visualization of subaqueous entities and subterranean structures. And 1949 was marked by a successful attempt to apply a pulsed technological flaw detector for visualization of calculi in bile ducts and extraneous bodies in soft tissues of humans, which triggered the development of medical ultrasound diagnostics.

Ultrasonic waves can propagate into opaque materials and return to the surface with important information about subsurface microstructure and properties. Only ultrasonic measurements are able to clarify the elastic properties of materials and biological tissue quantitatively. As ultrasonic measurements are considered to be harmless to a human body, ultrasonic diagnostic equipments have been favourably accepted in clinics, and they are continuing to be developed in various fields of medical diagnosis.

However, the advancement of ultrasonoscopy technique with a truly high resolving power started as late as 1970s with the advent of high-performance facilities for high-frequency ultrasound generation and detection.

One of the most popular high-frequency acoustic visualization techniques is based on the acoustic holography principle. In this method, the whole object to be investigated is immersed in a liquid and irradiated with a plane ultrasonic wave. Having passed through the object, the wave impinges on the free liquid boundary and generates a wave disturbance pattern at it, which is read by a scanning laser beam. This pattern which in essence is an acoustic hologram is processed to visualize the structure of the object being studied.

This method was embodied in a laser scanning acoustic microscope by A. Corpel and L. Kessler, "Zenith Corporation" (USA), in 1971. Operating frequencies of such ultrasonoscopes presently range from 10 to 200 MHz, which corresponds to resolution from 200 to 10 μm. However, a dramatic rise in the resolving power of laser scanning microscopes is impossible because of principal restrictions of physical nature. A higher resolution was achieved in an acoustic microscope with mechanical scanning devised by C. Quate and R. Lemons in 1974 at Stanford University. In this device, a sample is placed in the focal region of a converging ultrasonic beam generated by an acoustic lens with a large angular aperture.

The reflected signal received by a similar system of lenses is recorded, while the sample is moved mechanically relative to the focus, and thereby, an acoustic image is formed. The recorded signal primarily depends on the structure of that part of the object which finds itself in the focal region of the acoustic lens system, and therefore, the resolution of a scanning acoustic microscope is controlled by a focal spot diameter. Owing to simplicity of the acoustic system, operating frequencies of the microscope are as high as 3–4 GHz, which corresponds to the spatial resolution within the submicron range.

An essential advantage of scanning acoustic microscopy is the possibility of measuring viscoelastic parameters of local domains of test objects. A variety of ingenious procedures for measuring moduli of elasticity, viscosity coefficients, internal friction coefficients, local elastic anisotropy, etc. have been developed thus far at multiple research centers all over the world.

Acoustic microscopy has found a wide utility in solving multiple practical problems, namely, in controlling the quality of semiconducting and microelectronic products, magnetic information carriers and antireflection and protective coatings, in studying local properties of poly- and single crystal films, piezo- and photorecording materials, high-temperature superconducting samples, metals and alloys, ceramics and polymeric composite materials, biomedical entities, etc.

Beginning in the 1980s, microscopes of this type were produced in quantity by such companies as "Leitz" (Germany), "Olympus," "Honda Electronics," "Toshiba" (Japan), "Bruker" (France), "Sonix," "Sonoscan," and "Tessonics" (USA).

In USSR, the first laboratory mock-up of a scanning acoustic microscope was assembled at the Acoustics Department of Moscow State University in 1976 (V. E. Lyamov and S. I. Berezina). Thereupon, starting in 1978, the research along this line was underway at the Institute of Radio Engineering and Electronics, USSR Academy of Sciences (Yu. V. Gulyaev, A. I. Morozov, and M. A. Kulakov) and also at the Moscow Institute of Radio Engineering, Electronics and Automatics (L. D. Bakhrakh and S. A. Titov).

At the Institute of Chemical Physics, USSR Academy of Sciences, an investigation into high-resolution ultrasonics was begun in 1978 in the biointroscopy laboratory (R. G. Maev). To promote fundamental and applied research in this area, the Bureau of General Physics and Astronomy Department and Department of General and Technical Chemistry, USSR Academy of Sciences, passed a joint resolution, according to which the Acoustic Microscopy Center was founded in 1987 on the basis of the research staff of the Institute of Chemical Physics.

In 1996, this Center was evolved into the Center for investigation of modern materials within the framework of Emanuel Institute of Biochemical Physics, Russian Academy of Sciences. Organized and headed by R. G. Maev, the Center of Acoustic Microscopy has become a truly unique school of thought owing in no small part to Maev's colleagues, among which were outstanding experts in acoustics, such as L. A. Chernozatonskii, V. M. Levin, and S. A. Titov.

Maev's scientific school graduated a series of brilliant researchers, among which were O. V. Kolosov, K. I. Maslov, O. I. Lobkis, T. A. Senyushkina, E. Yu. Lagutenkova (Mrs. Maeva), P. D. Zinin, L. A. Matsiev, M. A. Pyshnyi, M. A. Bukhny, V. A. Zharov, and A. A. Denisov. The majority of them have become top-level experts in acoustic imaging and persist in exploring this captivating area of knowledge both in Russia and abroad.

The principles and methods of high-frequency ultrasonic visualization have been developed for as long as thirty years now. Current focus in high-resolution ultrasonic imaging, particularly for biomedical applications, is to develop procedures adding functionality in addition to the morphology contained in the images conventionally obtained. In this context, it has always been the goal of ultrasonic imaging to grade lesions when tissue had been detected as being suspicious for cancer. New strategies for advanced ultrasonic imaging applications include quantification on a rather molecular level (molecular imaging), determination of blood flow through vessels oriented arbitrarily in space and not aligned with the axis of the interrogating ultrasound beam. High-frequency ultrasonic imaging technology will cultivate new scientific and technological fields in a near future continuingly. For example, acoustic microscopy will contribute to the nanotechnology through the micro-NDE technology that is expected to be cultivated by acoustic microscopy. In 2004, the scientific community celebrated the 30th anniversary of the first scanning microscope and the 80th anniversary of its creator, Prof. C. F. Quate.

In 2005 the first monograph in Russian language "Acoustic Microscopy" by Prof. Roman Gr. Maev was published by Torus Press in Moscow, Russia. This book was intended as a monograph covering all the fundamental physical principles of acoustic microscopy and its numerous applications and created a lot of interest from researchers, engineers and students studying physics of solids, materia characterisation and, obviously, acoustic microscopy. I was asked by Dr. Maev to write an introduction for his book and was glad to do this.

One year later Dr. Maev began preparation of an English version of "Acoustic Microscopy" monography for Wiley and Sons Publishing House. It took about one more year for Dr. Maev to significantly revised material of "Russian version," to added some recent results and also added three more new chapters before the new

English version of the book was completed. And again, in that new case, I was asked by Dr. Maev to prepare an introductory for his new book and was pleased to do this again. So, the goal of this new book is to introduce recent advances in high-resolution acoustic imaging for material evaluation and nondestructive analyses and to fill lacunae left by earlier one and to convey some of the very exciting advances that are being made in acoustic microscopy.

It reviews the current status of research in the hottest areas of acoustic microscopy with emphasis on quantitative methods of studying microstructures, on acoustic diagnostics and on processing signals and images. The material is set forth at a rigorous scientific level and yet it is intelligible even to a nonspecialist.

I am sure that this new book will be interesting to broad circles of research workers, engineers and technicians, including multiple users of acoustic microscopy studying physics of solids, materials technology, microelectronics, biology, and medicine.

I am also convinced that this monograph will be useful as a textbook for students and postgraduates.

February 5, 2008

Yurii Vasil'evich Gulyaev
Academician, Professor
Doctor of Science in Physics and Mathematics
Member of the Presidium
of the Russian Academy of Sciences
Director of the Institute of Radioelectronics
of the Russian Academy of Sciences
Chairman, National Radio Society by Popov

Introductory Comments

Investigation and analysis of the internal microstructure of various entities ranging from crystals to biological tissues are the major challenges not only of modern science but also of diversified industries.

The analysis is carried out using a broad spectrum of physical methods, including optical, electron, field-ion, and tunneling microscopies. Each method has its own advantages and limitations, and pieces of information gained by them often supplement one another.

The idea of using acoustic fields for visualization of the internal structure of matter was advanced by the Russian scientist, Prof. S. Ya. Sokolov, as early as 1934. It was from this time on that ultrasonic testing, an efficient method of nondestructive control over materials, started progressing rapidly.

In 1949, a pulsed technological flaw detector was first used to reveal calculi in bile ducts and alien bodies in soft tissues, which marked the beginning of research in a new field of application for acoustic visualization techniques – medical ultrasonic diagnostics.

Improvement of the procedures for generation and reception of ultrasonic waves in the frequency range from hundreds of megahertz to several gigahertz opened up the way for the advent of acoustic visualization techniques with a spatial resolution as high as that of an optical microscope.

The new visualization method has come to be known as acoustic microscopy. It permits layer-by-layer imaging of the internal structure of test objects, such that the quality and resolution of the images are the same (and even higher) as those achieved with optical microscopy, and most of the test objects are absolutely opaque to electromagnetic optical waves.

High spatial resolution, in addition to the acoustic wave propensity to "penetrate" deep into opaque entities, is an extremely important advantage of acoustic microscopy.

The most outstanding distinction of acoustic microscopy compared with other tools of microstructure imaging is the possibility to detect **current magnitudes of acoustic fields** and not just their intensities, which owes its existence to the linearity of acoustic receivers. In other words, acoustic microscopy allows the measurement of scattered acoustic fields to within a "phase" of the ultrasound carrier frequency used. The feasibility of measuring current magnitudes of "acoustic fields" has given

Acoustic Microscopy. Roman Gr. Maev
Copyright © 2008 WILEY-VCH Verlag GmbH & Co. KGaA, Weinheim
ISBN: 978-3-527-40744-6

rise to the acoustic microscopy procedures which would be impossible to realize by means of optics or electron microscopy.

Excellent examples of two-dimensional (or three-dimensional) imaging of the internal "layer-by-layer" structure of a test object, which nowadays seems quite commonplace, are the systems of medical ultrasonic diagnostics and nondestructive control. By analyzing the ultrasonic response from a sample, the different interfaces and features of that sample can be verified, and flaws can be detected nondestructively. That was one of the critical reasons to claim acoustic microscopy as a valuable tool for the nondestructive inspection of materials and biomaterials. For these applications, flaws such as delimitations, cracks, voids, die tilt, underfill density and various distortions, are of interest.

However, these examples demonstrate certain technical advantages of acoustic microscopy. Of more importance for physics and technology is the possibility of straightforward and precise measurements of a series of microstructural parameters, such as viscosity, elasticity, dispersion constants, phase transition points, etc. Similar quantitative measurements by other microscopy methods are either completely impossible or possible only indirectly.

It is worth noting that despite apparent "geometric" correspondence between an acoustic image of some object and its optical counterpart, these two images are **quite different in their information content**, inasmuch as they result from interaction of the object with fields of totally different physical nature. This fact is often disregarded, especially by ultrasonic apparatus engineers striving for maximum "similarity" between optical and acoustic images. It is the possibility of deriving quantitative information about the most essential microstructural parameters from acoustic images that allows proper interpretation of these images and their comparison with optical images.

Acoustic microscopy first made it possible to explore the dynamics of various *in vivo* physicochemical and physicomechanical processes, which proved to be particularly important for biology by virtue of the noninvasive nature of the method and harmlessness of acoustic waves to living organisms.

Thus, acoustic microscopy has become not only a new imaging method extensively used in many areas of physics, biology, and technology, but also a new efficient tool of **quantitative characterization of microstructure of various species and materials**.

Now that nanotechnologies and new structural and biological materials are being developed, the unique capabilities of acoustic microscopy are difficult to overestimate.

The mid-1970s were marked by the realization of the fundamental principles of high-frequency acoustic visualization underlying scanning acoustic microscopes. They have been widely used for more than twenty years. Although considerable advances in theoretical justification of acoustic microscopy procedures and their practical implementation were achieved even at that time, many questions concerning the physical principles of acoustic microscopy and the interpretation of acoustic images of various types of objects remained unanswered. Many of these problems were successfully solved later, in particular, by the author, his colleagues, and his apprentices.

It is worth noting that modern acoustic microscopy is a synthesis of various physical and technical disciplines. On the one hand, it extensively uses classical physical-mathematical means of describing wave fields, based on diffraction, interference, and acoustic wave propagation in various nonhomogeneous media. On the other hand, acoustic microscopy also leans heavily on the methods of system analysis based on "locational" approaches allowing for the statistical nature of various noises and fluctuations and optimizing both the types of probing signals and the algorithms of their space and time processing.

And finally, modern acoustic microscopy has spurred the development of new processes for production of high-frequency piezoelectric materials exhibiting high sensitivity and broad frequency bands.

During the last decades, acoustic microscopy has become a common instrument for investigating the internal structure of materials and for industrial nondestructive evaluation. At the same time, quantitative acoustic imaging research was mostly applied through high resolution acoustic microscopy in material sciences and biomedical imaging. Further improvement of quantitative methods can be realized using other types of ultrasonic-media interactions and, accordingly, can provide additional specific information about the object. Realization of such methods will constitute a considerable expansion in the applicability of acoustic microscopy. In 2005, I prepared a first monograph in Russian, "Acoustic Microscopy", which was published by Torus Press in Moscow, Russia. This monograph was intended to cover both the fundamental physical principles of acoustic microscopy and its numerous applications. One year later, in 2006, I began preparation of an English version of the "Acoustic Microscopy" monograph for Wiley and Sons Publishing House. It has been delightful to observe the astonishing advances that have taken place in acoustic microscopy during the last five years since the monograph in Russian was written. I significantly revised most of the chapters of the "Russian version," added recent results to give an account of some of the key developments since then and also added three new chapters to introduce recent advances in high-resolution acoustic imaging for material evaluation and nondestructive analyses to fill lacunae left by the earlier one and to convey some of the very exciting advances that are being made in this field.

In closing, I pay homage to my teachers, colleagues, apprentices, and assistants without whom this book would never have made its appearance. First of all, I wish to thank Academician Yu. V. Gulyaev and corresponding member V. I. Pustovoit for their support and interest in my work, and Prof. C. F. Quate, the originator of the first scanning microscope at Stanford University, for fruitful discussions of the structure and certain chapters of the book. I am also indebted to my colleagues L. A. Denisova, S. A. Titov, V. A. Burov, B. O'Neil, I. Yu. Solodov and C. Miyasaka for their invaluable assistance in writing the book.

I am grateful to all my co-authors and apprentices with whom I have been working for the last 20 years and obtained many unique results included in the monograph. Undoubtedly, I would not be able to write this book without the support of my family, without the understanding and patience of my wife, Elena Maeva, and my children, Anna and Grigorii, who forgave my inattention to them and my preoccupation with the work on the book. Many thanks to you all.

Introduction

Investigation into the microstructure of entities of different nature, ranging from crystals to biological tissues, is one of the hottest areas of modern science and technology. A broad spectrum of physical tools, including optical, electron, field-ion, tunneling microscopies, etc., have been developed for this purpose. Each tool has its own advantages, and pieces of information gleaned by these methods complement one another.

In recent years, the trend has been to a new rapidly progressing physical method of studying microstructure – scanning acoustic microscopy. It allows one to obtain acoustic images, to measure local elastic and viscous properties of materials of different nature and to explore the dynamics of the physicochemical processes taking place in them. It has been turned to good account in many technical fields as an efficient means of nondestructive control over materials and products manufactured from them. In this method, the probing radiation is an acoustic wave in the ultrasonic or hypersonic range (frequencies from 10 MHz to 3 GHz). The resolving power of acoustic microscopes is commensurate with that of optical ones, and as the operating frequency of ultrasound is increased to tens of gigahertz, it closely approaches the resolution of electron microscopes.

The history of acoustic microscopy evolution is short, but it abounds with interesting events. The idea of using acoustic radiation for the visualization of the mechanical structure of materials of different nature was advanced by the Russian scientist – corresponding member of USSR Academy of Sciences S. Ya. Sokolov – back in 1934. It was he who introduced the term "ultrasonoscopy" [1–4]. From this time, ultrasonic testing, the method of nondestructive control over materials, started developing quite vigorously.

This method is based on partial reflection and scattering of ultrasonic waves by discontinuities or inhomogeneities in the test material, at which acoustic parameters experience change.

In 1947, a pulsed ultrasonic signal was first applied in clinical practice to detect calculi in bile ducts and alien bodies in soft tissue [5]. Unusually, the vast diagnostic potential of ultrasound has captured the imagination of many specialists, and from the mid-1960s, a variety of ultrasonic diagnostic devices operating at frequencies from 1 to 15 MHz and having a wide spectrum of clinical applications [6] were put into full-scale production. The further (anywhere near) essential increase in fre-

Acoustic Microscopy. Roman Gr. Maev
Copyright © 2008 WILEY-VCH Verlag GmbH & Co. KGaA, Weinheim
ISBN: 978-3-527-40744-6

quency and, hence, in resolving power of these devices was hampered by problems related to the generation of high-frequency ultrasonic waves.

In the late 1950s, Prof. K. N. Baranskii (USSR) and Profs. H. E. Bömmel and K. Dransfeld (Germany) independently devised a means of generating ultrasonic waves in the range from hundreds of megahertz to several gigahertz and thereby opened the way for the development of acoustic microscopy [7, 8].

Since the publication of these papers and up until the early 1970s, several research teams had striven to realize acoustic microscopy. The main bottleneck on the way to this goal was acoustic-field visualization. The majority of the proposed methods did not advance further because of their relatively low resolution, insufficient sensitivity, and long time of exposure.

Improvements in the procedure of hypersonic wave generation and reception, new processes for the production of highly efficient hypersonic transducers, and high-quality single crystal acoustic lines adapted to a commercial level, allowed two research teams to realize two different approaches to the visualization of high-frequency sound in the early 1970s. With the first approach proposed by Profs. A. Corpel and L. Kessler in 1971 ("Zenith Corporation," USA), a sound pattern is read by a surface-scanning laser beam by the methods typical of acoustic holography; namely, the object is immersed in water and irradiated with a plane ultrasonic wave forming a pattern at the free liquid boundary, which is read by a laser beam.

The practical aspects of this approach had advanced greatly by 1974, and in 1975 the "Sonoscan" Company (USA) instituted by L. Kessler started commercial production of this type of microscope operating at 100 MHz which corresponded to the sound wavelength of 15 µm in water and a resolution of optical and acoustic images of the order of 20–25 µm at a sensitivity of 10 W cm^{-1} [9]. However, the principles underlying this microscope restrict the possibility of improving its resolution in the future.

High resolution was achieved with the other approach to acoustic visualization – scanning acoustic microscopy – realized by C. Quate and R. Lemons in 1974 at Stanford University [10]. In this method, a focused ultrasonic beam formed by an acoustic lens, interacts with a sample. After the interaction, the signal is picked up by the same lens or by a lens confocal to it or by another special device for acoustic signal reception. The acoustic image is produced in the shape of a grating, while an acoustic beam scans across the sample.

The resolution of the microscope depends on its operating frequency. At room temperature and an operating frequency from tens of megahertz to several gigahertz, the acoustic microscope resolution ranges from hundreds of microns to tenth fractions of a micron. With superfluid helium as an immersion liquid, the operating frequency of a cryogenic acoustic microscope rises to tens of gigahertz, and its resolution to hundreds of angstroms.

An essential advantage of the method is the possibility of measuring local viscoelastic parameters of matter; namely, moduli of elasticity, viscosity coefficients, internal friction coefficient, local elastic anisotropy, etc.

This principle has since been widely used at many research centers all over the world to develop a variety of ingenious investigation techniques; efforts are underway to formulate the physical fundamentals and scientific principles of the method.

The method itself has gained recognition as a research tool and found wide use in solving a number of practical problems, such as quality control over semiconducting and microelectronic products, magnetic information carriers and anti-reflection and protective coatings, investigation into local properties of poly- and single-crystal films, piezo- and photorecording materials, high-temperature superconducting samples, metals and alloys, ceramics and polymeric composite materials, biomedical entities, etc. Beginning in the 1980s, microscopes of this type were produced in quantity by such companies as "Leitz" (Germany), "Olympus," "Honda Electronics," "Toshiba" (Japan), "Bruker" (France), "Sonoscan," and "Sonics" (USA).

Currently, in the world market, you can find scanning acoustic microscopes produced by the following high-tech companies: "Sonix" (USA), "Honda Electronics" (Japan), and "Tessonics" (Canada).

Among the challenges facing acoustic microscopy, the following are worth noting:

- The analysis of physical principles of formation of an output signal from an acoustic microscope.
- The formation of an acoustic image.
- Investigation into the nature of acoustic contrast and interpretation of acoustic images of entities differing in physicochemical properties and structure on the basis of theoretical concepts.
- Further experimental improvement and theoretical justification of quantitative methods and facilities for acoustic visualization, including visualization of distribution of micromechanical characteristics of objects.

Of prime practical importance is the purposeful development of acoustic microscopic procedures for studying a broad spectrum of materials: the mechanical microstructure of polymers, their mixtures and composite materials; the internal structure of ceramics and layered structures; various metals and alloys; the flaw structure of crystals and films and the structure of biological tissues and cells, etc. Advanced high-resolution acoustic microscopy will encourage the scientific community to use this unique research tool and methods for new broad applications. New strategies for various novel high-resolution ultrasonic imaging applications including complicated quantification analysis, are continuing to be developed in various fields of biophysics and medical diagnosis using advanced materials, and will contribute to nanotechnology through the micro-NDE technology that is expected to be cultivated by scanning acoustic microscopy.

1
Scanning Acoustic Microscopy. Physical Principles and Methods. Current Development

1.1
Basics of Acoustic Wave Propagation in Condensed Media

An acoustic wave is a physical phenomenon responsible for the transfer of dilatational and shear strains [11–16]. Generally, acoustic waves of three sorts can propagate along a certain direction, each of them spreading at its own velocity and imparting oscillatory motion in particles of the medium in its own direction – wave polarization. Only two sorts of waves propagate in an isotropic solid; namely, longitudinal and transverse ones. In a longitudinal wave, particles are displaced in the direction of wave motion while the overall strain transferred is a combination of the dilatational and shear strains. Transverse waves transfer shear strain alone, particles of the medium oscillate in an arbitrary direction in a plane normal to the propagation direction.

Distinguishing between longitudinal and transverse waves in a crystalline solid makes no sense because all the three acoustic waves spreading in a given direction transfer both the dilatational and shear strains [12, 16–21]. A wave in which longitudinal particle displacements dominate is called quasi-longitudinal; normally, it propagates at the highest velocity. The other two waves are called fast and slow transverse waves, in accordance with the value of their propagation velocity. A crystalline medium, even of high symmetry, exhibits a significant anisotropy of its acoustic properties, that is, its velocity heavily depends on the propagation direction with respect to the crystallographic axes. The phase and group acoustic wave velocities can appreciably differ both in their values and directions [12, 18, 21].

Longitudinal waves alone can propagate in liquids where no shear stresses exist. However, many liquids exhibit significant shear stresses at hypersonic frequencies, thereby approaching solids in their mechanical properties. Thus, high-frequency shear waves can propagate in them. This phenomenon was discovered in studying the Mandelstam–Brillouin scattering on the so-called anisotropy fluctuations in liquids which are, in essence, shear waves [16, 22, 23].

The propagation velocity and absorption coefficient of acoustic waves are the basic characteristics of their acoustic properties. According to the customary notions of fluid mechanics, the speed of sound is a parameter of the medium which is

Acoustic Microscopy. Roman Gr. Maev
Copyright © 2008 WILEY-VCH Verlag GmbH & Co. KGaA, Weinheim
ISBN: 978-3-527-40744-6

independent of frequency [13–16, 20]. Absorption of acoustic waves in a medium is governed by dissipative processes, namely, by viscosity and heat conduction, coefficients of which are assumed to be constants of the material. The absorption coefficient is proportional to viscosity and increases proportionally to the sound frequency v: $\alpha \propto v^2$ [13–16, 20]. Within this assumption, dispersion of the acoustic velocity shows up at very high frequencies at which the notion of continuity of the material becomes unjustified and the discrete nature of its structure manifests itself [24].

However, many materials exhibit frequency dispersion of their speed of sound and deviation from the quadratic frequency dependence of their acoustic absorption coefficient caused by the frequency dependence of the effective material viscosity [12–17, 20, 25–30]. The frequency dependence of the speed of sound and viscosity stems from the occurrence of two sorts of processes in the material induced by the acoustic wave; namely, resonant excitation of the internal degrees of freedom in the material and acoustic relaxation [13–16, 25, 27, 28].

Resonant excitation processes are inherent in heterogeneous media (oscillations of individual crystallites in polycrystals [31] or of air microbubbles in liquids), although they can be observed in homogeneous materials as well, e.g., when displacements are induced in crystals [16, 32].

Acoustic relaxation is a process of local equilibration of the material perturbed by an acoustic wave [13–16, 25, 27, 28]. If the relaxation time of a certain parameter τ is much greater than the wave period: $v\tau \gg 1$, its variation in an acoustic wave can be neglected, that is, the appropriate degrees of freedom are "frozen" in the wave.

If, however, the period of acoustic oscillations is commensurate with the relaxation time or is greater than τ, that is, $v\tau \leq 1$, the parameter at issue varies at an acoustic frequency due to instantaneous attainment of its equilibrium value. Inasmuch as relaxation processes are irreversible, some portion of the acoustic energy goes to heat. Hence, an additional dissipation mechanism arises in the material when the sound frequency decreases and the fluid viscosity increases. Concurrently, the elasticity modulus reduces because the relaxation processes proceed in such a way that the elastic stresses induced by material deformation diminish. As a result, dispersion of the acoustic velocity arises near frequency $v \sim v_r = 1/\tau_r$ called the relaxation frequency: as the frequency increases so does the speed of sound from its low-frequency value c_0 at $v\tau \ll 1$ to the high-frequency value c_∞ at $v\tau \gg 1$.

The frequency dependence of acoustic absorption in the relaxation domain deviates from the quadratic law because the material viscosity reduces from its low-frequency value η_0 to its high-frequency one η_∞ as the frequency increases. The above-mentioned shear stress arising in a liquid at high frequencies [25, 27] is an example of acoustic relaxation. Displacement of one fluid layer with respect to another brings about elastic shear stresses that lead within time $\Delta t \geq \tau_r$ to irreversible motion of the layers, that is, to their flow. Naturally, at frequencies $v\tau_r \gg 1$, the time allotted is too short for the displacement to occur and the liquid behaves as an elastic solid.

Apart from the shear relaxation, there are a number of other mechanisms of acoustic relaxation in liquids, e.g., those associated with liquid restructuring, energy redistribution among the internal degrees of freedom, and some other processes. The relaxation frequencies lie in the mega- and gigahertz ranges; the dispersion of speed of sound in these ranges can attain a few tens of percent. It is worth noting that water, which is the liquid most frequently used in acoustic microscopy, exhibits no dispersion of its speed of sound.

The basic process responsible for dissipation in solids is relaxation in the system of thermal phonons, the equilibrium state of which is perturbed by an acoustic wave (Akhiezer's mechanism) [26, 30]. Inasmuch as the relaxation time of a phonon system is short: $\tau_r \approx 10^{-11}$ s, the aforesaid mechanism fails to give rise to a frequency dependence of the speed of sound and viscosity up to a few tens of a gigahertz. Dispersion in this frequency range is governed by the fluid discreteness and absorption of sound is controlled by the direct three-wave interaction of the acoustic wave with thermal phonons (Landau–Rumer mechanism) [26, 30].

Apart from the phonon viscosity, some other mechanisms responsible for the absorption and dispersion of the acoustic velocity exist in crystals. In many crystals, interaction of an acoustic wave with crystalline structure defects (with dislocations in the first place) becomes significant at frequencies in the 10^6–10^{11} range [30, 32, 33].

Interaction with dislocations can be of a resonant nature, that is, it is most efficient when the frequency of the acoustic wave is close to the fundamental frequency of dislocation loops [32]. Interaction of oscillating dislocations with thermal phonons brings about dislocation absorption of acoustic waves. Naturally, the absorption and dispersion associated with dislocations depend on the degree of crystal perfection.

Semiconductors have a peculiar mechanism of absorption and dispersion of their speed of sound associated with the interaction of conduction electrons with an acoustic wave [16, 34–37]. This interaction is inherent in all sorts of semiconductor crystals, but is most prominent in piezosemiconductors in which the acoustic wave induces a variable electric field redistributing the free charge carriers in the material bulk due to the piezoelectric effect. As a result, there arises an induced wave of the electron density that spreads together with the acoustic wave.

Dissipation of the energy of directed electron motion results in effective ultrasound absorption. Interaction of an acoustic wave with free charge carriers is of a relaxation nature, that is, inhomogeneous electron distribution forms within some representative relaxation time period τ_e. At high frequencies, when $\nu\tau_e \gg 1$, the electron wave has no time to arise so that the presence of electrons virtually does not affect the behavior of the propagating acoustic wave. In many semiconductors (CdS, CdSe, ZnO, Zn, Te, and others) electron absorption dominates within a wide frequency range up to few gigahertz.

Dispersion of the speed of sound and absorption are also caused by interaction of ultrasound with other crystal subsystems, namely, spin waves, domain walls in ferromagnetics and ferroelectrics [15, 16, 19, 21, 24, 30, 38–40].

Apart from bulk acoustic waves, surface acoustic waves can also propagate in bounded solids; the amplitude of these latter waves diminishes exponentially as they move away from the boundary [12, 16, 20, 41–48]. The energy transferred by surface waves is concentrated in a surface layer whose thickness is of the order of the wavelength. The velocity of a surface wave must be less than that of bulk waves in bounded materials; otherwise, the surface wave would be re-emitted inwards into the solid in the form of bulk waves.

Depending on the nature of contacting materials, various types of surface wave can propagate along the interface between them [12, 41, 47, 48]. Rayleigh waves spreading along a solid–vacuum or solid–rarified medium interface [12, 41, 42, 45–48] are particularly important for reflection acoustic microscopy. The motion of the material particles in a Rayleigh wave is a superposition of two oscillations at the surface aligned with the wave propagation direction and perpendicular to it. The propagation velocity of Rayleigh waves is close to the speed of transverse acoustic waves, but is always somewhat lower than this.

A Rayleigh-type wave can also propagate along a solid–liquid or solid–solid interface [41, 46–49]. Propagation of Rayleigh waves along the boundary between solid half-space and a solid or liquid layer is of importance for acoustic microscopy practice.

The use of Stoneley–Scholte surface waves arising at the solid–liquid interface is a promising approach to developing a system for surface wave acoustic microscopy because absorption of these waves is sensitive to the mechanical structure of liquid or quasi-liquid (that is, a fluid with a small shear stress) fluid. In compliance with the structure of Stoneley–Scholte waves, their absorption is governed by energy dissipation in the crystal and liquid. Contribution of the internal friction effects in crystals is insignificant; therefore, the wave absorption is controlled solely by the viscoelastic properties of a liquid. A theoretical analysis demonstrates [50] that the absorption behavior depends on the ratio between the speed of sound in the liquid and the velocity of the transverse acoustic waves in the crystal.

Dissipation of the surface wave energy can be controlled both by absorption of the longitudinal component of the surface acoustic wave in liquid and by a viscous shear wave arising in the liquid near the interface.

Surface waves of another type are the Love waves that propagate along the boundary of the solid half-space on which the solid layer lies [41, 44, 47]. The material particles oscillate in the wave normal to the wave propagation direction within the layer plane; therefore, the wave should be qualified as transverse. The energy transferred by the wave is concentrated both in the layer and in the half-space zone adjacent to the interface. The velocity of Love waves depends on frequency and varies in the range between the velocities of transverse acoustic waves in the layer and half-space.

The existence of specific acoustic surface waves on some planes of piezoelectric crystals, called the Guliayev–Bluestein waves, is a distinguishing feature of these crystals. These waves are Love waves spreading over a free surface [49].

1.2
Physical Principles of Scanning Acoustic Microscopy

The lens system (Figure 1.1) is the heart of a scanning acoustic microscope (SAM). An ultrasonic wave is generated in the acoustic lens system by a transducer mounted at one of its ends. The wave spreads through an acoustic duct possessing a large acoustic impedance and is then focused with the aid of a spherical recess (lens) at the other duct end in an immersion material (liquid as a rule) filling the space between the lenses and objects examined. The immersion liquid provides a large refractive index for the acoustic lens and good acoustic contact between the duct and object. The focused beam interacts with the object, being partially reflected and scattered by the object and partially transmitted through it. If the reflected wave is detected, the microscope operates in the reflection mode. When the transmitted acoustic flux is recorded with the aid of the second lens, we have a transmission acoustic microscope.

It should be mentioned that a great number of modifications has been suggested within the basic principle underlying SAM that extends opportunities of the method. Thus, detection of acoustic radiation scattered by the object was arranged by rotating the lens through various angles with respect to the axis of the emitting lens [51]. This operating regime is essentially similar to the dark field regime of optical microscope operation and allows the effective resolution depth to be varied. A transmission regime with a single lens was realized in which a detector based on the acoustic-electric effect quadratic in the signal amplitude is used [52].

Additional information is obtained with the use of various nonlinear regimes in which the signal from the focal zone is recorded at harmonics of the input sig-

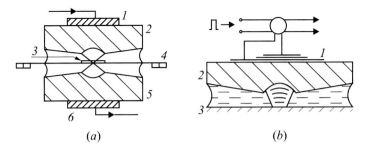

Figure 1.1 The lens system of a scanning acoustic microscope.
(a) Transmission microscope: 1 – piezotransducer, 2 – emitting lens, 3 – object, 4 – scanner, 5 – receiving lens, and 6 – receiving transducer. (b) Reflection microscope: 1 – piezotransducer, 2 – lens, and 3 – object.

nal [53] or at combined frequencies [54] rather than at the frequency of the input signal itself. In this case, both linear and nonlinear object properties contribute to the measurement results. An interference technique in the transmission microscope operation mode was suggested in [55] to measure local speeds of sound. The $V(z)$-characteristics method described in detail below is at present the most advanced technique for measuring local values of the velocity of Rayleigh waves spreading over the surface of samples studied. Of certain practical interest is also the possibility of using it in SAM surface acoustic waves; nowadays, some ideas have already been realized in acoustic microscopy.

At present, some promising approaches in enhancing the contrast and improving the quality of the image and SAM resolution have been proposed. One of these approaches is the use of appropriate immersion liquids, in particular, those possessing the lowest values of the speed of sound and transmission coefficient.

A set of cryogenic acoustic microscopes employed to study material surfaces was developed in Stanford University (USA). One of the instruments using liquid helium at $T = 0.1$ K [56] and operating at 8 GHz provided a 250–300-angstrom resolution, which is so far the highest level attained.

Finally, acoustic microscopy opportunities can be significantly extended by invoking the contemporary methods for computer image processing. This would obviously improve the image quality [57], enhance resolution, and offer the possibility of processing and documenting the data, and analyzing dynamic processes, etc. [58].

Resolution of the instrument and depth of radiation penetration are the most important characteristics of the method. They depend on the ultrasound frequency, parameters of the lens system, immersion material, and properties of the object. As the resolution increases, the depth of ultrasound penetration in the object reduces. Therefore, the ultrasound frequency is to be selected after taking into account the properties of the object and the problem to be solved and reasonably compromising between the penetration depth and resolution.

It is ultrasound radiation which permits one to obtain new information about the mechanical properties of a micro-object that cannot be provided by other microscopic methods.

Currently, we can formulate the following basic points of acoustic microscopy development:

- The development of physical grounds and the search for new foundations of acoustic microscopy.
- The study of the physical basics of acoustic imaging.
- The development of automated methods for recovering and analyzing the acoustic structure of objects.
- Investigations into the principles of acoustic microtomography.
- The development of acoustic microscopy techniques as applied to contemporary scientific and technical problems.
- Engineering developments.
- The design and manufacture of industrial samples of new acoustic microscopes.

1.3
Acoustic Imaging Principles and Quantitative Methods of Acoustic Microscopy

Obviously, of the above investigation lines, the first two are the most interesting and important; of particular importance is understanding the physical principles of acoustic imaging, in the first place the problems of image contrast and artifacts. Acoustic images cannot be interpreted without clear understanding of the physical mechanisms of their formation and the nature of acoustic contrast. Knowledge of these mechanisms allows one to perform quantitative measurements and to quantitatively characterize the materials studied. Measuring the output signal of an acoustic microscope, its amplitude and phase, and comparing them with the amplitude and phase of the reference signal in liquid, one obtains information about the speed of sound, acoustic impedance, attenuation, and geometric characteristics of the sample (thickness, curvature, and slope angle of the surface).

We start with considering how the output signal of an acoustic lens forms in a general case. The signal is detected with a piezoelectric transducer whose response is linear. In order for electric signal to be generated by the transducer, the incident wave front must be parallel to its surface, that is, the acoustic radiation refracted at the lens surface must impinge on the transducer surface normally to it. In other words, all the beams must pass through the lens focus. Furthermore, they all must arrive at the transducer in the same phase. Otherwise, signals from various beams would interfere and the resulting signal would be attenuated. Consider first how the signal forms in a reflection microscope (Figure 1.2). If the object boundary is at the lens focus (Figure 1.2a), the output signal is controlled by the integral (over all incidence angles from $\theta = 0$ to $\theta = \theta_m$, where θ_m is half the aperture angle) refraction index. If the lens is moved away from the object (Figure 1.2b), the cone

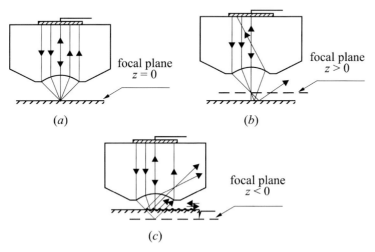

Figure 1.2 Schematic of the output signal of a reflection microscope: (a) $z = 0$, (b) $z > 0$, and (c) $z < 0$; z is the deviation from the focal plane of the lens.

of beams received by the transducer narrows rapidly and the level of the signal from the transducer drops drastically.

The $V(z)$ curve depicting the output signal amplitude as a function of distance from the focal plane of the lens shows a rapid decrease in the signal amplitude with small shallow oscillations as z increases in the $z > 0$ range. The oscillations are caused by a difference between the phases of differently directed beams. A dissimilar behavior of the $V(z)$ dependence can be observed at negative z values when the lens approaches the object (Figure 1.2c).

In solid objects, a Rayleigh wave propagates over the sample surface. If its velocity exceeds the sound speed in the immersion liquid, a surface wave would arise which is re-emitted back into the liquid. Waves of this type are called leaky Rayleigh waves. They arise at any position of the lens with respect to the object. However, they are detected solely at $z < 0$. Figure 1.2(c) demonstrates how it occurs. The output signal shows up as a superposition of the signal generated by the mirror-reflected paraxial beam and the signal generated by the leaky surface wave. The phase difference between these signals depends on the distance z; as a result of their interference, the $V(z)$ dependence exhibits a regular set of maximums and minimums [59, 60]. The distance between neighboring maximums and minimums is unambiguously related to the velocity of the Rayleigh wave spreading over the sample surface (Figure 1.3). The onset of a leaky Rayleigh wave is extremely important in acoustic imaging in the reflection mode. In particular, it brings about interference fringes near sharp inhomogeneities, at curved surfaces, and so on. It also gives rise to the effect of acoustic contrast inversion caused by small lens displacements. Figure 1.2(c) demonstrates how the interference fringes arise near a surface defect. Reflection from inhomogeneities produces not only the forward wave but the backward leaky Rayleigh wave as well, the latter wave can also be detected by the transducer. Inasmuch as its phase depends on the position of the lens axis with respect to the inhomogeneity, lens scanning produces interference fringes that follow the inhomogeneity contour.

The use of linear lenses that excite Rayleigh waves spreading in one direction turned out to be extremely informative in reflection microscopy. Rotating the lens with respect to the sample and thereby varying the propagation direction, one can measure local anisotropic properties of the material surface studied. This method was suggested and successfully developed in the works of Japanese scientists J. Kushibiki and N. Chubachi. At present, local velocities of Rayleigh surface waves are measured by this method with an accuracy of 10^{-4} [61].

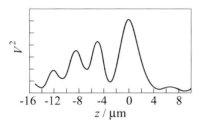

Figure 1.3 Representative pattern of the amplitude of output microscope signal V vs distance z between the object and lens focal plane.

What produces the image in an acoustic microscope operating in the transmission regime? We start with a discussion of the refraction phenomena. Let us consider transmission of a focused beam through a thin plate. The beam passing through a plate is refracted twice. As a result, its direction after it exits from the plate is parallel to that of the incident beam; however, the exit beam is shifted with respect to the incident one. The shift value depends on the incidence angle.

The rays, after passing a plate, form a diverging beam when the speed of sound in the plate exceeds the speed of sound in the immersion liquid; rays which are close to the lens axis make a paraxial focus while the rays incident at greater angles on the plate surface are focused at different points of the acoustic axis. As a result, the cone of rays received by the transducer narrows after passing the lens and the output signal diminishes. Variations of the signal due to the refraction effects depend on the ratio of the speeds of sound in the sample and liquid. Similarly, the cone of rays recorded by the receiving lens is narrowed due to phase aberrations. Phase aberrations stem from the fact that rays incident at various angles upon the object travel in it through various distances. Therefore, they have a different shape and can mutually attenuate the signals they induce in the output piezoelectric transducer. The values of phase aberrations are also controlled by the difference between the speeds of sound in the liquid and sample. Therefore, phase aberrations cause additional contrast of acoustic images.

To measure the local mechanical properties of samples quantitatively, we have developed a method of $A(z)$-characteristics [62, 63]. An $A(z)$ curve is the dependence of the output signal A of the receiving lens on the distance z between the lenses. A model $A(z)$ curve is displayed in Figure 1.4. We place an object between the lenses. As follows from the aforesaid, the focal point of the received acoustic beam is shifted due to refraction in the plate. Accordingly, the maximum of the

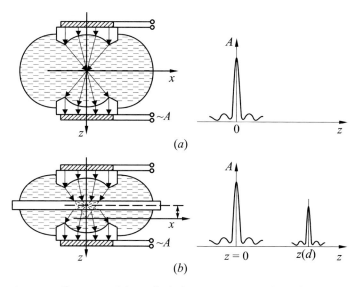

Figure 1.4 Illustration of the method of $A(z)$-curves: (a) without object and (b) with object.

$A(z)$ curve is displaced toward smaller z, if the speed of sound in the plate exceeds the speed of sound in the immersion liquid, and toward greater z, if the ratio of the acoustic velocities reverses.

The shift of the $A(z)$ maximum is proportional to the difference between the speeds of sound in the immersion liquid c_0 and in the object c, and to the sample thickness $\Delta Z = d(c/c_0 - 1)$. A local value of the speed of sound in a zone 5 or 10 µm in size can be assessed by measuring ΔZ. The ratio between the maximum amplitudes yields the coefficient of sound transmission through the plate. Knowing the object density and the value of the velocity c, the local ultrasound attenuation coefficient can be evaluated using the known impedance and transmission coefficient. The method of $A(z)$ characteristics was employed to measure the ultrasound velocity in various polymer films [63].

It should be mentioned that refraction in solids gives rise to shear waves. Being refracted at the bottom plate surface, these shear waves generate an additional converging or diverging acoustic beam with a different focus position [64]. The $A(z)$ dependence in this case has two maximums. The main maximum is shifted more and pertains to the longitudinal waves traversing the plate. The additional maximum arises due to the transverse waves. It is shifted to a lesser extent because its shift depends on the velocity of transverse waves in the material. Measuring the two shifts by the method of $A(z)$ characteristics, one can assess local distributions of the velocities of both longitudinal and transverse acoustic waves in the material [65]. Bifurcation of the maximums of the $A(z)$ curve was observed in studies of a dry gelatin film 7 microns thick at the same parameters of the setup (450-megahertz frequency, lens aperture of 30°) using mercury as an immersion liquid. Measurement of the two maximums permitted the longitudinal and shear wave velocities in the sample to be calculated and the vanishing of the shear modulus to be traced in samples swollen with water [66].

Another interesting example is the generation of a Lamb leaky wave with a set of modes in an individual layer or multi-layer system with dissimilar impedance and low absorption. This wave can significantly contribute to the received signal of the microscope operating in the reflection or transmission mode.

All the above examples can be related to the field of purely physical investigations; therefore, it is expedient to dwell, at least briefly, on the practical value of particular applications.

1.4
Methodological Limitations of Acoustic Microscopy

The acoustic microscopy technique is fairly sensitive to the presence of various inhomogeneities in a sample and to breaks of material continuity because mismatch of acoustic impedances at the boundaries brings about intense reflections. At present, acoustic microscopy allows one to reveal the following defects: failure of adhesion, exfoliation, microcracks, alien inclusions, deviations from the preset layer thickness in multi-layer systems and coatings, technological deviations in sizes, orientation and distribution of grains.

Based on the aforesaid, we believe that the following points are important in developing the methods of reflection and transmission acoustic microscopy and are promising in studies of surface and subsurface structures of diverse materials:

- The topography of surfaces, including height of steps, width of cracks and the pattern of mechanical stress fields around them, the curvature radii of bulges or concavities, wedge angles, etc.
- The morphology of smooth surfaces with inhomogeneous distribution of acoustic properties, including characterization of individual components of granulated and laminated structures, acoustic images of internal planes, structures, grains, and analysis of thin-filmed heterogeneous objects.
- The measurement of local values of the propagation velocity and attenuation of Rayleigh waves in materials employing acoustic microscopy techniques in which spherical and cylindrical lenses are used.
- The study of distributions of local anisotropic elastic properties in crystals and other materials.
- Quantitative measurements of the mechanical properties by acoustic microscopy techniques, including local measurements of piezoelectric, photoelectric, and high-temperature superconducting properties of films.
- Studies of dynamic phenomena associated with reconstruction of the material properties induced by varying physical factors (temperature, ultraviolet (UV), infrared (IR), and superhigh frequency impacts) and also by mechanical, chemical, and pharmacological impacts.

2
Acoustic Field Structure in a Lens System of a Scanning Acoustic Microscope

2.1
Calculation of the Focal Area Structure with Due Regard for Aberrations and Absorption in a Medium

To gain an understanding of acoustic field structure in SAM, it is essential that the effect of aberrations and absorption in a medium on the acoustic pressure distribution in the focal area of a concave spherical lens is analyzed theoretically. A converging wave front formed by this lens departs from a regular spherical shape, and the departure is referred to as an aberration.

In acoustic microscopy, aberration is reduced with crystals such that the sound velocity in them is higher than in an immersion liquid. Nevertheless, the use of wide-aperture lenses as a means of improving the resolving power calls for research on aberrations of such systems.

Acoustic pressure is calculated assuming that the linear dimensions of a lens far exceed the acoustic wavelength in a liquid. The angle between the direction toward the observation point and a normal to the lens surface is small, because the sound velocity in the lens, c_1, far exceeds that in the immersion liquid, c_{im}. The vibrational velocity potential Ψ at the observation point is a superposition of the fields induced by point sources having equal amplitudes and different phases and located on the lens surface S [67]:

$$\Psi = \frac{V_0}{2\pi} \iint_S \frac{\exp(-ikr - i\varphi)}{r} dS$$

where V_0 is the vibrational velocity on the lens surface, φ is the phase distribution formed on the lens surface as a plane wave impinges on it, and r is the distance to the observation point.

The above-derived field distribution along the acoustic axis and in the focal plane of the lens can be compared with the familiar expressions for a spherical focusing transducer. From the viewpoint of geometrical acoustics, all rays converge at the center of curvature of a spherical radiator. Because of aberrations, the rays intersect the acoustic axis of the lens on interval $[F_e, F_p]$, where F_e denotes the edge focus at which the rays from the lens edge arrive, and F_p is the paraxial focus to which

Acoustic Microscopy. Roman Gr. Maev
Copyright © 2008 WILEY-VCH Verlag GmbH & Co. KGaA, Weinheim
ISBN: 978-3-527-40744-6

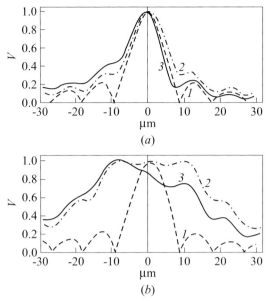

Figure 2.1 Normalized distribution of the acoustic pressure amplitude (V) along the acoustic axis of the lens. Ultrasonic frequency is 500 MHz. (a) Lens with a small aberration ($R = 630$ μm, $\theta_m = \pi/3$, $k = 2$ μm^{-1}, $N = 5$, $\beta = 1.33$, $\alpha = 0.01$ μm^{-1}, and $\gamma = 0.63$); (b) lens with a large aberration ($R = 1800$ μm, $\theta_m = \pi/3$, $k = 2$ μm^{-1}, $N = 5$, $\beta = 3.8$, $\alpha = 0.003$ μm^{-1}, and $\gamma = 0.54$): 1 – spherical radiator; 2 – nonabsorbing lens; and 3 – absorbing lens.

the rays from the lens center converge. The separation between these foci, referred to as maximum longitudinal aberration, equals

$$\delta_m = F_p - F_e = \frac{R(1 - \cos\theta_m)}{N^2 - 1}$$

where R is the radius of curvature of the lens, θ_m is its aperture angle, and $N = c_l/c_{im}$ is the ratio of the sound velocities in the lens and in the immersion liquid.

Inasmuch as the converging wave front is not strictly spherical, the phase may not be constant on any surface drawn from any point on the acoustic axis through the lens edges. The phase difference is a minimum when the center of curvature of the surface resides at the wave focus, i.e., at the point where the acoustic pressure amplitude is maximum at small aberrations. At the wave focal point, the in-phase waves from the point sources at the edges and center of the lens come together.

Aberrations are of little significance, when the maximum phase difference β on the spherical surface with its center at the wave focus satisfies the condition:

$$\beta \approx \frac{kR(1 - \cos\theta_m)^2}{8N^2} < \frac{\pi}{2}$$

where k is the wave number of the liquid.

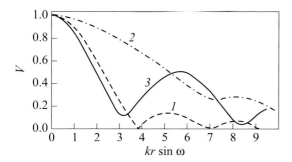

Figure 2.2 Normalized distribution of the acoustic pressure amplitude (V) in the focal plane of the lens. Ultrasonic frequency 500 MHz ($R = 1900$ μm, $\theta_m = \pi/3$, $k = 2$ μm^{-1}, $N = 5$, $\beta = 4$, and $\omega = 500$); $\sin(\omega) \sim \sin(\theta_m)(1 - \cos(\theta_m)/N)$; 1 – spherical scatterer; 2 – paraxial focal plane; 3 – wave focal plane; ω is the wave front aperture; and r is the coordinate in the focal plane.

The pressure distribution along the acoustic axis of the lens differs from the distribution for a spherical radiator in that its principal maximum is broader, its subordinate maxima are greater in amplitude, and the pressure is nonzero at all points on the axis (Figure 2.1).

At small aberrations, the principal-maximum width on the acoustic axis, Δ, is governed solely by the diffraction broadening and is independent of the lens dimensions (Figure 2.1a). In our case, Δ can be expressed as:

$$\Delta = \lambda \left(1 + \frac{N^2}{N^2 - 1} \operatorname{ctg}\left[\frac{\theta_m}{2} \right] \right)$$

where λ is the acoustic wavelength in the liquid.

At large aberrations ($\beta > \pi/2$), the principal-maximum width primarily depends of the geometrical beam spread which exceeds the diffraction broadening in this case and equals to δ_m (Figure 2.1b).

When $\beta > \pi/2$ in the focal plane passing through the paraxial focus, the principal maximum broadens, and the amplitude decays almost monotonically (Figure 2.2). This is because the main contribution to the amplitude distribution in this plane comes from the rays arriving from the central region of the lens and tilted at a small angle to the acoustic axis. In the plane intersecting the axis at the wave focus, the principal maximum narrows down, and the amplitude is an oscillating function (see Figure 2.2).

At the high frequencies used in acoustic microscopy, one has to take into account absorption in the immersion liquid which, in addition to general attenuation of the signal, leads to a nonuniform amplitude distribution along the converging front. This is because the path of peripheral rays in the liquid is shorter than that of paraxial rays, and hence they are absorbed to a lesser extent.

The effect of absorption on the acoustic field structure in the vicinity of the focal area can be expressed in terms of the parameter γ:

$$\gamma = \ln\left(\frac{P_e}{P_p}\right) = \alpha R \frac{1 - \cos\theta_m}{N}$$

where P_e and P_p are the acoustic pressures at the edge and center of the wave front, and α is the absorption coefficient of the immersion liquid. Parameter γ accounts for the nonuniformity of the amplitude along the wave front due to absorption. When absorption is significant ($\gamma > 0.5$), the pressure distribution along the acoustic axis becomes asymmetric about the focus (see Figure 2.1). In the focal plane, absorption causes narrowing of the principal maximum and an increase in the amplitude of subordinate maxima at small aberrations.

2.2
The Field of a Spherical Focusing Transducer with an Arbitrary Aperture Angle

Converging wave fronts are calculated using two basic approaches. With the first approach [68], the Helmholtz–Huygens integral is taken over the wave-front surface. The linear dimensions of this surface are assumed to be great compared to the sound wavelength in the medium, which makes it possible to approximate the integration surface by a large array of almost planar sections with dimensions exceeding the wavelength. The resultant field is a sum of the fields of these planar sections radiating independently of one another in space and coherent in time. In the context of this method, O'Neyl [69] derived the exact expression for the field on the acoustic axis of a spherical focusing transducer, valid at small aperture angles and large, compared to the wavelength, linear dimensions of the transducer.

The other computing method [70] involves a changeover from a converging wave front to a planar surface and from the true boundary conditions on the transducer surface to the boundary conditions in the plane, so that the field is calculated by integration over the planar surface. The advantage of this method lies in the fact that, for a plane with a potential or a vibrational velocity preset in it, there is an exact integral solution of the Helmholtz equation, whereas for an open spherical front, the solution has yet to be derived. Using this method, Lucas and Muir [71] obtained the field distribution for a spherical focusing transducer with a small aperture angle. The results derived by O'Neyl [69] and Lucas and Muir [71] are identical in the vicinity of the center of the curvature of the transducer [72].

When calculating the field [69], it was assumed that none of the waves issuing from the planar sections into which the transducer surface was broken were diffracted by the other sections, which was true only for transducers with small aperture angles. On the other hand, the aperture angle was supposed to be small in [71] merely for convenience of the field calculations. Therefore, the field of a transducer with an arbitrary aperture angle is calculated here using the transfer of the boundary conditions into the plane adjacent to the rear surface of the transducer.

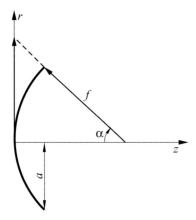

Figure 2.3 Geometry of a spherical focusing transducer: f is the radius of the curvature of the transducer, a is the transducer radius, and α is the aperture angle.

Figure 2.3 depicts a spherical focusing transducer with radius of curvature f, radius a, and aperture angle α. The normal component of the vibrational velocity on the transducer surface is constant and equals V_0.

It is reasonable to switch to cylindrical coordinates (r, z), where z is the coordinate along the acoustic axis of the transducer reckoned from the transducer surface, and r is the coordinate in the plane perpendicular to the acoustic axis. The solution of the first boundary problem for the Helmholtz equation in an infinite plane (with the vibrational velocity potential preset in it), expressed in terms of cylindrical coordinates, takes the form [70]:

$$\Psi(r, z) = \frac{-ikz}{2\pi} \int_0^\infty \Psi(x, 0) x \, dx \int_0^{2\pi} \frac{e^{ikR}}{R^2} \left(1 + \frac{i}{kR}\right) d\varphi \qquad (2.1)$$

Here, $\Psi(r, z)$ is the vibrational velocity potential at an arbitrary point (r, z); k is the wave number of the medium; $\Psi(x, 0)$ is the vibrational velocity potential in the $z = 0$ plane; x is the coordinate in this plane; R is the separation between points $(x, 0)$ and (r, z):

$$R = \sqrt{z^2 + x^2 + r^2 + 2xr \cos\varphi}$$

and φ is the angle between the directions toward points x and r in the plane perpendicular to the acoustic axis.

The true boundary conditions on the transducer surface (vibrational velocity equals to V_0 on the transducer and zero outside it) are now to be transformed into the boundary conditions in the $z = 0$ plane. Assuming that the linear dimensions of the transducer far exceed the sound wavelength in the medium, the field in this plane can be calculated in the geometrical–acoustics approximation. Within this approximation, the acoustic field of a spherical transducer is identical to the field of a point source at its center of curvature. The potential of this field equals $(A/\rho) \exp(ik\rho)$, where ρ is the distance from the center of curvature of the transducer and A is the normalizing factor to be derived from the boundary conditions: $\delta\Psi/\delta\rho = -V_0$ on the transducer surface. If $kf \gg 1$, the potential in the plane is

$$\Psi(x,0) = \begin{cases} \dfrac{V_0 f \exp[ik(f - \sqrt{f^2 + x^2})]}{ik\sqrt{f^2 + x^2}} & \text{at } x < f \operatorname{tg}\alpha \\ 0 & \text{at } x > f \operatorname{tg}\alpha \end{cases} \quad (2.2)$$

Thus, the problem reduces to finding the field of a planar transducer of radius $b = f \operatorname{tg}\alpha$ on an infinite absolutely soft screen with phase- and amplitude-modulated potential.

At $r = 0$, the integrand in Equation (2.1) is independent of φ, and the potential on the acoustic axis equals

$$\Psi(0,z) = -V_0 f z e^{ikf} \int_0^b \frac{e^{ik(\sqrt{z^2+x^2}-\sqrt{f^2+x^2})}}{(z^2+x^2)\sqrt{f^2+x^2}} \left(1 + \frac{i}{k\sqrt{z^2+x^2}}\right) x\, dx$$

Considering that the integrand is an exact differential, we obtain:

$$\Psi(0,z) = \frac{V_0 f}{ik(z-f)} \left\{ \frac{z(z-f)e^{ik(\sqrt{z^2+b^2}-\sqrt{f^2+b^2}+f)}}{\sqrt{z^2+b^2}(\sqrt{z^2+b^2} - \sqrt{f^2+b^2})} - e^{ikz} \right\} \quad (2.3)$$

Let us now determine the range of validity of Equation (2.3). For the boundary conditions to hold, it is required that $d\Psi/dz$ in the limit of $z = 0$ be equal to the vibrational velocity on the transducer surface. Calculation of this limit for Equation (2.3) yields:

$$\left. \frac{\partial \Psi}{\partial z} \right|_{z=0} \approx V_0 \left(1 + \frac{1}{ika} \exp\left(ikf \left(1 + \operatorname{tg}\alpha - \frac{1}{\cos\alpha} \right) \right) \right) \quad (2.4)$$

Equation (2.4) implies that the necessary condition for the validity of distribution (2.3) throughout the acoustic axis is the smallness of the acoustic wavelength compared to the transducer radius, while the similar check of the boundary conditions for the O'Neyl expression shows that the necessary condition is the smallness of the aperture angle of the transducer [69]. A limitation on the transducer size stemming from Equation (2.4) has to do with the geometrical acoustics approximation used to calculate the field in the $z = 0$ plane. If the first term of Equation (2.3) is expanded in a power series of small parameter $|z/f - 1| \ll 1$ in the vicinity of the center of curvature, (2.3) reduces to the expressions derived by O'Neyl [69], Lucas and Muir [71].

In the focal plane ($z = f$) at $kf \gg 1$, Equation (2.1) yields:

$$\Psi(r,f) = \frac{V_0 f^2}{2\pi} \int_0^b \frac{\exp(ik(f - \sqrt{f^2 + x^2}))}{\sqrt{f^2 + x^2}} x\, dx$$

$$\times \int_0^{2\pi} \frac{\exp(ik\sqrt{f^2 + x^2 + r^2 + 2xr\cos\varphi})}{f^2 + x^2 + r^2 + 2xr\cos\varphi} d\varphi$$

2.2 The Field of a Spherical Focusing Transducer with an Arbitrary Aperture Angle

Consider the field at small, compared to the transducer dimensions, distances from the acoustic axis, i.e., at $r \ll f$. Let us expand the radical appearing in the phase multiplier in the integral over φ, restricting the expansion to the terms linear in r/f,

$$\sqrt{f^2 + x^2 + r^2 + 2xr\cos\varphi} \approx \sqrt{f^2 + x^2} + \frac{xr\cos\varphi}{\sqrt{f^2 + x^2}}$$

and disregard the r-dependence of the amplitude multiplier. Taking advantage of the integral representation of the Bessel function, we obtain the expression for the potential in the focal plane identical to the O'Neyl equation derived on the assumption that the aperture angle of the transducer is small.

Thus, as revealed by the calculations, the O'Neyl expressions apply in the vicinity of the center of curvature of the transducer with an arbitrary aperture angle and produce incorrect results away from the center of curvature.

For a hemispherical transducer, it is convenient to use another representation of the solution of the Helmholtz equation [71]

$$\Psi(r, z) = \int_0^\infty \exp\left(iz\sqrt{k^2 - s^2}\right) J_0(rs) s \, ds \int_0^\infty \Psi(x, 0) J_0(sx) x \, dx \quad (2.5)$$

where J_0 is the zero-order Bessel function. It follows from Equation (2.2) that, at $\alpha = \pi/2$, the "shadow" region is missing in the $z = 0$ plane. Sequential calculation of the integrals over x and s [73] in (2.5) yields ahead of the focus ($z < f$);

$$\Psi(r, z) = \frac{V_0 f}{ik} e^{ikf} \left\{ \frac{e^{-ik\sqrt{q^2+r^2}}}{\sqrt{q^2+r^2}} \right.$$

$$- \frac{1}{\sqrt{q^2+r^2}} \sum_{m=0}^{\infty} (-1)^m \varepsilon_m J_{2m}(kr) \left(\frac{r}{\sqrt{q^2+r^2} - q} \right)^{2m}$$

$$\left. - \frac{1}{\sqrt{p^2+r^2}} \sum_{m=0}^{\infty} (-1)^m \varepsilon_m J_{2m}(kr) \left(\frac{r}{\sqrt{p^2+r^2} + p} \right)^{2m} \right\} \quad (2.6a)$$

behind the focus ($z > f$);

$$\Psi(r, z) = \frac{V_0 f}{ik} e^{ikf} \left\{ -\frac{e^{-ik\sqrt{q^2+r^2}}}{\sqrt{q^2+r^2}} \right.$$

$$+ \frac{1}{\sqrt{q^2+r^2}} \sum_{m=0}^{\infty} (-1)^m \varepsilon_m J_{2m}(kr) \left(\frac{\sqrt{q^2+r^2} - q}{r} \right)^{2m}$$

$$\left. + \frac{1}{\sqrt{p^2+r^2}} \sum_{m=0}^{\infty} (-1)^m \varepsilon_m J_{2m}(kr) \left(\frac{\sqrt{p^2+r^2} + p}{r} \right)^{2m} \right\} \quad (2.6b)$$

where $q = z - f$, $p = z + f$, $\sqrt{q^2 + r^2}$ is the distance from the center of curvature, $\varepsilon_m = 1$ at $m = 0$ and $\varepsilon_m = 2$ at $m > 0$, and J_{2m} is the Bessel function.

In Equation (2.6), the field structure is represented in the form of three terms: the first describes the point source placed at the center of curvature of the transducer (geometrical–acoustics approximation), while the second and the third are the corrections for diffraction. In the vicinity of the center of curvature of the transducer, the first and the third terms are commensurate, which results in a finite value of the field at the center of curvature. The third term of Equation (2.6), responsible for the damped cylindrical modes with $s > k$, is much less than the two preceding terms in the vicinity of the center of curvature and becomes commensurate with the second term as the $z = 0$ plane is approached, owing to which the boundary conditions (2.2) in this plane are satisfied.

Inasmuch as the above-derived approximate solution (2.1) and (2.2) provides us with the field distribution at any point, it is possible to check whether the original boundary conditions are met and to calculate the vibrational velocity distribution over all the transducer surface and not just at a single point on the acoustic axis, as in the case of Equation (2.4). Since the vibrational velocity on the surface of a transducer with an arbitrary aperture angle can be found only numerically, let us perform the calculations for a hemispherical transducer ($\alpha = \pi/2$). The normal component of the vibrational velocity on the transducer surface equals

$$V_n(\theta) = \frac{\partial \Psi}{\partial n} = \frac{\partial \Psi}{\partial z} \cos\theta - \frac{\partial \Psi}{\partial r} \sin\theta \tag{2.7}$$

where n is the normal to the transducer surface, θ is the slope of n to the acoustic axis and $\partial\Psi/\partial z$ and $\partial\Psi/\partial r$ are the axial and radial components of the vibrational velocity on the transducer surface. Substituting Ψ given by (2.6a) into Equation (2.7) yields the following expression for the relative deviation $\delta(\theta)$ of the vibrational velocity from the true boundary conditions:

$$\delta(\theta) = \frac{V_n(\theta)}{V_0} - 1 \approx 2i e^{ikf} \sum_{m=0}^{\infty} (-1)^m \left(\operatorname{tg}\frac{\theta}{2}\right)^{2m+1} J_{2m+1}(kf \sin\theta) \tag{2.8}$$

(the terms of the order of $1/(kf)$ are neglected).

Note that, at the edge of the transducer, the boundary conditions fail in so far as $\delta(\pi/2) = \exp(2ikf) - 1$, and the relative deviation becomes of the order of unity in magnitude. At $\theta < \pi/2$, the series converges quickly and therefore it is safe to retain just a few terms of the series in (2.8). In view of the asymptotic behavior of Bessel functions at a large magnitude of the argument, the relative deviation becomes $(kf)^{-1/2}$ at $\theta < \pi/2$.

It is easily shown that the range of angles $\Delta\theta$ within which $\delta(\theta)$ varies from the value of the order of unity at the edge of the transducer to the value of the order of $O(kf)^{-1/2}$ at $\theta < \pi/2$, is also proportional to $(kf)^{-1/2}$. At $kf \gg 1$, the boundary conditions do not hold only in a narrow range of angles $\Delta\theta \approx (kf)^{-1/2}$, the contribution of which can be neglected. It is probably safe to assume that the above estimates are also true for a transducer with an arbitrary aperture angle.

Thus, the field distribution is derived for a spherical transducer with an arbitrary aperture angle and linear dimensions far exceeding the acoustic wavelength in the medium. This solution is valid at any point in space, except for the regions adjacent to the edges of the transducer.

2.3
Analysis of Acoustic Field Spatial Structure with a Spherical Acoustic Transducer

Measurement and visualization of acoustic fields still remain topical problems [74]. Gavrilov and co-authors [75, 76] proposed that a spherical focusing transducer should be used in reconstructing the field structure as a receiver of radiation with a wavelength much shorter than the transducer size. The method was invoked to study the fields of planar and spherical focusing transducers. In [77], the field of a planar transducer was investigated with a receiving lens of an acoustic microscope at a frequency of about 500 MHz.

To justify the method theoretically, the authors of [75, 76] derived the integral expression for the output signal and considered the formation of this signal in acoustic fields of various types. The authors came to conclude that the condition for the feasibility of the method was that the characteristic size of the amplitude (or phase) inhomogeneity of the fields in question is greater than $\lambda/\sin\alpha$ (where λ and α denote the acoustic wavelength and aperture angle of a spherical transducer, respectively).

However, a simple example demonstrates that this condition does not suffice to measure local amplitudes and phases of an arbitrary acoustic field with a spherical transducer. Let a plane wave $P_0 \exp(2\pi i z/\lambda - i\omega t)$ propagate parallel to the acoustic z-axis of the receiver. The output signal V is calculated through direct integration of the pressure over the transducer surface. Its amplitude $|V|$ is actually proportional to the pressure amplitude P_0:

$$|V| = \frac{2\pi f}{k} \left| \exp\left(\frac{2\pi i f}{\lambda}\right) - \exp\left(\frac{2\pi i f \cos\alpha}{\lambda}\right) \right| P_0 \qquad (2.9)$$

with the proportionality factor dependent on the radius of curvature f and the aperture angle of the receiver α as a quickly oscillating function, $|\sin[pf(1-\cos\alpha)/\lambda]|$. Inasmuch as $f/\lambda \gg 1$, this factor, and hence the signal at the receiver, can be reduced to zero by a slight variation of the frequency or transducer parameters. In these conditions, the relation between the output signal and the incident-field parameters becomes unstable.

Contrary to Equation (2.9), it stems from [75, 76] that the factor relating the signal V to the plane-wave amplitude P_0 must be a smooth function of frequency ω and transducer parameters f and α. This disagreement is associated with the approximate expressions used in calculations in [75, 76] (Debye approximation [67]).

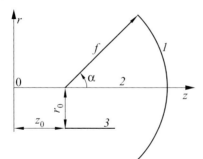

Figure 2.4 Geometry of the system: 1 – receiving transducer, 2 – its acoustic axis, and 3 – radiator axis.

This implies that the above-mentioned approximation is inaccurate for certain types of fields. In this context, it would be pertinent to revert to a theoretical justification of the method in order to establish all types of acoustic fields, the local parameters of which can be measured by the method.

The analysis is performed within the framework of the concept of a spatial spectrum of an acoustic field. Assume that an acoustic wave generated by an arbitrary radiator impinges on a spherical transducer (Figure 2.4). Consideration is restricted to the case of an axisymmetric radiator with the acoustic axis running parallel to that of the radiator and separated from it by r_0. Let us set the cylindrical frame of reference (z, r, φ), such that the z-axis is aligned with the receiving-transducers axis. Incident-field potential $\Psi(z, r, \varphi)$ is represented as a superposition of cylindrical waves $J_0(s\rho)$ with the amplitudes forming a spatial spectrum $q(z, s)$ [78]:

$$\Psi(z, r, \varphi) = \int_0^\infty q(z, s) J_0(s\rho) s \, ds \tag{2.10}$$

where $\rho = \sqrt{r^2 + r_0^2 - 2rr_0 \cos\varphi}$ is the distance from the radiator axis to the observation point with coordinates (z, r, φ), and $J_0(s\rho)$ is the zero-order Bessel function. The voltage at the transducer, V, is expressed in terms of the overall pressure on the transducer surface generated by the incident field

$$V = \iint \Psi(z, r, \varphi) \, dS \tag{2.11}$$

The transducer sensitivity is assumed here to be uniform all over the surface; the proportionality factor in (2.11) is omitted. In the spherical frame of reference centered at the focus of the receiving transducer, the points on its surface are given in the form $z = z_0 + f \cos\theta$, $r = f \sin\theta$ at fixed φ, where z_0 is the coordinate of the receiving-transducer's focus on the z-axis, $0 < \theta < \alpha$ is the slope of the radius vector of a point on the transducer surface with respect to the acoustic axis, and $dS = f^2 \sin\theta \, d\theta \, d\varphi$. The spatial spectrum $q(z, s)$ is expressed in terms of the incident-radiation spectrum $q(z_0, s)$ in the focal plane of the receiver as $q(z, s) = q(z_0, s) \exp(if \cos\theta \sqrt{k^2 - s^2})$, where k is the wave number. Equation (2.10) is subsequently substituted in (2.11). Changing the order of integration and taking into account that the integral over φ equals $2\pi J_0(sr_0)J_0(sf\sin\theta)$ [79, p. 206], we obtain the expression for the output signal V in terms of $q(z_0, s)$:

$$V = \int_0^\infty q(z_0, s) J_0(r_0 s) F(s) s\, ds$$

Function

$$F(s) = 2\pi f^2 \int_0^\alpha \exp\left(if\sqrt{k^2-s^2}\sin\theta\right) J_0(sf\sin\theta) \sin\theta\, d\theta$$

is the function of the transducer sensitivity to various components of the incident radiation.

Signal V is proportional to potential $\Psi(z_0, r_0)$ at the center of curvature of the receiving transducer, only if $F(s)$ is independent of s. In this case, a focusing transducer is similar to a point receiver. However, in reality $F(s) \neq \text{const}$. When analyzing $F(s)$, consideration was restricted to the case of a hemispherical transducer ($\alpha = \pi/2$), which made it possible to derive an analytical expression for $F(s)$ almost without the loss of generality. To within a proportionality factor $\exp(ikf)2pf/(ik)$, $F(s)$ can be represented as a sum of three terms [73]:

$$F(s) = 1 - \Theta(s-k) - \exp(-ikf)\Delta F(s)$$

where $\Theta(s-k)$ is the Heaviside theta-function, and ΔF equals:

$$\Delta F = \begin{cases} J_0(sf) + 2\sum_{n=1}^\infty (-1)^n \left[\dfrac{s}{k+\sqrt{k^2-s^2}}\right]^{2n} J_{2n}(sf) & s < k \\ 2i\sum_{n=1}^\infty (-1)^n \sin\left(n\arcsin\left(\dfrac{k}{s}\right)\right) J_n(s,f) & s > k \end{cases} \quad (2.12)$$

At $s = 0$, $\Delta F(0) = 1$; at $s = k$, $\Delta F(k-0) = \cos(kf)$ and $\Delta F(k+0) = i\sin(kf)$.

The plot of $|F(s)|$ at $kf = 100$ is shown in Figure 2.5. The function $F(s)$ is basically localized in the range $0 < s < k$. Over this interval, small-amplitude oscillations (of the order of $(kf)^{-1/2}$) are superimposed on the average value equal to unity.

The oscillation amplitude, however, becomes commensurate with unity in the vicinity of $s = 0$ and $s = k$ (over the intervals of the order of f^{-1}). At $s > k$, $|F(s)| \approx (kf)^{-1/2} \ll 1$. Thus, $F(s)$ is a rectangular function on which an oscillating perturbation is superimposed. In keeping with the aforesaid, the output signal is given as a sum of three terms, the term equal to the incident-field potential at the center of curvature of the transducer, $\Psi(z_0, r_0)$, and the two corrections, ΔV_1 and ΔV_2:

$$V = \Psi(z_0, r_0) - \Delta V_1 - \exp(-ikf)\Delta V_2$$

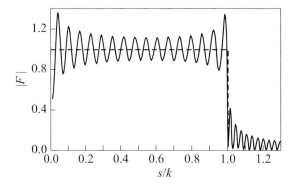

Figure 2.5 The plot of $|F(s)|$ ($kf = 100$, $\alpha = \pi/2$).

where

$$\Delta V_1 = \int_k^\infty q(z_0, s) J_0(r_0 s) s\, ds$$

$$\Delta V_2 = \int_0^\infty q(z_0, s) J_0(r_0 s) \Delta F(s) s\, ds$$

Correction ΔV_1 arises whenever the spectrum involves the components with $s > k$, while ΔV_2 owes its existence to the oscillating part of $F(s)$.

A focusing system can serve as a point receiver if the following inequalities

$$|V_{1,2}| \ll |\Psi(z_0, r_0)| \tag{2.13}$$

imposing certain conditions on the spatial distribution of the field studied are good. Inequality $|\Delta V_1| \ll |\Psi|$ implies that the spatial spectrum of the incident radiation is predominantly localized in the range $s \leq k$. Accordingly, the characteristic size a of the field inhomogeneity in the plane perpendicular to its axis of symmetry must be greater than the acoustic wavelength, $ka \gg 1$.

The criterion for inequality $|\Delta V_2| \ll |\Psi|$ is somewhat harder to formulate. Correction ΔV_2 is associated with a steep increase in the amplitude of oscillations of the sensitivity function $F(s)$ in the vicinity of $s = 0$ and $s = k$. If the spatial spectrum of the incident radiation, $q(z_0, s)$, is predominantly localized in the neighborhood of these points, the correction is of the order of the leading term $\Psi(z_0, r_0)$. For these fields, a focusing system cannot be used as a point receiver.

These limitations are essential for the fields with very smooth amplitude and phase distributions in the transverse plane.

If the characteristic size of the distributions, a, exceeds the focusing-system dimensions (e.g., radius of curvature f), the spatial spectrum is localized in the neighborhood of $s = 0$ within a narrow range of values less than or of the order $1/f$. In this region, $\Delta F \sim 1$, and hence, in this case, $|\Delta V_2| \approx |\Psi|$. This result can be derived analytically. When the spectrum is localized in the vicinity of $s = 0$, the sensitivity function, in accordance with Equation (2.12), is approximated by $\Delta F \approx J_0(sf)$, and correction ΔV_2 takes the form:

$$\Delta V_2 \approx \int_0^\infty q(z_0, s) J_0(fs) J_0(r_0 s) s \, ds \tag{2.14}$$

When the field inhomogeneity far exceeds the radius of curvature of the transducer ($a \gg f$), the expressions become much simpler, and $J_0(sf) \approx 1$ and $\Delta V_2 \approx \Psi(z_0, r_0)$. The signal at the receiving transducer is found to be proportional to the field potential $\Psi(z_0, r_0)$ at the receiving-system focus: $V = [1 - \exp(-ikf)]\Psi(z_0, r_0)$; however, in this case, the proportionality factor represents a quickly oscillating function of the frequency and receiving-transducer parameters. Signal V changes significantly even under a slight variation of the transducer parameters, which renders the results of measurements uncertain. When $a \approx f$, ΔV_2 and $\Psi(z_0, r_0)$ are also commensurate; however, they depend differently on r_0. By virtue of this fact, signal V ceases to be at all proportional to the field value at the receiving-transducer focus.

By way of illustration, let us consider the field of a collimated acoustic beam, the amplitude of which is distributed in the focal plane of a receiving transducer according to the law:

$$\Psi(r) = \frac{1}{1 + r^2/a^2}$$

The spatial spectrum of this distribution expressible in terms of the MacDonald function K_0, $q(s) = a^2 K_0(as)$, exhibits a maximum at $s = 0$ and decays exponentially at $s > 1/a$. If a, characterizing the beam width, exceeds the wavelength ($ka \gg 1$), the spectrum is localized in the vicinity of $s = 0$.

Correction

$$\Delta V_1 = \Psi(r_0)\left[ka J_0(kr_0) K_1(ka) - kr_0 J_1(kr_0) K_0(ka)\right] \tag{2.15}$$

is exponentially small: $\Delta V_1/\Psi \approx \exp(-ka)$. However, correction ΔV_2 can make a tangible contribution to the signal magnitude. Using Equation (2.14) for ΔV_2, the signal at the receiving transducer can be expressed as a function of distance r_0 as

$$V = \frac{a^2}{a^2 + r^2} - \frac{a^2 \exp(-ikf)}{\sqrt{[a^2 + (f + r_0)^2][a^2 + (f - r_0)^2]}} \tag{2.16}$$

At $a \approx f$, both terms of this equation are commensurate; however, they depend differently on r_0. The actual field distribution over the beam cross-section and the dependence of signal V on the location of the receiving-transducer focus with respect to the beam axis are displayed in Figure 2.6 at $a = f$ with kf equal to a multiple of 2π. Correction ΔV_2 in this case is seen to cause a significant difference between the spatial dependence of the recorded signal and the actual amplitude distribution in the beam. Thus, using a focusing receiver to measure the local structure of acoustic fields is worthwhile when the characteristic size of the transverse

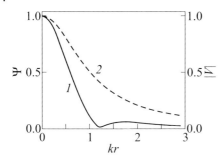

Figure 2.6 The calculated signal at a receiving transducer impacted by the field $\Psi(r) = (1 + r^2/a^2)^{-1}$: 1 – $\Psi(r)$, and 2 – output signal $|V|$.

inhomogeneity of the incident field in the focal plane of the receiver exceeds the wavelength and/or is perceptibly less than the receiving-transducer dimensions.

With the above-formulated criteria in mind, we can now turn to the possibility of using spherical transducers to study the structure of particular types of fields. As intimated in [75, 76], spherical transducers are appropriate for exploring the structure of the focal area of a focusing radiator with the characteristic inhomogeneity size commensurate with the wavelength of the radiation received.

For a transducer with aperture angle β, the spatial radiation spectrum is similar to $F(s)$ as shown in Figure 2.5; however, it is localized in the range between 0 and $k\sin\beta$. This range is much wider than $1/f$, and therefore ΔV_2 is small. Hence, when calculating the output signal, it is justified to use the Debye approximation (angular spectra of the radiator and receiver are represented in the form of bars) and the above-mentioned theoretical analysis of a focusing transducer carried out by Gavrilov and co-workers is also valid.

Let us now consider the possibility of studying the planar-transducer field with a focusing receiver. For a planar transducer mounted on an absolutely rigid screen, the spatial radiation spectrum is available from [80]:

$$q(z,s) = \frac{kaJ_1(sa)}{s} \exp\left(iz\sqrt{k^2 - s^2}\right)\sqrt{k^2 - s^2}$$

The normal component of the vibrational velocity V_z on the transducer surface is assumed here to be equal to ik to ensure that the average value of the potential is normalized to unity. Discussion is restricted to the case where the transducer focus is scanned along the radiator axis. The field potential on the planar-transducer axis equals

$$\Psi(z) = \exp(ikz) - \exp\left(ik\sqrt{z^2 + a^2}\right)$$

the absolute value of the potential in the near zone (at $z < z_n = ka^2/(2\pi)$) is an oscillating function with the amplitude equal to 2; in the far zone ($z \gg z_n$), the potential decreases monotonically, $|\Psi(z)| \approx ka^2/(2z)$. A spherical transducer is appropriate in those regions of the near and far zones, where $|\Delta V_{1,2}| \ll 1$ and $|\Delta V_{1,2}| \ll ka^2/(2z)$, respectively.

Correction ΔV_1 associated with the fact that the spatial radiator spectrum is unbounded equals

$$\Delta V_1(z_0) = -ika \int_k^\infty \frac{\exp(-z_0\sqrt{s^2-k^2})}{\sqrt{s^2-k^2}} J_0(r_0 s) J_1(as)\, ds \qquad (2.17)$$

In the immediate vicinity of the radiator, it is close in magnitude to the potential itself, $\Delta V_1(0) = -i\sin(ka)$ and $|\Delta V_1(0)| \sim |\Psi(0)|$. To estimate ΔV_1 at $z_0 \neq 0$, one needs to switch to variable $t = \sqrt{s^2-k^2}$ in the integrand of Equation (2.17) and develop $J_1(a\sqrt{t^2+k^2})$ as a series in terms of Bessel functions according to the addition theorem. Calculating Equation (2.17), we obtain

$$\Delta V_1(z_0) = -2i \sum_{n=0}^\infty \left[\frac{\sqrt{z_0^2+a^2}-z_0}{a}\right]^{2n+1} (-1)^n J_{2n+1}(ka)$$

Taking into account that the transducer size far exceeds the wavelength, ΔV_1 can be assessed using the asymptotic expansion of $J_{2n+1}(ka)$. Summing up the series with respect to n for each power of the asymptotic expansion, we obtain $\Delta V_1(z_0)$ in the form of the expansion in powers of $\xi = \sqrt{ka}/(kz_0)$, with the leading term equal to $i\xi\sqrt{2/\pi}\sin(ka - \pi/4)$. In the far field, inequality $|\Delta V_1(z_0)| \ll |\Psi(z_0)|$ is good automatically: $|\Delta V_1/\Psi| \sim 1/(ka) \ll 1$; in the near field, this condition is identical to $\xi < 1$ and holds at $z > z_g = \sqrt{a/k}$. The boundary of the region of reliable measurements is much closer to the transducer than the near-zone boundary, $z_g/z_n \sim (ka)^{-3/2} \ll 1$.

Correction ΔV_2 governs the upper bound of z_0 values at which a spherical transducer can serve as a point receiver. At small z_0 values, most of the spatial spectrum is confined in the s range from 0 to $1/a$; for ΔV_2 to be small, this range must exceed $1/f$ and, accordingly, the radiator size a must be less than the receiver size: $a < f$. Suppose that this inequality is good, and $|\Delta V_2(0)| \ll 1$; as z_0 increases, spectrum $q(0,s)$ is multiplied by fast-oscillating factor $\exp(iz\sqrt{k^2-s^2})$. The main contribution to integral (2.14) is produced by interval $\Delta s \sim \sqrt{k/z_0}$ in the vicinity of the stationary-phase point neighboring zero. In the far field at $z > z_n$, this interval becomes narrower than $1/a$, i.e., the spatial spectrum seemingly narrows down.

In actual space, this corresponds to an increase in the characteristic transverse size of the beam. As z_0 increases further, the range Δs, within which most of the spatial spectrum is confined, decreases and can become of the order of or less than $1/f$. Accordingly, inequality $|\Delta V_2| \ll |\Psi|$ ceases to be true. To assess the boundary value of z_0, the spectrum is substituted in Equation (2.14), $J(sf)$ is replaced by its asymptotic expansion at large values of the argument, a is assumed to be $a \ll f$, and $J_1(as)$ is supposed to vary only slightly. Calculations of ΔV_2 by the stationary-phase procedure yield $\Delta V_2 \approx (a/f)J_1(kaf/z)$. In the near zone, $\Delta V_2 \approx (a/f)^{3/2} \ll 1$, while in the far zone, inequality $|\Delta V_2/\Psi| \ll 1$ is good only at $z < z_l = kaf$. Inasmuch as $z_l = 2\pi(f/a)z_n \gg z_n$, the boundary of the region

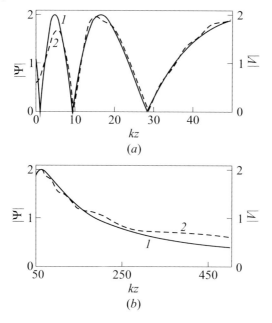

Figure 2.7 The calculated signal at a spherical transducer scanned along the acoustic axis of a planar transducer in the near (a) and far (b) fields: 1 – field distribution $|\Psi|$ on the axis, and 2 – output signal $|V|$ at a receiver.

of reliable measurements of the planar-radiator field with a focusing receiver lies deep in the far zone.

Thus, the planar-radiator field can be studied with the aid of a spherical transducer on the condition that the receiver is much larger than the radiator. The region to be studied must be neither too close to the radiator surface nor too deep in the far zone: $(a/k)^{1/2} \equiv z_g < z < z_l \equiv kaf$.

The foregoing is illustrated below by the results of numerical calculation of the output signal of a focusing transducer ($kf = 50$, $\alpha = \pi/2$) with the center of curvature scanned along the acoustic axis of a planar transducer with $ka = 20$. The distributions of the transducer field potential and the output signal of the spherical receiver in the near and far radiation zones, respectively, are shown in Figure 2.7. These distributions match almost exactly at $kz > 5$ and $kz < 200$ (the near-zone boundary for the planar transducer was $kz_n = 64$).

In closing, brief mention should be made of a transducer with arbitrary aperture angle α. The analytical expression for $F(s)$ in this case is somewhat more intricate; however, the function retains its shape: as previously, it represents a unit step of width $\Delta s = k \sin \alpha$, to which oscillating function $\overline{\Delta F}$ is added. The oscillation amplitude of $\overline{\Delta F}$ is of the order of unity in the vicinity of $s = 0$ and $s = k \sin \alpha$ and of the order of $(kf)^{-1/2}$ at the other s values. Correspondingly, the voltage at the transducer is given by the expression similar to Equation (2.15); however, the lower limit of integral (2.16) for ΔV_1 equals to $k \sin \alpha$. Inequality $|\Delta V_1| \ll |\Psi|$ holds, if the spatial spectrum of the radiation source is primarily confined in the range $\Delta s < k \sin \alpha$ (i.e., $ka \sin \alpha > 1$), while $|\Delta V_1| \ll |\Psi|$ is good, if the range within which most of the spatial spectrum is confined is wider than interval $1/f \sin \alpha$, over which $\overline{\Delta F}$ is of the order of unity (i.e., $a < f \sin \alpha$). Therefore, conditions $ka > 1$ and $a < f$, valid in the case of a hemispherical transducer, are replaced with $ka \sin \alpha > 1$ and $a < f \sin \alpha$ in the case of a transducer with arbitrary aperture angle α.

Thus, it is demonstrated that a spherical transducer is applicable for studying only the acoustic fields with the spatial spectra satisfying conditions (2.13). For these conditions to be good, the characteristic size of the incident-field inhomogeneity must be greater than the acoustic wavelength and less than the linear dimensions of the receiving transducer.

2.4
Experimental Study of the Focal Area Structure of a Transmission Acoustic Microscope

Formation of the output signal of a transmission acoustic microscope under conditions of relative displacement of the lenses was analyzed theoretically in [81, 82]. In particular, it was shown that the signal at the receiving transducer was equal to the field potential generated by the radiating lens at the receiving-lens focus, as long as the receiving-lens aberrations were insignificant. This conclusion is of prime importance when experimentally checking the acoustic field structure in the vicinity of the focus, which constitutes a formidable problem of acoustic microscopy because of the small size of the focal area.

Experimental examination of a confocal system in the transmission acoustic microscope (TAM) mode was performed using a laboratory TAM setup with an operating frequency of 420 MHz. Acoustic lenses with radius of curvature R of approximately 500 μm and an aperture angle of $\theta \approx 45°$ were ground out in the end faces of two single-crystal cylindrical acoustic lines fabricated of Al_2O_3 with the axis parallel to the C_3-axis of the crystals (acoustic lines were 20 mm long and 6 mm across). Mounted on the end faces opposite to the lenses were CdS film piezoelectric transducers 2 mm in diameter. The electric circuit of the microscope allowed for measurement of the amplitude of its output signal within a dynamic-range of 50 dB.

The lens system geometry (mutual arrangement of the acoustic lenses subjected to plane-parallel displacement) was varied using a three-coordinate two-stage lens system positioner with rough mechanical (displacement amplitude ±5 mm and accuracy ±5 μm) and exact piezoelectric (displacement ±20 μm) driving; and the displacement (defocusing) vector ($\bar{r} = (x, y, z)$) components were measured by three mutually perpendicular induction displacement pickups (DT-310, "Messtechnik," FRG) to 0.1 μm over the ±0.5 mm range. The output signal from the pickups, proportional to the displacement of one of the lenses, was applied to the coordinate port of a display.

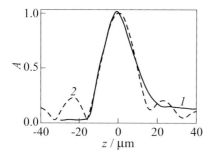

Figure 2.8 Normalized signal from the microscope under conditions of displacement of the lenses along the acoustic axis z (*1*) and the field distribution along the acoustic axis of the radiating lens (*2*).

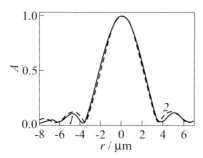

Figure 2.9 Normalized signal from the microscope under conditions of transverse displacement of lenses (*1*) and field distribution in the focal area (*2*).

The one-dimensional dependence of the output signal of the microscope obtained under conditions of displacement of the lenses along acoustic axis z is compared with the theoretical field distribution along the acoustic axis of the radiating lens in Figure 2.8. The curves are seen to be in good agreement within the limits of the principal maximum, which is roughly 20 µm in size. The reasons for the lack of pronounced subordinate maxima are that (i) the actual lens apodization (decrease in the amplitude of peripheral rays compared to paraxial ones) is greater than the one predicted by the theory, and (ii) the refractive lens surface deviates from a perfectly spherical shape. The latter trouble arises from severe technological difficulties in manufacturing such small lenses.

The output signal of the microscope in the case of relative displacement of the lenses in the focal area and the field distribution in the focal area of the radiating lens are shown in Figure 2.9. The theoretical and experimental curves are seen to agree fairly well within the boundaries of the principal maximum, which is approximately 7 µm in size.

Thus it is demonstrated that, under conditions of relative displacement of the lens system of a TAM, its output signal is proportional to the structure of the field formed by the radiating lens. Because of the small size of the focal area and essential inhomogeneity of the amplitude inside it, this structure is almost impossible to measure by other methods.

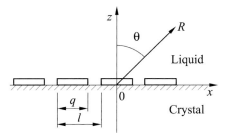

Figure 2.10 Interdigital transducers and the frame of reference.

2.5
Formation of a Focused Beam of Bulk Acoustic Waves by a Planar System of Transducers

This section of the chapter considers a system of interdigital transducers (IDTs) constituting a plane lens of an acoustic microscope. Such a lens furnishes a means of generating a linearly focused beam of bulk waves in an immersion liquid and can be used in much the same way as a well-known cylindrical lens in studying the anisotropy of the mechanical properties of surfaces [83, 84].

Operation of the system is based on the IDT's ability to initiate a bulk wave in the medium adjacent to the surface [85]. Our prime interest here is with bulk waves in an immersion liquid – which is usually water in acoustic microscopy. Before proceeding to focusing systems, it is expedient to calculate the acoustic field of a single IDT in a liquid contiguous with the surface on which the transducer is positioned. We next consider a number of phased IDT systems and calculate the acoustic field structure in the focal area for each IDT.

Location of the transducer at the interface between the media is shown in Figure 2.10. The support occupies the lower half space $z < 0$ and is made of a piezoelectric crystal with density ρ, elastic constant tensor C_{ijkl}, piezoelectric constant tensor e_{ijk} and permittivity ε_{ij}. Half-space $z > 0$ is filled with dielectric liquid having density ρ_{liq}, sound velocity c_{liq}, and permittivity ε_0. Electrodes in the shape of infinitely long bands of width q are applied on the $z = 0$ surface of the piezoelectric with period l parallel to the Y-axis. Electrodes are assumed to be infinitely thin and perfectly conductive.

Initiation of acoustic waves by an IDT at the piezocrystal/liquid interface was covered quite comprehensively by Guzhev and co-authors in [86]. Therefore, it is appropriate here only to present all the necessary mathematics to be used in the subsequent discussion and to note that the calculations are performed within the approximation of low piezoelectric coupling. The support is taken to be the YZ cut of $LiNbO_3$ that is widely used in devices based on surface acoustic waves (SAWs).

In the so chosen frame of reference with the z-axis that is normal to the support surface and the x-axis parallel to the Z-axis of crystallographic orientation of $LiNbO_3$, the equation of motion of the crystal takes the form

2 Acoustic Field Structure in a Lens System of a SAM

$$\rho \frac{\partial^2 U_x}{\partial t^2} = C_{33} \frac{\partial^2 U_x}{\partial x^2} + C_{44} \frac{\partial^2 U_x}{\partial z^2} + (C_{13} + C_{44}) \frac{\partial^2 U_z}{\partial x \partial z} - e_{33} \frac{\partial E_x}{\partial x} - e_{15} \frac{\partial E_z}{\partial z}$$

$$\rho \frac{\partial^2 U_z}{\partial t^2} = C_{11} \frac{\partial^2 U_z}{\partial z^2} + C_{44} \frac{\partial^2 U_z}{\partial x^2} + (C_{13} + C_{44}) \frac{\partial^2 U_x}{\partial x \partial z}$$

$$- e_{22} \frac{\partial E_z}{\partial z} - e_{15} \frac{\partial E_z}{\partial x} - e_{31} \frac{\partial E_z}{\partial z} \quad (2.18)$$

where U is the displacement in the crystal and E is the external electric field. To solve (2.18), one needs to take the Fourier transform for coordinate x. The Fourier components of the electric field, $\bar{E}_x(k, z)$ and $\bar{E}_z(k, z)$, can be written as [87]:

$$\bar{E}_x(k, z) = \bar{E}(k) \exp(m|k|z)$$

$$\bar{E}_z(k, z) = -i \sin(k) m \bar{E}(k) \exp(m|k|z)$$

$$z < 0$$

where

$$\bar{E}(k) = i \frac{\pi}{2K(s')} N F(k) V_0 \exp(i\omega t)$$

$$F(k) = \sum_{n=0}^{\infty} P_n(\cos \eta) \left\{ \frac{\sin[(2n+1+kl/\pi)N\pi]}{(2n+1+kl/\pi)N\pi} - \frac{\sin[(2n+1-kl/\pi)N\pi]}{(2n+1-kl/\pi)N\pi} \right\}$$

Here, N is the number of pairs of electrodes in the transducer, $K(s')$ is the complete elliptic integral of the first kind with supplementary modulus $s' \sqrt{1-s^2}$ as the argument, $s = \sin(\eta/2)$, $\eta = \pi q/2$, $m = \sqrt{\varepsilon_{23}/\varepsilon_{22}}$, $P_n(\cos \eta)$ are the Legendre polynomials, and $V_0 \exp(-i\omega t)$ is the voltage at the transducer. The solution of (2.18) has the form:

$$\bar{U}_x(k, z) = A_1 \exp(\alpha_1 k z) + A_2 \exp(\alpha_2 k z) + \frac{a}{k} \bar{E}(k) \exp(mkz)$$

$$\bar{U}_z(k, z) = i \left[p_1 A_1 \exp(\alpha_1 k z) + p_2 A_2 \exp(\alpha_2 k z) + \frac{b}{k} \bar{E}(k) \exp(mkz) \right] \quad (2.19)$$

where amplitudes A_1 and A_2 of the eigensolutions are unknown. Coefficients for acoustic-vibration penetration into the crystal, α_1 and α_2, are given by expressions

$$\alpha_{1,2} = \left\{ \frac{1}{2C_{11}C_{44}} \left[-B \pm \sqrt{B^2 - 4C_{11}C_{44}\left(\frac{\rho\omega^2}{k^2} - C_{33}\right)\left(\frac{\rho\omega^2}{k^2} - C_{44}\right)} \right] \right\}^{1/2}$$

$$B = C_{11}\left(\frac{\rho\omega^2}{k^2} - C_{33}\right) + C_{44}\left(\frac{\rho\omega^2}{k^2} - C_{44}\right) + (C_{13} + C_{44})^2$$

2.5 Formation of a Focused Beam of Bulk Acoustic Waves by a Planar System of Transducers

where ω is the acoustic-vibration frequency. Coefficients p_i, $i = 1, 2$, govern the trajectories of motion of particles in partial acoustic waves:

$$p_i = \frac{C_{44}\alpha_i^2 + \rho\omega^2/k^2 - C_{33}}{\alpha_i(C_{13} + C_{44})}$$

while coefficients a and b, the efficiency of forced-motion initiation by the electric field are:

$$a = \frac{1}{d}\left[i\left(m^2 C_{11} + \frac{\rho\omega^2}{k^2} - C_{44}\right)(e_{33} - m^2 e_{15})\right.$$

$$\left. + m^2(C_{13} + C_{44})(e_{15} - me_{22} - ie_{31})\right]$$

$$b = \frac{1}{d}\left[\left(m^2 C_{44} + \frac{\rho\omega^2}{k^2} - C_{33}\right)m(e_{15} - me_{22} - ie_{31})\right.$$

$$\left. - im(C_{13} + C_{44})(e_{33} - m^2 e_{15})\right]$$

The liquid occupying half-space $z > 0$ is supposed to be ideal, and its equation of motion, as well as the variable pressure in it, can be written in terms of displacement potential φ such that $u^{\text{liq}} = \text{grad}\, \varphi$:

$$\Delta\varphi + \frac{\omega^2}{c_{\text{liq}}^2}\varphi = 0 \tag{2.20}$$

$$p = -\omega^2 \rho_{\text{liq}} \varphi \tag{2.21}$$

where c_{liq} is the sound velocity in the liquid.

The external electric field fails to initiate any forced motion of the liquid, and therefore, the solution of Equation (2.20) is an acoustic wave with unknown amplitude A_3:

$$\bar{\varphi}(k, z) = \frac{A_3}{k}\exp\left(-z\sqrt{k^2 - \frac{\omega^2}{c_{\text{liq}}^2}}\right) \tag{2.22}$$

Motion in the crystal and liquid is consistent with the boundary conditions at $z = 0$, namely the continuity of the normal displacements and stresses and the lack of tangential stresses on the crystal surface. Taking into account Equations (2.19), (2.21), and (2.22), the boundary conditions can be written as:

$$ip_1 A_1 + ip_2 A_2 + \sqrt{1 - \frac{\omega^2}{k^2 c_{\text{liq}}^2}} A_3 = -mb \frac{\overline{E}(k)}{k}$$

$$(C_{13} + C_{11}\alpha_1 p_1) A_1 + (C_{13} + C_{11}\alpha_2 p_2) A_2 - \frac{i\rho_{\text{liq}}\omega^2}{k^2} A_3$$
$$= -(C_{13}a + mC_{11}b) \frac{\overline{E}(k)}{k} \qquad (2.23)$$

$$(\alpha_1 - p_1) A_1 + (\alpha_2 - p_2) A_2 = m(b - a) \frac{\overline{E}(k)}{k}$$

Unknown amplitudes A_i ($i = 1, 2, 3$) are to be derived from (2.23) using a routine procedure for solving sets of inhomogeneous linear equations:

$$A_i = \frac{\Delta_i}{\Delta} \frac{\overline{E}(k)}{k}$$

where

$$\Delta = \begin{vmatrix} ip_1 & ip_2 & \sqrt{1 - \frac{\omega^2}{k^2 c_{\text{liq}}^2}} \\ C_{13} + C_{11}\alpha_1 p_1 & C_{13} + C_{11}\alpha_2 p_2 & -\frac{i\rho_{\text{liq}}\omega^2}{k^2} \\ \alpha_1 - p_1 & \alpha_2 - p_2 & 0 \end{vmatrix} \qquad (2.24)$$

is the principal determinant of (2.23), and of all Δ_i, Δ_3 is the most interesting, because it determines wave amplitude A_3 in the liquid:

$$\Delta_3 = \begin{vmatrix} ip_1 & ip_2 & -mb \\ C_{13} + C_{11}\alpha_1 p_1 & C_{13} + C_{11}\alpha_2 p_2 & -(C_{13}a + mC_{11}b) \\ \alpha_1 - p_1 & \alpha_2 - p_2 & m(b - a) \end{vmatrix}$$

Taking the inverse Fourier transform, we obtain the displacement potential in the liquid:

$$\varphi(x, z, t) = \frac{1}{2\pi} \int_{-\infty}^{\infty} \frac{\Delta_3}{\Delta} \frac{\overline{E}(k)}{k} \exp\left(-z \sqrt{k^2 - \frac{\omega^2}{c_{\text{liq}}^2}} + ikx - i\omega t\right) dk \qquad (2.25)$$

To evaluate integral (2.25), we switch from coordinates (x, z) to polar coordinates (R, θ): $x = R \sin\theta$, $z = R \cos\theta$. In terms of the new coordinates, (2.25) takes the form:

$$\varphi(R, \theta) = \int_{-\infty}^{\infty} \beta(k) \exp\bigl(i R \gamma(k)\bigr) dk \qquad (2.26)$$

where

$$\beta(k) = \frac{\Delta_3}{\Delta} \frac{\overline{E}(k)}{2\pi k}$$

$$\gamma(k) = \sqrt{k_{\text{liq}}^2 - k^2} \cos\theta - k \sin\theta$$

$$k_{\text{liq}} = \frac{\omega}{c_{\text{liq}}}$$

2.5 Formation of a Focused Beam of Bulk Acoustic Waves by a Planar System of Transducers

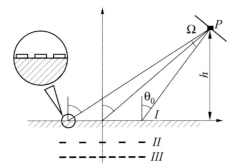

Figure 2.11 Phased IDT system: P – focus, $h = 100\lambda$, $\theta_0 = 45°$, $\Omega = 17°$ (I – 3 IDTs, II – 5 IDTs, III – 9 IDTs).

(multiplier $\exp(-i\omega t)$ is omitted). At a fairly long distance from the transducer, i.e., at large R values, integral (2.26) can be estimated by the method of stationary phase. The stationarity condition $d\gamma(k)/dk = 0$ is met at $k = k_0 = \pm k \sin\theta$. The fact that k_0 can assume two values is associated with the fact that, by virtue of symmetry, the transducer radiates bulk waves on either side of the vertical z-axis. Since at both k_0 values the governing equations are identical, consideration here is restricted to the case of $k_0 = k_{\text{liq}} \sin\theta$ corresponding to positive θ. At this value, $\gamma(k_0) = k_0$ and $\gamma''(k_0) = -(k_{\text{liq}} \cos^2\theta)^{-1}$. It is worth noting that function $\beta(k)$ also incorporates an oscillating multiplier: $\beta(k) = D(k)\sin(klN)/(klN)$, where $D(k)$ is a slowly varying function of k. At distances R far exceeding transducer size $2lN$ ($R\gamma(k) \gg klN$), $\sin(klN)$ varies slowly compared to $\exp[iR\gamma(k)]$ and therefore it can be omitted from the phase multiplier. The resulting expression for integral (2.26) is:

$$\varphi(R,\theta) = \sqrt{\frac{2\pi k_{\text{liq}} \cos^2\theta}{k}} \beta(k_0) \exp\left(ik_{\text{liq}}R - \frac{i\pi}{4}\right)$$

$$= \sqrt{\frac{1}{2\pi k_{\text{liq}} R}} \operatorname{tg}\theta \left.\frac{\Delta_3}{\Delta}\right|_{k_0} \overline{E}(k_0) \exp\left(ik_{\text{liq}}R - \frac{i\pi}{4}\right) \quad (2.27)$$

The last cofactor in the foregoing expression governs the wave phase, and $\overline{E}(k_0)$ the radiation pattern. The principal lobe of the radiation pattern has a maximum at an angle of $\theta = \arcsin(c_{\text{liq}}/(2lf))$ (f is the frequency) and changes the slope, as the period l of the IDT electrodes is varied.

We now proceed to the analysis of phased systems incorporating several transducers. Consider a system of three transducers (Figure 2.11, system I) arranged so that the directions of the maxima of their radiation patterns intersect at a chosen point (focus) P, and the waves radiated by IDT add together in phase. For definiteness, the radiation-maximum angle of the central transducer is taken equal to $45°$ and the distance from the focus at point P to the interface between the media – to $h = 100\lambda$ (λ is the wavelength in the liquid). The phasing condition is met if the distances R_i, $i = -1, 0, 1$ from the centers of the transducers to the focus differ by a whole number of wavelengths. Assume that $R_{-1} - R_0 = R_0 - R_1 = 20\lambda$.

2 Acoustic Field Structure in a Lens System of a SAM

Table 2.1 Coordinates of IDT centers a_i along the x-axis for the water–lithium niobate interface ($f = 100$ MHz and $\lambda = 15$ μm).

a_i	−26.71	−20.28	−13.70	−6.95	9.0	7.28	14.73	22.65	31.13
System I	a_{-1}				a_0				a_1
System II	a_{-2}		a_{-1}		a_0		a_1		a_2
System III	a_{-4}	a_{-3}	a_{-2}	a_{-1}	a_0	a_1	a_2	a_3	a_4

Table 2.2 Periods l_i of the electrodes in the transducers for the water–lithium niobate interface ($f = 100$ MHz and $\lambda = 15$ μm).

l_i, μm	19.1	19.5	19.9	20.5	21.2	22.1	23.1	24.5	26.4
System I	l_{-1}				l_0				l_1
System II	l_{-2}		l_{-1}		l_0		l_1		l_2
System III	l_{-4}	l_{-3}	l_{-2}	l_{-1}	l_0	l_1	l_2	l_3	l_4

With these assumptions, the doubled aperture angle of the system, i.e., the angle accommodating the IDT system, as viewed from point P, equals $\Omega = 17°$.

Placing one more transducer between the neighboring IDTs of system I, so that $R_{i-1} - R_i = 10\lambda$, we obtain a phased system of 5 IDTs (Figure 2.11, system II). System II, complemented in the same manner, evolves into a system of 9 IDTs, such that $R_{i-1} - R_i = 5\lambda$ (Figure 2.11, system III). The periods l of the electrodes in the transducers are chosen to comply with the condition for intersection of the directions of IDT radiation maxima at the focal point P.

The acoustic pressure in the liquid produced by an IDT system is represented as a sum of the pressures from individual transducers, and taking into account Equation (2.21), we obtain:

$$p = \sum_i p_i = -\omega^2 \rho_{\text{liq}} \sum_i \varphi_i \qquad (2.28)$$

Thus, the acoustic pressure of the beam generated by the system can be calculated at any point in the liquid using Equations (2.27) and (2.28). The pressure in the focal area of systems I, II and III depicted in Figure 2.11 was calculated numerically. When performing the calculations, the parameters of the ith transducer in (2.24) were set equal to $R_i = \sqrt{(x-a_i)^2 + z}$, $\sin\theta_i = (x-a_i)R_i$, and $\cos\theta_i = z/R_i$ (where a_i is the coordinate of the center of the ith transducer along the x-axis), based on the geometry of the problem. The values of a_i and l_i used in calculations at frequency $f = 100$ MHz and $\lambda = 15$ μm are listed in Tables 2.1 and 2.2.

The distributions of the pressure modulus $|p|$ along the $z = 200\lambda - x$ straight line perpendicular to R_0 are shown in Figure 2.12 for systems I, II, and III of transducers having three electrodes each. As is obvious from the plots, the system

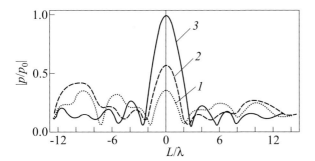

Figure 2.12 Distribution of acoustic pressure modulus $|p/p_0|$ along the $z = 200\lambda - x$ line: 1 – 3 IDTs, 2 – 5 IDTs, 3 – 9 IDTs; L is the distance from focus P and p_0 is the pressure at the focal point of the system of 9 IDTs.

of three transducers is actually multifocal, and in the system of nine IDTs, the principal maximum is pronounced and exceeds the first subordinate maximum by 12.5 dB.

As the transducers are added to the system, the separation between the first minima governing the focal-waist size slightly increases: from 4λ to 6λ. The systems in question are long-focus: the focus dimension in the longitudinal direction along the $z = x$ line at a level of 3 dB is approximately 40λ for system *III*.

These structures are similar in phasing principle to a Fresnel lens, and the possibility of displacing the focal waist by varying the frequency still exists for them, as was demonstrated experimentally in [85]. They, however, differ from an ordinary Fresnel lens [88] in that each of their transducers radiates directional waves, so that almost all the radiation energy is concentrated within the confines of the principal lobe, which makes it possible to achieve a considerable gain in the pressure at the focal point.

The focus structure can be somewhat improved by narrowing the IDT radiation patterns, which, in turn, is accomplished by increasing the number of electrodes in each transducer. The pressure distributions along the $z = 200\lambda - x$ line for the system of nine IDTs with $N = 1.5, 2.5$, and 3.5 are illustrated in Figure 2.13. The ratio of the pressures at the principal and first subordinate maxima increases and equals 4.43, 4.58, and 4.80, respectively; the pressure decays faster with distance from the focus in a system incorporating more electrodes. It should, however, be remembered that an increase in the number of electrodes is limited by the separation between the transducers.

Thus, a phased IDT system allows the generation of a focused beam of acoustic waves in a liquid. Considering the availability of photolithographic means of engraving surface electrode structures, there can be little doubt that such systems can compete with a cylindrical lens in studies of materials with an acoustic microscope, and the above-described calculations can serve as a theoretical basis for the analysis of these systems.

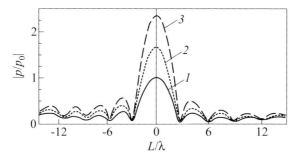

Figure 2.13 Distribution of acoustic pressure modulus $|p/p_0|$ along the $z = 200\lambda - x$ line for the system of 9 IDTs (*III*) at different numbers of electrodes in the transducers: $1 - N = 1.5$; $2 - N = 2.5$; and $3 - N = 3.5$; L is the distance from focus P and p_0 is the pressure at the focal point of the system with $N = 1.5$.

2.6
About the Possibility of Using Scholte–Stoneley Waves for Surface Waves' Acoustic Microscopy

Visualizing surface flaws with the largest characteristic dimension perpendicular to the surface (i.e., narrow and deep cracks), thin liquid monolayers, etc., by means of scanning acoustic microscopy presents certain problems. It has been suggested that, in the aforesaid and some other cases, SAM is operated in the mode of surface acoustic waves. Theoretical and experimental principles of realization of SAW-based devices intended for visualization and characterization of objects are set forth in [89–99].

Of much interest is the SAM operation mode associated with the possibility of generating and receiving surface leaky Rayleigh waves at the object–immersion liquid interface. Information gained under these operating conditions provides an insight into the elastic properties of the test object surface, given the characteristics of the immersion liquid, which is assumed isotropic and homogeneous. However, the situation is conceivable where the object of investigation is a liquid or a quasi-liquid medium, and a crystal (which is generally solid) serves as a sound duct with known mechanical characteristics. In this case, it is also possible to use leaky Rayleigh waves, inasmuch as their velocity and absorption depend on the liquid medium properties.

However, acoustic-energy dissipation in a pseudo-Rayleigh wave is insignificant compared to re-emission losses into the liquid. Therefore, pseudo-Rayleigh waves are not very sensitive to a variation in the liquid parameters.

Stoneley–Scholte surface waves appear to hold greater promise for SAW-based acoustic microscopy because their absorption is sensitive to the mechanical structure and properties of liquid or quasi-liquid media (i.e., to systems with a low shear strength). Displacement amplitudes of particles in pseudo-Rayleigh (left) and

2.6 About the Possibility of Using Scholte–Stoneley Waves in SAW Microscopy

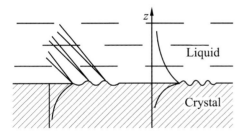

Figure 2.14 Displacement amplitudes of particles in pseudo-Rayleigh (left) and Stoneley (right) waves at the liquid–solid interface.

Stoneley–Scholte (right) waves at the solid/liquid interface are illustrated schematically in Figure 2.14.

If liquid and solid differ dramatically in density and elastic constants, i.e., if $\rho/\rho_{liq} \ll 1$ and $(c_{liq}/c)^2 \ll 1$, the velocity of SAW in question closely approximates the longitudinal wave velocity in liquid:

$$c_{St} = c_{liq}(1 - \beta) = c_{liq}\left[1 - \frac{1}{8}\left(\frac{\rho_{liq}}{\rho} \frac{c_{liq}^2}{c_L^2 - c_T^2}\right)\right] \quad (2.29)$$

where c_{liq} is the longitudinal wave velocity in liquid; and c_L and c_T are the longitudinal and transverse wave velocities in solid. Wave localization depth in the interfacial layers is expressed by the following equations:

- liquid:

$$z_{loc}^{liq} = \frac{1}{\sqrt{k^2 - k_{liq}^2}} = \frac{\lambda_{liq}}{2\pi\sqrt{2\beta + \beta^2}}$$

- solid:

$$z_{loc}^{sol} = \frac{1}{\sqrt{k^2 - k_T^2}} = \frac{\lambda_{liq}}{2\pi\sqrt{(1+\beta)^2 - c_{liq}^2/c_T^2}}$$

where β is the coefficient from Equation (2.29) and k is the wave number.

By virtue of the structure of Stoneley–Scholte waves (Figure 2.14), their absorption is controlled by energy dissipation in the crystal and liquid. Inasmuch as internal friction effects are insignificant in crystals, wave absorption is totally governed by the viscoelastic characteristics of the liquid. As revealed by the theoretical analysis, wave absorption changes with the ratio between the sound velocity in the liquid and the transverse wave velocity in the crystal. Surface wave energy dissipation is governed by two mechanisms: (i) absorption of the longitudinal component of SAW in the liquid and (ii) generation of a viscous shear wave in the liquid in the vicinity of the interface. The calculated dependence of the absorption coefficient α on the velocity c_T at fixed values of the other parameters is illustrated in Figure 2.15. Water

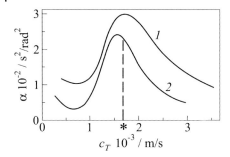

Figure 2.15 Absorption coefficient α of a Stoneley wave vs. transverse wave velocity c_T in a solid: 1 — $\rho_s = 4.5\,\text{g cm}^{-3}$ and 2 — $\rho_s = 9\,\text{g cm}^{-3}$; * corresponds to $c_{\text{liq}} \sim \tau_R$.

was selected for the liquid medium. At velocities $|c_T - c_{\text{liq}}| \ll c_{\text{liq}}$, i.e., when the rate of shear closely approximates the sound velocity in water, $c_{\text{liq}} = 1.5 \times 10^5\,\text{cm s}^{-1}$, there is a pronounced absorption peak. In this velocity range, the dependence of decrement α on frequency f involves a radical; α is governed solely by the shear viscosity of the liquid η and is independent of its bulk viscosity ξ: $\alpha \sim \eta^{1/2} f^{3/2}$. This is true for a water–$\text{Bi}_{12}\text{GeO}_{20}$ system. Theoretical estimates are in good agreement with the measured coefficient of Stoneley–Scholte wave absorption in this system [86].

When the velocities in liquid and solid differ dramatically ($|c_T - c_{\text{liq}}| \sim c_T$), absorption of Stoneley–Scholte waves primarily depends on dissipation of the longitudinal component of SAW in water. The decrement appears to be proportional to the squared frequency and a combination of the shear (η) and bulk (ξ) viscosities: $\alpha \sim (7/6\eta + \xi) f^2$. This is true, when Stoneley–Scholte waves propagate along the interface between water and crystals like lithium niobate or quartz. The measured absorption coefficients for these pairs agree fairly well with their theoretical counterparts.

From the standpoint of using Stoneley–Scholte SAWs in acoustic microscopy, both absorption modes hold much interest, because shear viscosity measured in the former case is quite sensitive to structural modifications, while absorption measured in the latter case furnishes information about a thicker interfacial layer. Note that a prerequisite for realization of one or another mode is the proper choice of a crystal for a sound duct.

Measurement of the absorption of Stoneley–Scholte waves can be an efficient method of acoustic characterization of shear viscosity of liquids and biological tissues in the frequency range from 10 to 100 MHz. In particular, modification of the method allows layer-by-layer measurement of the shear frequency of a surface wave. This possibility stems from the wave-frequency dependence on the depth of Stoneley–Scholte wave penetration into a liquid.

In order to develop SAW-based SAM facilities and procedures, one has first to tackle the question of generation and reception of Stoneley–Scholte SAWs.

The possibility of efficient initiation of high-frequency Stoneley–Scholte waves in piezoelectric crystals and their reception by interdigital transducers, was demonstrated experimentally. The interface between water and crystals of bismuth germanate ($\text{Bi}_{12}\text{GeO}_{20}$) and lithium niobate ($\text{LiNbO}_3$) with three identical IDTs on each was chosen to study Stoneley–Scholte SAW initiation and to determine

2.6 About the Possibility of Using Scholte–Stoneley Waves in SAW Microscopy

Table 2.3 Amplitudes and propagation velocities of various excitation waves at the interface between water and a (001) 100 cut $Bi_{12}GeO_{20}$ crystal.

Propagation velocity 10^5 cm s^{-1}	Amplitude arb. units
$c_{St} = 1.478$	1
$c_R = 1.633 + 0.03i$	0.11
$c_{liq} = 1.495$	0.17
$c_T = 1.68$	0.008
$c_L = 2.75$	0.002

the propagation characteristics. In a water–bismuth germanate system, Stoneley–Scholte SAW generation was observed over a frequency span from 15 to 30 MHz.

The measured losses to double transformation were 30 dB for an IDT with 18 pairs of electrodes and 20 dB for an IDT with 28 pairs [86, 100].

As revealed by calculations of the initiation process for a water–bismuth germanate system, the optimum conditions are achieved with a transducer having 28 pairs of electrodes. In addition to a Stoneley–Scholte wave, the IDT also initiates a leaky Rayleigh wave and bulk waves in the adjacent media. A leaky wave decays by an exponential law, as it propagates along the interface, while a longitudinal wave in water, as well as longitudinal and shear waves in the crystal, decays by the power law, $(kx)^{-3/2}$, where k is the wave vector and x is the coordinate along the direction of wave propagation. Wave amplitudes calculated at a distance of $10\lambda_{liq}$ (λ_{liq} is the sound wavelength in liquid) from the transducer, are summarized in Table 2.3. The estimates presented suggest that the bulk of acoustic power is radiated by the transducer in the form of Stoneley–Scholte waves.

Stoneley–Scholte wave velocities measured by the group delay time method in both systems were found to be close to the longitudinal sound velocity in water. The measured velocity in a water–bismuth germanate system, equal to 1.478×10^5 cm s^{-1}, is identical to the theoretical estimate. In a water–LiNbO$_3$ system, the theoretical estimate of the difference between the Stoneley–Scholte wave velocity and the sound velocity in liquid comes to only 10^{-5}, which is above the accuracy of measurement.

Absorption of a Stoneley–Scholte wave was measured by the pulsed method using a set of three transducers (one for initiation and two for reception). Absorption of a wave propagating along the water–$Bi_{12}GeO_{20}$ interface at a frequency of 22.9 MHz turned out to be 4.5 dB cm^{-1}, which was four times that of longitudinal sound in water. Absorption of a Stoneley–Scholte wave in a water–LiNbO$_3$ system is equal 3 dB cm^{-1} and is close to the longitudinal wave absorption in water at the same frequency.

The results obtained were accounted for within the framework of the theoretical analysis of Stoneley–Scholte wave absorption in a liquid/solid system. Inasmuch as the internal friction effects in crystals are insignificant, Stoneley–Scholte wave absorption depends on the liquid viscosity. Dissipation of the surface wave energy is governed by two mechanisms; namely, by absorption of the longitudinal wave component and by formation of a viscous shear wave in a liquid in the vicinity of the interface.

The pattern of surface wave absorption changes with the ratio between the sound velocity in a liquid c_{liq} and the shear wave velocity in a crystal in a given direction (c_T). If the transverse wave velocity is less than or commensurate with the sound velocity in a liquid, $c_T \leq c_{\text{liq}}$, the absorption is totally governed by energy dissipation in a viscous shear wave (abnormal absorption of Stoneley–Scholte waves). At these values of the elastic parameters of the media, decrement α depends solely on the shear viscosity of liquid (η), is independent of its bulk viscosity (ξ), and incorporates a square-root of f:

$$\alpha \sim \eta^{1/2} f^{3/2}$$

The decrement heavily depends on the ratio between sound velocity c_{liq} and Rayleigh velocity c_R: at $c_{\text{liq}} \sim c_R$, there is a clearly defined absorption peak (Figure 2.15). The absorption maximum is associated with the Stoneley–Scholte wave structure. In the case of $c_{\text{liq}} > c_R$, the energy flux is predominantly localized in a solid and depends on the amplitudes of longitudinal and transverse waves in a solid, while dissipation takes place in a liquid where the amplitude of a viscous shear wave grows, as c_R approaches c_{liq}. Accordingly, the dissipated energy also increases.

To the right of the maximum the Rayleigh wave velocity exceeds the sound velocity in liquid and the energy flux in a Stoneley–Scholte wave is localized inside the liquid layer and depends on the amplitude of the longitudinal wave component.

The energy dissipation rate is predetermined by the amplitude of a viscous shear wave. Wave amplitudes being fixed, the depth of longitudinal-component penetration, as well as the integral energy flux, increases with c_R. As a consequence, decrement α decreases at $c_R > c_{\text{liq}}$ with increasing c_R (Figure 2.15).

Abnormal absorption of Stoneley–Scholte waves was studied in the water–$Bi_{12}GeO_{20}$ system ($c_{\text{liq}} = 1.49 \times 10^5$ cm s^{-1} and $c_R = 1.68 \times 10^5$ cm s^{-1}), because theoretical estimates obtained for this system within the framework of the above-developed concepts check well with the measurements.

When the velocities in liquid and solid differ dramatically, absorption of Stoneley–Scholte waves is primarily governed by dissipation of the longitudinal SAW component in water. The decrement was found to be proportional to the squared frequency f and a combination of the shear (η) and bulk (ξ) viscosities,

$$\alpha \sim \left(\frac{7}{6}\eta + \xi\right) f^2$$

Such is the case when Stoneley–Scholte waves propagate along the interface between water and crystals like lithium niobate or quartz. The measured absorption coefficients for these systems closely approximate their theoretical counterparts.

Thus, to develop a surface acoustic microscope, it is justified to use the totality of facilities and procedures devised for the excitation and reception of SAWs on

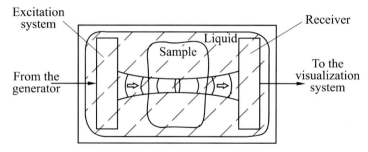

Figure 2.16 Surface acoustic microscope.

the free surface of crystals. A surface acoustic microscope can be based on the principles of both A- and B-scanning and, in either case, it is necessary to generate a focused beam of Stoneley–Scholte waves. To solve the problem of SAW focusing, it is common practice to use ring IDTs and loop-type multivoid couplers [101], as well as phased transducer arrays [102], which offer a means of controlling the focal spot position. A schematic sketch of a surface acoustic microscope is illustrated in Figure 2.16. As in the case of a conventional SAM, a focused SAW travels through the test object, and the signal arriving at the receiver carries information about absorption and variation of the wave phase in the part of the object corresponding to the focus location.

Electron scanning with a focal spot along the interface plane provides orderly information about viscoelastic characteristics of test objects which eventually enters a visualization system.

3
Output Signal Formation in a Transmission Raster Acoustic Microscope

3.1
Outline of the Problem

The nature of acoustic contrast characterizing the correlation between the amplitude and phase of the output signal in an acoustic microscope and local values of the acoustic sample parameters (density, elasticity, viscosity) is the central problem of acoustic microscopy [103, 104]. There are two approaches to analyzing formation of the output signal of an acoustic microscope. In the first [59, 105–107], a converging beam is considered as a set of plane waves and its interaction with a sample is described, employing the methods of Fourier optics [78]. The output signal is expressed as an integral function of the reflection coefficient at the immersion liquid – sample interface (reflection regime) or of the coefficient of acoustic wave transmission through the sample (transmission regime). The reflection and transmission coefficients carry the basic information about acoustic sample properties. This approach makes it possible to construct a rigorous quantitative theory of output signal formation for objects with flat boundaries [108]. Because of mathematical difficulties, application of this approach is generally confined to numerical calculations [59, 60, 105–112].

A ray approach based on geometrical acoustics [60, 113–117] seems to be more descriptive. A focused beam is considered, in this approach, as a set of rays interacting with the object. Incident rays are reflected and refracted at the object boundaries in a normal way; some waves incident on the interface at certain angles induce side waves in the sample [41], leaky surface waves [118], or waveguide modes (of the Sezawa or Lamb type), and so on [119–122]. These waves are re-emitted in the immersion liquid as they propagate over the sample surface and take part in formation of the output signal of a microscope. Formation of the output signal in reflection SAM has been discussed in many papers [59, 60, 105, 106, 108–115, 117]. Studies of this problem for transmission microscopes are very scarce. Formation of the output signal in a microscope was considered solely for an infinitesimally thin object with no account taken of the realistic expression for the transmission coefficient [105]. Accordingly, mechanisms of the acoustic contrast in the transmission operation mode have not been analyzed. Only a few factors governing the contrast can be obtained based on general reasoning: namely, reflection at the object–liquid interface, attenuation of the ultrasound wave in the sample, aberrations caused by refraction effects at the sample boundaries, re-reflection in the sample and resonance effects, existence of leaky waveguide modes and their participation in formation of the output signal, and so on. Development of notions of the output signal formation in SAM serves also as a theoretical basis for devising techniques for measurements of local physical-mechanical characteristics of the objects studied.

The present chapter analyzes theoretical models of formation of the output signal and discusses quantitative methods for measurements by transmission acoustic microscopy. A general formula for the output signal is derived, principles of signal formation in samples with a small shear modulus are formulated, and a new technique for measuring local elastic properties and acoustic wave absorption in such samples is proposed which uses the dependence of the output microscope signal A on the distance z between the receiving and transmitting lenses.

3.2
Transmission Acoustic Microscope: Formation of the Output Signal as a Function of Local Properties of Flat Objects. General Concepts

The first SAM made by Lemons and Quate [10] was realized in the transmission mode. Since that time, various independent groups, including ours, have designed, built and used in their research, transmission raster acoustic microscopes (TRAM) for many years. Research results obtained using those systems show serious advantages compared with the reflection mode when investigating biological samples, polymer materials, or thin films of various types [66, 104]. The TRAM proves to be effective for investigation of multilayer heterogenous films, for example, multilayer photo films, films with special electromagnetic properties like magnetic films, films with high-temperature superconductive properties in particular, etc. Another indisputable advantage of TRAM is the visualization of inner structures of optically opaque objects.

It may also be reasonable to point out that there is a serious difference between transmission and reflection microscopes, related to the principles of output acoustic signal formation. Reflection microscopes enable us to investigate the surface structures of highly absorbing samples or samples distorting acoustic waves; while transmission microscopes make it possible to study samples having a rough surface. Using the amplitude and phase of the signal, TRAM is capable of determining, separately, elastic and viscous properties of the object under study [104]. This makes it possible to carry out quantitative measurements of these properties. The complementary character of the features and application fields of reflection and transmission acoustic microscopes suggests that instruments operating in these modes can fulfil the contemporary scientific and technical demands.

A focused beam traversing a thin plate is refracted twice and the rays escape the plate parallel to the directions of the initial rays, though shifted with respect to them (see Figure 1.4). The shift value depends on the beam incidence angle.

As a result, rays incident at various angles are focused at different points of the acoustic axis; hence, the focus of a converging beam is shifted and smeared after it passes through the plate. The cone of rays recorded by the receiving transducer narrows due to its smearing; therefore, the output signal level decreases. The magnitude of variation of the output signal caused by refraction is controlled by the ratio between the speed of sound in the immersion liquid and in the sample. This effect is one of the sources of acoustic contrast in transmission microscopy.

Two more physical phenomena brought about by the interaction of a focused beam with a plate affect the magnitude of the output signal in a transmission microscope: these are reflection at the sample boundaries and acoustic wave attenuation within the sample. The contribution of reflection to signal variations depends on the difference between the acoustic impedances of the immersion liquid and sample; the effect of wave attenuation is controlled by the acoustic energy dissipated in the sample and by the presence and efficiency of scattering processes.

Variations in the output signal are dictated by the integral reflected, absorbed, or dispersed energy of all waves incorporated in the acoustic beam and incident on the sample surface at various angles. The contribution of any factor (refraction, reflection, and attenuation) to output signal formation and, hence, to the acoustic contrast, depends on the physical properties of the sample and is to be analyzed individually for each class of objects. We concentrate our attention mostly on samples possessing a low shear modulus (biological samples, polymer films, etc.) because they offer a possibility of developing a simple physical model and their study by transmission acoustic microscopy is of great practical importance.

The effects associated with refraction of a focused beam can be used to measure local viscoelastic properties of samples.

We suggested a new technique for such measurements that uses the dependence of the output signal A of a microscope on the distance z between the receiving and transmitting lenses. The technique rests on the above-mentioned shift of the focus of a converging beam due to its refraction (see Figure 1.4). With no object, the output signal in the lens system of a transmission microscope is at a maximum when the lens focuses are identical: the signal drops, executing regular diffraction oscillations, as the lenses are displaced. If an object in the form of a plane-parallel plate is placed in the way of a focused beam, rays incident on the plate at different angles θ are focused, after passing the plate, on the acoustic axis at a point shifted with respect of the initial position of the emitting-lens focus through distance Δz:

$$\Delta z = d \left(1 - \frac{\operatorname{tg} \alpha}{\operatorname{tg} \theta} \right) \tag{3.1}$$

where α is the refraction angle in the object. At small angles θ, tangent can be replaced by sine (paraxial approximation), leaving Δz independent of θ:

$$\Delta z_{\text{par}} = d\left(1 - \frac{c_L}{c}\right) \tag{3.2}$$

Here, c_L is the longitudinal sound velocity in the object, and c is the speed of sound in the immersion liquid. Therefore, one can assume that the angular aperture of a sound beam is small, the transmitted beam is focused at a new position, and the shift of the focus is specified by formula (3.2) [64, 123].

A Δz value assessed by the $A(z)$ curve can be used to evaluate the local speed of sound in the sample by formula (3.2), and the maximums of the transmission coefficient can be calculated from the measured ratio of $A(z)$. A local value of the ultrasound absorption in the sample is estimated using the transmission coefficient and known reflection coefficient. The feasibility of the measurement of acoustic velocities and absorption by the above technique was proved in experiments with polymer films, sections of collagen fibers, etc. [124].

However, the aperture of beams used in acoustic microscopes is usually wide, so that the beams experience significant aberration upon traversing the studied object. The points where refracted rays incident at various angles θ converge are positioned at different sites of the acoustic axis. The length of the segment covered by these points is comparable with their shift Δz. Therefore, the interpretation of experimental $A(z)$ curves calls for a more sophisticated theory of formation of the output acoustic signal and its dependence on the distance between the lenses. To realize this program, a general expression for the output signal A of the microscope was derived, which in the integral form related the output signal to the transmittance through the plate T.

3.3
General Representation of the Output Signal of the Transmission Acoustic Microscope

The formation of the output signal of an acoustic microscope can conveniently be analyzed using notions of the spatial radiation spectrum borrowed from Fourier optics [78]. We introduce a frame of reference with the origin at point O_1 which is the focus of the emitting lens. The field generated by the emitting lens in plane $z = 0$ is expanded in a two-dimensional (2D) Fourier integral and presented in the form of a spatial spectrum $U_1(\vec{k})$ which represents the amplitude of a planar wave with a wave vector

$$\vec{k} = \left(k_x, k_y, \left(\frac{\omega^2}{c^2} - k_x^2 - k_y^2\right)^{1/2}\right)$$

where ω is the acoustic wave frequency.

When acoustic radiation propagates in a homogeneous material through a distance z, the spectrum is transformed due to the phase multiplier $\exp(ik_z z)$ and when it traverses a plane-parallel object of thickness d, its variation is governed by the transmission coefficient multiplier $T(k_x, k_y, d)$. Therefore, the spatial spectrum $U(k_x, k_y, z)$ in the focal plane of the receiving lens shifted through distance z with respect to the focal plane of the emitting lens reads:

$$U(k_x, k_y, z) = T(k_x, k_y, d) U(k_x, k_y) \exp\bigl(ik_z(z - d)\bigr) \qquad (3.3)$$

The receiving lens is impinged by a set of plane waves with various k_x and k_y; each of these waves induces a signal at the receiving transducer. The signal phase depends on the position of the receiving lens with respect to the emitting lens. We introduce a sensitivity function of the receiving lens $U_2(k_x, k_y)$. It is a signal induced on the receiving transducer by a planar wave with a wave vector $\vec{k} = (k_x, k_y, k_z)$, unit amplitude, and phase equal to zero at the focus of the receiving lens. If the axis of the receiving lens is shifted relative to the axis of the emitting lens by vector $\vec{\rho} = (\rho_x, \rho_y, 0)$, the signal generated by the receiving transducer equals $U_2(k_x, k_y) \exp(i\vec{k}\vec{\rho})$. With the use of these relations, we write the microscope output signal as a superposition of signals produced by individual waves contained in the spatial radiation spectrum:

$$A(\vec{\rho}, z, d) = \iint_\infty dk_x dk_y\, U_1(k_x, k_y) T(k_x, k_y, d) \exp\bigl(ik_z(z-d) + i\vec{k}\vec{\rho}\bigr) \qquad (3.4)$$

This depends both on the mutual arrangement of lenses (shifts in their focal planes and axes ρ) and on the thickness d and acoustic properties of the object. If there is no shift in the acoustic axes ($\vec{\rho} = 0$) and the sample is either isotropic or is axially symmetric with respect to the z-axis, 2D integration with respect to k_x and k_y in Equation (3.4) is replaced by one-dimensional integration with respect to the angle θ between wave vector $\vec{k}$ and axis z:

$$A(z, d) = \int d\theta\, U_1(\theta) U_2(\theta) T(\theta, d) \exp\bigl(ik(z-d)\cos\theta\bigr) \sin\theta \qquad (3.5)$$

where $k = \omega/c$. When calculating the integral in Equation (3.5), one can confine oneself to positive values θ and neglect the contribution of inhomogeneous waves (waves with imaginary values of k_z) to the output signal of the microscope [125]. The form of functions $U_1(\theta)$ and $U_2(\theta)$ has been repeatedly discussed in the literature [59, 107]; it depends on the characteristics of the lens system and is independent of the sample properties. The quantity $P(\theta) = U_1(\theta) U_2(\theta)$ is usually called the aperture function of a microscope. Following [107], we use the simplest approximation of $P(\theta)$. Let us assume that the field produced by the focusing lens is a set of planar waves lying within the angular lens aperture θ:

$$P(\theta) = \begin{cases} B & \text{if } \theta \leq \theta_m \\ 0 & \text{if } \theta > \theta_m \end{cases} \qquad (3.6)$$

where B is a constant. Under this assumption, the output signal is controlled by the form of the angular dependence of the transmission coefficient $T(\theta, d)$ within the angular aperture θ_m of the lens system:

$$A(z, d) = B \int_0^{\theta_m} T(\theta, d) \exp(ik(z - d)\cos\theta) \sin\theta \, d\theta \tag{3.7}$$

The angular dependence of the transmission coefficient $T(\theta, d)$ is normally very complicated because of possible resonance effects. We restrict our consideration to a simple case of objects with a small shear modulus and longitudinal acoustic impedance, which is close to that of the immersion liquid. It turns out that, for such objects, one can develop a consistent quantitative theory of formation of the output signal A and of its dependence on z, $A(z)$.

3.4
Formation of the $A(z)$ Dependence for Objects with a Small Shear Modulus

Above we considered samples in which the velocity of shear waves c_T is appreciably lower than the velocity of longitudinal sound waves. This is true for such materials as biological tissues, polymeric and biopolymeric systems [64, 123, 124], and so on. Inasmuch as the absorption coefficient of sound is inversely proportional to the appropriate propagation velocity cubed, one can expect considerable absorption of shear waves α_T, which appreciably exceeds the absorption of longitudinal waves α_L. Therefore, the contribution of shear waves to the transmission coefficient can be neglected under the assumption that $c_L/c_T \gg 1$ and $\alpha_T d \gg 1$ [41]. If we additionally assume that $\alpha_L d \ll 1$, the following expression can be used for the transmission coefficient $T(\theta, d)$ [126]:

$$T(\theta, \varphi(\theta)) = i\tau \left(\frac{1}{\mathrm{ctg}\,\delta - i\tau} + \frac{1}{\mathrm{tg}\,\delta + i\tau} \right) \tag{3.8}$$

where $\tau = (\rho_l c/\cos\theta)/(\rho_T c_L/\cos\alpha)$ is the ratio of the longitudinal acoustic impedances, $\varphi(\theta) = \omega d \cos\alpha/(2c_L)$ is the phase delay of a wave caused by its traversing the plate, α is the refraction angle, ρ_l and ρ_T are the densities of the immersion liquid and sample, and ω is the ultrasound frequency. Furthermore, we assume that densities and velocities of sound in contacting media have similar values: $\rho_l \approx \rho_T$ and $c \approx c_L$, so that $\cos\alpha \approx \cos\theta$, and τ is close to unity. It is this case that is realized in biological samples and many polymeric systems, provided water is used as an immersion liquid. Then, as to the accuracy of $(\tau^2 - 1)/(2\tau) \ll 1$, the transmission coefficient equals the phase multiplier that corresponds to the appropriate change in phase of the incident wave travelling through the object studied:

$$T(\theta, d) = \exp(2i\delta) = \exp\left(ikd \sqrt{\frac{c^2}{c_L^2} - \sin^2\theta} \right) \tag{3.9}$$

where function $A(z)$ is expressed by the integral of an exponential function with an exponent that varies in a complicated way:

$$A(z,d) = B \int_0^{\theta_m} \exp(i\Psi(\theta, z, d)) \sin\theta \, d\theta \qquad (3.10)$$

Here, $\Psi(\theta, z, d) = kd\sqrt{c^2/c_L^2 - \sin^2\theta} + k(z-d)\cos\theta$. We derive the explicit $A(z)$ dependence for an important limiting case when the aperture θ_m of the acoustic lens is so small that $\cos\theta$ is close to unity. Expansion of phase $\Psi(\theta)$ in (3.10) in terms of the small parameter $x = 1 - \cos\theta$, yields

$$\Psi(\theta, z, d) \approx \Psi_0 + k\left(d\left(1 - \frac{c_L}{c}\right) - z\right)x + \frac{1}{2}kd\left(\frac{c_L^2}{c^2} - 1\right)x^2 \frac{c_L}{c} \qquad (3.11)$$

where $\Psi_0 = k[z - d(1 - c/c_L)]$. At fairly small apertures – angles θ_m up to $\theta_m \approx 30°-40°$ ($x_m = 1 - \cos\theta \approx 0.13-0.3$) – only a few initial terms of the expansion can be retained. The number of retained terms depends on the smallness of the aperture and the thickness of the object. Only the first term can be retained in the expansion when the lens aperture $\theta_m \leq 15°-20°$ and $x_m < 0.1$. In this case,

$$A(z,d) = Bx_m \exp(i\Psi_0 + i\Gamma)\frac{\sin\Gamma}{\Gamma} \qquad (3.12)$$

where $\Gamma = (kd/2)(1 - c_L/c - z/d)x_m$.

Most often, the only recorded parameter is the amplitude of the output microscope signal $|A(z)|$; in the case considered, it reads:

$$|A| = Bx_m \left|\frac{\sin(\pi x_m(d/\lambda)(1 - c_L/c - z/d))}{\pi x_m(d/\lambda)(1 - c_L/c - z/d)}\right| \qquad (3.13)$$

It is similar to the $|A(z)|$ function in the immersion liquid with no object; however, it is shifted along the z-axis. The $|A(z)|$ maximum is quite prominent and the periodicity of the minimums and lateral maximums depends on the ultrasound wavelength λ in the immersion liquid and lens aperture θ_m. The principal maximum is shifted with respect to the $|A(z)|$ maximum in the absence of an object by a value

$$\Delta z_{par} = d\left(1 - \frac{c_L}{c}\right) \qquad (3.14)$$

The above results precisely correspond to the paraxial approximation considered previously, based on the ray notions. If ultrasound attenuation in the sample is ignored, the value of the principal maximum is independent of the sample thickness d (no signal attenuation due to aberrations occurs in the paraxial approximation).

3 Output Signal Formation in a Transmission Raster Acoustic Microscope

At small apertures, the paraxial approximation depicts the $A(z)$ dependence for plates, the thickness of which satisfies the condition

$$d \ll \frac{\lambda}{2}\left(1 - \frac{c_L}{c}\right) x_m^2$$

In this case, plates can be of an appreciable thickness. Thus, at $x_m \sim 0.1$ ($\theta_m \sim 25°$) and $(c_L/c - 1) \sim 0.2$, Equation (3.13) adequately depicts the behavior of function $A(z)$ up to thickness d values of the order of a few tens of wavelengths. However, the basic quantity to be measured; namely, the shift of the principal maximum, is assessed in the paraxial approximation with a systematic error of the order of x_m. More precise determination of the dependence between the position of the principal maximum and acoustic properties of the object studied calls for taking into account the next expansion term in expansion (3.11) quadratic in x, particularly for lenses with angular apertures $\theta_m \sim 20°-40°$.

In Equation (3.11) retaining the terms linear and quadratic in x allows the $A(z)$ function to be expressed via the Fresnel integral [127]:

$$A(z) = B \exp\left(\frac{i\tilde{\Psi}_0}{b}\right) \int_{a_1}^{a_1+bx_m} \exp\left(\pm \frac{i\pi s^2}{2}\right) ds \qquad (3.15)$$

where $a_1 = a(1 - c_L/c - z/d)$ with

$$a = \sqrt{\frac{kdc}{\pi c_L(1 - c_L^2/c^2)}} \sin\left(1 - \frac{c_L}{c}\right)$$

$$\tilde{\Psi}_0 = \Psi_0 - \frac{0.5kd(1 - c_L/c - z/d)}{c_L(1 - c_L^2/c^2)/c}$$

$$b = \sqrt{\frac{kd(1 - c_L^2/c^2)c_L}{c/\pi}}$$

The plus or minus sign in the integrand exponential function is taken when the $(1 - c_L/c)$ value is positive or negative, respectively.

The use of the quadratic approximation for the phase imposes quite a mild restriction on the object thickness:

$$d \ll \frac{\lambda}{2x_m^3|c_L/c - 1|} \qquad (3.16)$$

Thus, at $x_m \sim 0.24$ ($\theta_m \sim 40°$) and $|c_L/c - 1| \sim 0.2$, the thickness of a sample for which the aforesaid approximation holds must be less than 160λ, and at $x_m \sim 0.1$ ($\theta_m \sim 25°$) it must be much less than 2500λ.

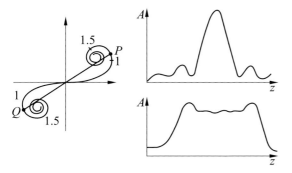

Figure 3.1 Analysis of the behavior of $|A(z)|$ dependence with the aid of the Cornu spiral.

As a function of z, the output signal amplitude is expressed through sine and cosine Fresnel integrals

$$C(y) = \int_0^y \cos\left(\frac{\pi t^2}{2}\right) dt \quad \text{and} \quad S(y) = \int_0^y \sin\left(\frac{\pi t^2}{2}\right) dt$$

of arguments $y_P(z) = a_1 + bx_m$ and $y_Q(z) = a_1$:

$$|A(z,d)| = \frac{|B|}{b}\left\{[C(y_P) - C(y_Q)]^2 + [S(y_P) - S(y_Q)]^2\right\}^{1/2} \qquad (3.17)$$

It is convenient to analyze the behavior of the $|A(z)|$ dependence with the aid of geometric plots based on the Cornu spiral (Figure 3.1). Plotting (with account taken of their signs) the values of arguments $y_P(z)$ and $y_Q(z)$, we obtain points P and Q, respectively. The PQ segment length is calculated as a square-root of the sum of the Fresnel integrals squared and, hence, equals the signal amplitude $|A|$ normalized to quantity $|B|/b$. The distance between points P and Q along the Cornu spiral $bx_m = \sqrt{kd(1 - c_L^2/c^2)c_L/c/\pi}(1 - \cos\theta_m)$ is independent of coordinate x. Only the position of the PQ segment, rather than its length, changes as z varies.

Function $|A(z)|$ is symmetric with respect to the PQ value, for which points P and Q are positioned at equal distances from the beginning of the Cornu spiral on its opposite sides [125]. This condition yields

$$z_0 = d\left\{\left(1 - \frac{c_L}{c}\right) + \left(1 - \frac{c_L^2}{c^2}\right)(1 - \cos\theta_m)\frac{c_L}{2c}\right\} \qquad (3.18)$$

that is, the position of the symmetry point of curve $|A(z)|$ depends on the plate thickness d and ratio c_L/c of the speeds of sound in the object and immersion liquid.

Below, we consider in more detail the form of curve $|A(z)|$ at various values of thickness d of the plate studied. At a moderate thickness value, when $bx_m < 2.44$ and, hence,

$$d < \frac{3\lambda}{x_m^2|1 - c_L/c|} \tag{3.19}$$

the PQ segment length is at a maximum when the points P and Q are positioned symmetrically with respect to the spiral beginning (Figure 3.1). As z deviates from z_0, the distance between the ends of segment PQ remains constant; however, the segment length diminishes, oscillating periodically as PQ coils on the right-hand spiral branch. As a result, if the sample thickness meets condition (3.19), the $|A(z)|$ curve behaves similarly to the graph of function $|\sin(z - z_0)/(z - z_0)|$: it exhibits a prominent principal maximum located at the symmetry point $z = z_0$ with side maximums decaying fairly rapidly. Thus, the $A(z)$ dependence for such samples, as in the case when paraxial approximation is used, is similar to a diffraction curve, but with a location of the principal maximum corrected according to formula (3.18). The $|A(z)|$ minimums for those samples are nonzero; as d increases so do their values, while the principal maximum value, which is independent of the sample thickness at its very small values, starts decreasing with increasing z and the maximum itself widens.

When samples are thicker, that is, when $d \gg 3\lambda/(x_m^2|1-c_L/c|)$, inequality (3.16) still holds and the approximation quadratic in x for phase $\Psi(x, z)$ provides quite good accuracy. The length of the Cornu spiral between points P and Q is so long that one of its end portions is heavily wound on one of its coils. The output signal is the highest at z values for which one of the PQ ends is wound on one of the coils while the other is either located between the spiral coils or is wound on the opposite coil. Accordingly, a wide zone in which the output signal is fairly great arises on the $|A(z)|$ curve on both sides of point $z = z_0$. The boundaries of this zone z_1 and z_2 are found from the condition that either the left-hand end of segment PQ finds itself in the beginning of the right-hand Cornu spiral coil: $y_Q(z_1) \approx 2$, or its right-hand end is positioned at the beginning of the left-hand coil: $y_P(z_2) \approx -2$. The width of this zone is evaluated by equation

$$\Delta z = z_2 - z_1 \approx (bx_m + 4)\frac{d}{a} \approx d\left(1 - \frac{c_L^2}{c^2}\right)(1 - \cos\theta_m)\frac{c_L}{c}$$

and approximately equals the distance between the points where the paraxial and extreme rays of the focused beam merge. This distance is calculated by formula (3.1). The output signal amplitude in this zone experiences shallow oscillations. Beyond this zone, the signal drops rapidly. $A(z)$ curves of this type are observed solely when the samples are very thick. Thus, at $\theta_m \sim 20°$, $x_m \sim 0.06$, and $|c_L/c - 1| \sim 0.2$, the sample thickness needed significantly exceeds 4000λ. Examining such a thick sample is unrealistic: first, because of purely construction peculiarities of transmission microscopes and, second, due to the intense absorption of the focused beam in the sample.

3.4 Formation of the A(z) Dependence for Objects with a Small Shear Modulus

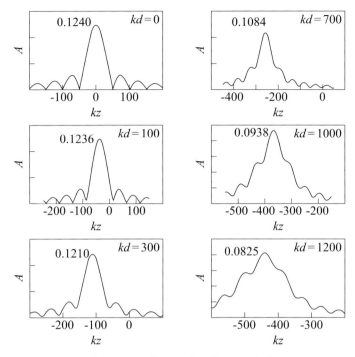

Figure 3.2 $A(kz)$ dependences for samples of various thickness.

The behavior of $A(z)$ curves discussed above, which follows from graphical interpretation based on Cornu spirals, is confirmed by our numerical calculations.

Figure 3.2 and, in more detail, Figures 6.5 and 6.6 (in Chapter 6) display the results of numerical calculations of $A(z)$ curves for samples of various thickness. The graphs show that the shift in the principal curve maximum increases linearly with the sample thickness d. However, inasmuch as the $A(z)$ method and data furnished by it seem to be quite informative, we intend to develop future methods for the efficient processing of the data in order to achieve high accuracy of the quantitative information about the properties of various materials.

4
Quantitative Acoustic Microscopy Based on Lateral Mechanical Scanning

4.1
Methods of Quantitative Ultrasonic Microscopy with Mechanical Scanning: Review

Several acoustic microscopy methods have been developed for quantitative investigation of a flat specimen immersed in a liquid. In most of these methods, the output voltage $V(z)$ of the focused transducer is recorded as a function of the distance between the focus and the surface of the specimen [60, 61, 128–133]. The phase velocity and propagation attenuation of Leaky Surface Acoustic Waves (LSAW), as well as a reflectance function for the specimen–water interface, can be obtained from the recorded $V(z)$ data.

A point-focus-beam acoustic lens was used in the first implementation of the $V(z)$ technique for an isotropic specimens study [60]. For characterization of anisotropic materials, a line-focus-beam acoustic lens was proposed [61]. A Lamb wave lens and directional lenses with noncircular shaped transducers [130] were developed to obtain enhanced sensitivity for particular ranges of incidence or orientation angles. Also, various electrical exciting waveforms and processing electronics were employed for $V(z)$ data acquisition. The conventional quantitative scanning acoustic microscopy (QSAM) system works in a tone burst mode and the amplitude of $V(z)$ is only used for analysis. The amplitude and the phase of the output voltage was recorded to reconstruct the reflectance function by the Fourier transform of the complex $V(z)$ [128]. A continuous-wave Doppler system produces a complex $V(z)$ for a single, particular frequency [131]. Conversely, the $V(z, t)$ waveform acquired in the pulse mode represents the properties of the specimen over a wide frequency band [132, 133].

With all these techniques, the accuracy of the measured LSAW parameters and the angular resolution of the reconstruction of the reflectance function increase with an increasing amount of $V(z)$ data [128, 134]. The maximal defocusing distance is limited in the $V(z)$ configuration by the geometry of the acoustic lens. Furthermore, the wave velocity in water is used as reference in the $V(z)$ schemes. Because of the temperature dependence of the wave velocity in water, the error in the temperature measurement significantly affects the accuracy of the measured velocity and attenuation of LSAW [129, 135, 136].

Acoustic Microscopy. Roman Gr. Maev
Copyright © 2008 WILEY-VCH Verlag GmbH & Co. KGaA, Weinheim
ISBN: 978-3-527-40744-6

Several ultrasonic systems employing two transducers have also been developed for quantitative material characterization. In an ultrasonic microspectrometer, spherical-planar-pair lenses were used for measurement of the reflection coefficient over a wide frequency range [137, 138]. An angular spectrum of the reflected wave is detected in this system by tilting the spherical-planar-pair lenses as a unit. The size of the planar transducer is large enough to provide sufficient angular selectivity and, because of geometrical restrictions, the measurements cannot be carried out at small and large angles of incidence.

A QSAM system employing separated transmitting and receiving point-focus transducers has been developed to study anisotropic propagation of LSAW. During the experiments [139, 140], the foci were located on the surface of the specimen at a fixed distance, and the specimen was rotated to obtain the group velocity of surface waves as a function of the propagation angle. In systems [141–144], the focus of the transmitter was located on the liquid–solid interface, but the recording of the scattered acoustic field was carried out by 2D scanning of the receiving transducer. The recorded spatial distribution represented the angular spectrum of the scattered wave associated with the reflection or transmission coefficients. Thus, using this technique, the angular resolution was restricted by the spatial resolution of the receiver and the distance from the point source to the plane of the data acquisition.

Lateral scanning of the receiving transducer along the surface of a specimen was used for Lamb wave measurement and study of the properties of the materials in plate form [145–147]. A 2D Fourier transformation of the acquired time-spatial data represented the wavenumber dispersion curves. Because of the high directivity of the transducers employed in the experiments, the single scan data correspond to a narrow range of angles of incidence. To obtain dispersion curves for the entire angular range, the data acquisition and processing should be repeated many times for different angular orientations of the transducers.

The QSAM technique, based on lateral scanning of the receiving transducer, can be considered as an alternative to the $V(z)$ method. In this chapter, a theory of the $V(x, t)$ formation and inversion is presented, and angular resolution and temperature stability of the $V(x, t)$ and $V(z, t)$ schemes are compared. The theory is illustrated by experimental results obtained using a time-resolved line-focused system with lateral scanning of the receiving transducer [148, 149].

4.2
Ray Models of $V(z)$ and $V(x)$ QSAM Systems

In the $V(z)$ system (Figure 4.1a), the LSAW is generated by a ray incident on the liquid–solid interface at the critical angle θ_R. The surface wave propagates along the interface re-radiating ("leaking") back to the liquid at the same angle θ_R. If the transducer is moved toward the specimen, one of the re-radiated rays (ray R) is effectively received by the transducer. Only a leaky wave whose critical angle is less

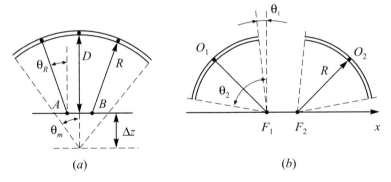

Figure 4.1 Ray models of the $V(z)$ (a) and $V(x)$ (b) systems.

than the half-aperture angle of the transducer $\theta_R < \theta_m$ can be excited and detected in this scheme.

The time delay Δt between the responses to the ray R and the directly reflected ray D is related to the velocity of the LSAW, c_R [150, 151]:

$$c_R = \left[\frac{\Delta t}{c \Delta z} - \frac{1}{4}\left(\frac{\Delta t}{\Delta z}\right)^2 \right]^{-0.5}$$

where c is the sound velocity in the liquid, and Δz is the defocusing distance. Obviously, the accuracy of the LSAW measurement increases with increasing Δz. The maximum value of Δz is limited by the focal distance F of the lens and the half-aperture angle θ_m: $\Delta z < F\cos(\theta_m)$. Usually, the maximum value of F is limited by the sound attenuation in the liquid. On the other hand, it is possible to obtain better accuracy by decreasing θ_m, but this is not desirable because of the reduction in the critical angle range.

The measured values of the velocity and attenuation of the LSAW depend on the velocity and attenuation of the ultrasound in the liquid because the propagation distances in the liquid of rays R and D change with the position of the transducer along the z-axis. The fractional error in the LSAW velocity can be expressed for tone burst [129] and pulse [152] modes by:

$$\frac{\delta c_R}{c_R} = \frac{1 - \cos\theta_R}{\sin^2\theta_R} \frac{\delta c}{c}$$

The fractional error in the velocity in the liquid, $\delta c/c$, is determined mostly by the temperature measurement error δT. We define η_z, the temperature coefficient of the velocity measurement error:

$$\eta_z = \frac{\delta c_R}{c_R \delta T} = \frac{1 - \cos\theta_R}{\sin^2\theta_R} \frac{\delta c}{c \delta T} \tag{4.1}$$

For the typical case of water at 20 °C as the immersion liquid and $c_R \approx$ 3000 m s^{-1}, the coefficient can be estimated to be $\eta_z \approx 10^{-3}$ (°C)$^{-1}$.

In the $V(x)$ scheme, the tilted transmitting and receiving transducers are used in a pitch-catch arrangement, and the receiver is translated along the interface in the x direction (Figure 4.1b). The velocity of the nondispersive LSAW is simply the ratio of the travel distance Δx and the relative time delay Δt of the leaky wave R:

$$c_R = \frac{\Delta x}{\Delta t} \tag{4.2}$$

Because of the geometry of the $V(x)$ system, the range of angles of incidence (θ_1, θ_2) can be made close to the values $(0, \pi/2)$, and there is no restriction on the maximum scanning distance Δx in the direction of the LSAW propagation.

For an ideal uniform temperature distribution in the immersion liquid, the time of flight in the liquid $\Delta t^* = (O_1 F_1 + F_2 O_2)/c$, is constant during data acquisition. Also, Δt^* remains constant if the temperature distribution along the rays $O_1 F_1$ and $F_2 O_2$ does not change during the scan of the receiving transducer. Under this sufficient condition, the relative time delay Δt is only associated with the LSAW velocity, and the temperature coefficient η_x of the c_R measurement error is negligible. It is difficult to estimate theoretically the value of η_x, but the results of the experimental study of the temperature stability of the $V(x)$ scheme will be presented below.

4.3
Wave Theory of $V(z)$ and $V(x)$ QSAM Systems

Let us consider an arbitrary ultrasonic system with two transducers (Figure 4.2). They are the transmitting (T) and receiving (R) transducers, and the flat laterally homogeneous specimen is immersed into the coupling liquid. The focus of the transmitting transducer is at the origin of the Cartesian coordinate system (x, y, z), and the surface of the specimen is aligned in the plane $z = z_0$. The focus of the receiving transducer is located in the plane $z = z_1$. Let us determine the output voltage of the system as a function of time t, the position of the receiving transducer (x, y, z_1), and the position of the specimen z_0.

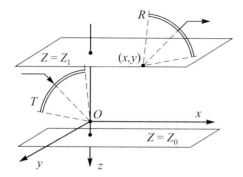

Figure 4.2 Configuration of an arbitrary QSAM system; T and R are the transmitting and receiving transducers, respectively. The specimen–liquid interface is aligned in the plane $z = z_0$.

Let $h_1(x, y, t)$ be a scalar field of the probing wave generated by the transmitting transducer in the focal plane $z = 0$. The spectrum $H_1(k_x, k_y, \omega)$ of this wave field is determined by the three-dimensional (3D) Fourier transform:

$$H_1(k_x, k_y, \omega)_1 = \iiint_{-\infty}^{\infty} h_1(x, y, t) \exp(j(\omega t - k_x x - k_y y)) \, dx \, dy \, dt$$

Here, ω is the frequency and (k_x, k_y, k_z) are the components of the wave vector. We suppose that the transducer function $H_1 = 0$ for $k_x^2 + k_y^2 > k^2$, where $k = \omega/c$ is the wavenumber. This wave propagates towards the specimen, and the reflected wave propagates back to the receiving transducer. Because of their travel in the liquid, the spectral components acquire phase shifts which may be expressed by the propagation function $\exp(jk_z(2z_0 - z_1))$ [153]. Also, let $R(k_x, k_y, \omega)$ be a reflectance function of the liquid–specimen interface. Thus, the spectrum of the reflected wave in the focal plane of the receiving transducer $z = z_1$ can be written as follows:

$$H_r(k_x, k_y, \omega) = H_1(k_x, k_y, \omega) R(k_x, k_y, \omega) \exp(jk_z(2z_0 - z_1)) \tag{4.3}$$

The acoustic field $h_r(x, y, t)$ associated with the spectrum (4.3) can be obtained by the inverse Fourier transform:

$$h_r(x, y, t) = F^{-1}_{k_x, k_y, \omega}[H_r(k_x, k_y, \omega)]$$

We define the impulse response of the receiver $h_2(x, y, t)$ as the output voltage of the transducer R produced by a point source $\delta(x, y, z_1, t)$, where $\delta(\,)$ is the Dirac delta function. Consequently, the transfer function of the receiver $H_2(k_x, k_y, \omega)$ can be obtained by the Fourier transform:

$$H_2(k_x, k_y, \omega) = F_{x,y,t}[h_2(x, y, t)]$$

Assuming the linearity and the spatio-temporal invariance of the receiving scanning system, the output voltage $V(x, y, t, z_0, z_1)$, recorded as a function of the scanning coordinates of the receiving transducer (x, y) and time t, can be expressed as a convolution over the variables x, y, t:

$$V(x, y, t, z_0, z_1) = h_r(x, y, t) \otimes h_2(x, y, t)$$

According to the convolution theorem, the output data can be written as:

$$V(x, y, t, z_0, z_1) = F^{-1}_{k_x, k_y, \omega}[H_1(k_x, k_y, \omega) H_2(k_x, k_y, \omega) R(k_x, k_y, \omega)$$
$$\times \exp(jk_z(2z_0 - z_1))] \tag{4.4}$$

Let us now consider a system with two line-focused transducers whose foci are parallel to the y-axis. Assuming the ideal directionality of the lens system in the x direction, we can define the transfer function $H_x(k_x, \omega)$ of the whole system:

$$H_1(k_x, k_y, \omega) H_2(k_x, k_y, \omega) = H_x(k_x, \omega)(2\pi)\delta(k_y)$$

In this case, Equation (4.4) can be simplified, and by letting $R(k_x, \omega) = R(k_x, 0, \omega)$, we have for the 2D output $V(x, t)$:

$$V(x, t) = F^{-1}_{k_x, \omega}\left[H_x(k_x, \omega) R(k_x, \omega) \exp\left(j\sqrt{k^2 - k_x^2}(2z_0 - z_1)\right)\right] \quad (4.5)$$

If the foci of the transducers are located at the surface of the specimen by adjustment of the vertical positions of the specimen and receiving transducer, $z_0 = 0$ and $z_1 = 0$, the inconvenient phase factor $\exp(j\sqrt{k^2 - k_x^2}(2z_0 - z_1))$ in (4.5) equals unity. Finally, we find that the output $V(x, t)$ and the product of the reflectance coefficient R and system transfer function H are related by the 2D Fourier transform:

$$S_x(k_x, \omega) = F_{x,t}[V(x, t)] = H_x(k_x, \omega) R(k_x, \omega) \quad (4.6)$$

To compare the $V(x, t)$ and $V(z, t)$ formation, let us also consider a measurement configuration with one transducer. In this case, it is necessary to let $x = y = 0$ and $z_1 = 0$ in Equation (4.4). By changing the variables, we obtain:

$$V(z_0, t) = (2\pi)^{-3} \iiint\limits_{-\infty}^{\infty} H_1(k_z, \varphi, \omega) H_2(k_z, \varphi, \omega) R(k_z, \varphi, \omega)$$

$$\times \exp(j(k_z 2z_0 - \omega t)) k_z \, dk_z \, d\varphi \, d\omega$$

where φ is the azimuth angle. Assuming that the transducer has the line focus oriented along the y-axis, we define the system transfer function H_z:

$$H_1(k_z, \varphi, \omega) H_2(k_z, \varphi, \omega) k_z = H_z(k_z, \omega)(2\pi)\delta(\varphi)$$

Denoting $z = 2z_0$ and $R(k_z, \omega) = R(k_z, 0, \omega)$, the spectrum of the output voltage can be expressed as a function of time t and scanning coordinate z:

$$S_z(k_z, \omega) = F_{z,t}[V(z, t)] = H_z(k_z, \omega) R(k_z, \omega) \quad (4.7)$$

Thus, the spatio-temporal data $V(z, t)$ and the product of the reflectance coefficient and the system transfer function are Fourier conjugated. Equation (4.7) is in agreement with the relationship between $V(z)$ and the reflectance function derived for a narrow band mode [128].

Comparing the $V(x, t)$ and $V(z, t)$ schemes, it should be noted that the first Fourier transform (4.6) is associated with the pair of variables (k_x, x) while the transform (4.7) is associated with (k_z, z). Traveling along the liquid–specimen interface, the transducer reads the spatial frequency $k_{xR}/(2\pi)$ of the leaky wave, where k_{xR} is the x-component of the wave vector of the LSAW. The velocity c_R of the LSAW is related as follows to the measured spatial frequency:

$$c_R(\omega) = \frac{\omega}{k_{xR}(\omega)} \quad (4.8)$$

The spatial frequency $k_{zR}/(2\pi)$ is detected during the scan of the transducer along the z-axis in the $V(z, t)$ scheme. In this case, the leaky wave velocity can be calculated as follows:

$$c_R(\omega) = \frac{\omega}{\sqrt{(\omega/c)^2 - k_{zR}^2(\omega)}}$$

Thus, to determine c_R from the $V(z, t)$ data, the sound velocity c in the liquid should be known, whereas Equation (4.8) does not demonstrate an explicit dependence on c.

4.4
Angular Resolution of QSAM Systems

The system transfer functions H_x and H_z are band-width-limited in the frequency domain because of the finite bandwidth of the transducers and the electronics. Also, the reflectance coefficient R and the transfer functions H_x and H_z vanish outside of the regions $|k_x| \leq k$ and $0 \leq k_z \leq k$. Thus, $S_x(k_x, \omega)$ and $S_z(k_x, \omega)$ are band-width-limited functions in the 2D frequency domain. It follows from Fourier-transform theory that, in this case, $V(x, t)$ and $V(z, t)$ must be infinite in the spatio-temporal domain.

In practice, $V(z, t)$ and $V(x, t)$ can only be acquired within the finite 2D data window. The width of the time window depends on the size of the storage memory and can be large enough to record all distinguishable wavelets. Thus, the spectrum distortion due to data truncation in the time domain can be neglected. The width of the spatial window is determined by the travel distance of the transducer. This distance is limited by the geometry of the system and can significantly reduce the spectral resolution.

Let $p(\alpha)$ be a window function: $p(\alpha) = 1$ at $|\alpha| \leq L_\alpha/2$; and $p(\alpha) = 0$ elsewhere. Here, $\alpha = x, z$ denotes the appropriate spatial coordinate and L_α is the displacement along the respective axis. According to the convolution theorem, for the spectrum of the truncated data we have:

$$W_\alpha(k_\alpha, \omega) = F[V(\alpha, t)p(\alpha)] = S_\alpha(k_\alpha, \omega) \otimes A_\alpha(k_\alpha)$$

where A_α is the Fourier transform of the window function:

$$A_\alpha(k_\alpha) = L_\alpha \operatorname{sinc}\left(\frac{k_\alpha L_\alpha}{2}\right)$$

As a result of the convolution between S and A, the measured spectrum W is distorted. The smoothing is most evident in the vicinity of the critical angles, where the reflectance function exhibits abrupt changes, and the oscillating side lobes of A

can produce ripples around these regions. The spectral resolution can be estimated from the width of the main lobe of A in the k-domain:

$$\Delta k_\alpha \approx \frac{4\pi}{L_\alpha}$$

To suppress the side lobes of A, the truncated data are usually weighted by an apodization function [154]. However, the apodization causes a slight increase in the main lobe width, resulting in some further loss of spectral resolution.

We now compare the angular resolutions of the $V(x,t)$ and $V(z,t)$ schemes. Noting that

$$k_x = k\sin\theta \qquad k_z = k\cos\theta$$

where θ is the angle of incidence of the spectral component, we can estimate the angular resolutions of the $V(x)$ and $V(z)$ schemes, respectively:

$$\Delta\theta_x = \frac{4\pi}{L_x k \cos\theta} \qquad \Delta\theta_z = \frac{4\pi}{L_z k \sin\theta} \qquad (4.9)$$

A similar formula for the angular resolution $\Delta\theta_z$ was obtained earlier in [128]. It was shown that the distortion of the inverted reflectance function is serious at large incidence angles and small data windows. Because of the convolution with the function A, the magnitude of the reflectance may exhibit a sharp dip in the vicinity of the critical angle and an abrupt phase jump can be significantly deformed. The uncertainty in the critical angle causes an error in measurement of the LSAW phase velocities which increases with decreasing travel distance L_z and angle of incidence θ.

A similar distortion of the experimental spectrum $S_x(k_x,\omega)$ is associated with the truncation of the $V(x,t)$ data. The angular resolution $\Delta\theta_x$ is also inversely proportional to L_x but, unlike in the case of L_z, the scanning distance along the liquid–specimen interface is not restricted by the geometry of the transducer and the data window L_x can extend by any amount in the direction of the leaky wave propagation. Furthermore, $\Delta\theta_x$ increases with decreasing angle of incidence θ, and the angular resolution $\Delta\theta_x$ is higher than $\Delta\theta_z$ at $\theta < \pi/4$, if the data windows are equal ($L_x = L_z$).

Let us now express the angular resolution as a function of maximum travel distance of the recorded LSAW. If the data windows are symmetrical with respect to the origin of the coordinates $x = 0, z = 0$, then, in the case of the $V(x)$ system, this distance is half of the width of the data window: $L_{\text{SAW}(x)} = L_x/2$. In the case of the $V(z)$ system, it follows from Figure 4.1(a) that the travel distance $L_{\text{SAW}(z)} = AB$ depends on the angle of incidence θ: $L_{\text{SAW}(z)} = L_z \tan\theta/2$. Using (4.9), we can derive the following expressions for the angular resolutions:

$$\Delta\theta_x = \frac{\lambda}{L_{\text{SAW}(x)} \cos\theta} \qquad \Delta\theta_z = \frac{\lambda}{L_{\text{SAW}(z)} \cos\theta}$$

where λ is the wavelength of the ultrasound in the liquid at a given frequency. Accordingly, the angular resolutions in the $V(x)$ and $V(z)$ schemes depend equally on the size of the region where the leaky wave is propagated. It should be noted that $L_{\text{SAW}(x)}$ is constant for all LSAW velocities whereas $L_{\text{SAW}(z)}$ decreases with decreasing critical angle, degrading the resolution $\Delta\theta_z$ for high-speed waves.

4.5
Application of the $V(x)$ QSAM System to LSAW Measurement

Several materials with known properties were tested using the wide-aperture, line-focused $V(x)$ system in the 20-megahertz frequency range. The $V(x, t)$ data recorded for lead are shown in Figure 4.3 as a grayscale image. The output voltage reaches a maximum at the focused position of the transducers ($x = 0$) and decays with defocusing.

The time of flight of the wave D depends linearly on the distance x. According to (4.2), the slope of the response D in Figure 4.4(a) corresponds to a velocity $c_D =$

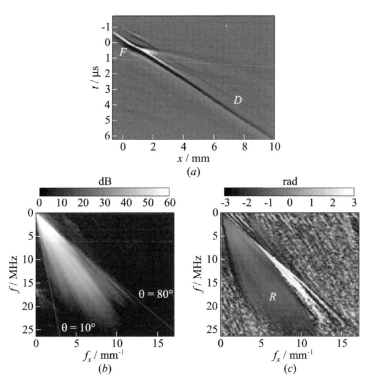

Figure 4.3 The $V(x, t)$ waveform recorded for lead (a), and the magnitude (b) and phase (c) of the spectrum of $V(x, t)$. Markers F and D correspond to the focused and defocused positions of the transducers, respectively.

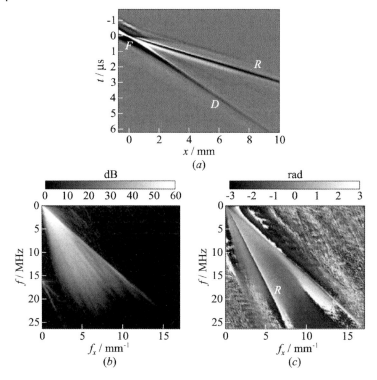

Figure 4.4 Waveform $V(x, t)$ (a), magnitude (b), and phase (c) of the spectrum $W(f_x, f)$ for fused quartz: R indicates the leaky Rayleigh wave, F and D correspond to the focused and defocused positions of the transducers, respectively.

1535 m s^{-1}, which is higher than the velocity of sound in water $c = 1491$ m s^{-1} at the temperature of the experiment $T = 23 \pm 0.2\,°\mathrm{C}$. The angle of incidence of the wave D can be approximately estimated as $\arcsin(c/c_D) = 76°$. This value is close to the maximum geometrical angle of incidence of the system $\theta = 80°$. Therefore, we can suggest that the wave D is generated by the edge area of the transmitting transducer. Grazing the specimen, the wave is reflected from the specimen–liquid interface and detected by the edge area of the receiving transducer. Due to the large angle of incidence, the time of flight of the response D depends practically linearly on the scanning coordinate x [149].

The spectrum of $V(x, t)$ of the water–lead interface, approximating the experimental transfer function of the system H_x, is shown in Figures 4.4(b) and (c). The amplitude and phase of the calculated spectrum $H_x(f_x, f)$ are presented as grayscale images, where $f = \omega/(2\pi)$ is the frequency and $f_x = k_x/(2\pi)$ is the spatial frequency. In the (f_x, f) domain, the transfer function occupies a temporal frequency bandwidth from 1–2 MHz up to 20–25 MHz, and its spatial frequency bandwidth is restricted by minimum ($\theta = 10°$) and maximum ($\theta = 80°$) angles of

Table 4.1 Published densities and velocities of materials used in the experiment [155].

Material	Density g cm^{-3}	Wave velocity	
		Longitudinal	Shear
Water (23 °C) [39]	0.998	1491.2	–
Fused quartz	2.2	5970	3765
Brass naval	8.42	4430	2120
Polystyrene	1.056	2340	1150

incidence of the system. Within this area, the amplitude and the phase of $H_x(f_x, f)$ are smooth and slowly varying functions.

Figure 4.4 shows $V(x, t)$ and $W(f_x, f)$ for fused quartz. In addition to the system response D, the Rayleigh wave R is present in the $V(x, t)$ data. To measure the velocity of the leaky Rayleigh wave, the time delay of the negative peak of wave R was determined as a function of x. Measurements were repeated several times at various positions of the specimen to estimate the reproducibility. The average value of the LSAW velocity c_R and its standard deviation δc_R were found to be 3428 and 0.37 m s^{-1}, respectively. The experimental c_R is in a good agreement with the leaky Rayleigh wave velocity of 3429.3 m s^{-1} which was calculated on the basis of the published data (Table 4.1), and the relative accuracy of the measurement obtained for fused quartz is estimated to be 0.03%.

The phase (Figure 4.4c) exhibits an abrupt change R caused by a phase jump of the reflectance function in the vicinity of the critical angle of the leaky Rayleigh wave. The critical spatial frequency f_{xR} was estimated as a value at which the derivative of the unwrapped phase had a minimum and the c_R was then calculated as a function of frequency using Equation (4.8). Assuming that the leaky Rayleigh wave propagates along the water-fused quartz interface without dispersion, the velocity of the LSAW was obtained by averaging the $c_R(f)$ within the frequency range from 4 to 20 MHz.

The velocities of the LSAW were calculated for each experimental spectrum $W(f_x, f)$, and the average value c_R was found to be 3433 m s^{-1} with a standard deviation $\delta c_R = 1.3$ m s^{-1}. The relative accuracy of the measurement is about 0.1% in this case. Consequently, at least for fused quartz, the reproducibility of direct processing in the (x, t) domain is higher than the reproducibility of processing in the (f_x, f) domain, although the same raw data set was used in the analysis.

Figure 4.5 shows $V(x, t)$ and $W(f_x, f)$ for a brass foil 0.2 mm thick. The foil was mounted on a rigid flat substrate and the edges of the sample were sealed, and consequently, the bottom interface of the foil was loaded by air only. Several modes of leaky Lamb waves could propagate along the layer. Because of the strong dispersion of Lamb waves, the recorded $V(x, t)$ waveform is very complicated (Figure 4.5a), but the spectrum $W(f_x, f)$ clearly demonstrates a set of dispersion curves. To make a comparison, a reflection coefficient for a layer loaded by water from one side was calculated according to [156]. The amplitude of the calculated reflectance function (not shown) was equal to unity, and the phase is presented in Figure 4.6

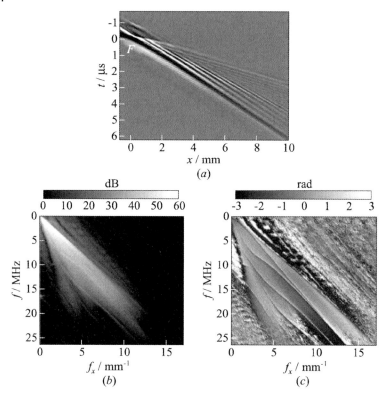

Figure 4.5 Waveform $V(x,t)$ (a), magnitude (b), and phase (c) of the spectrum $W(f_x, f)$ for 0.2-millimeter brass foil; F corresponds to the focused position of the transducers.

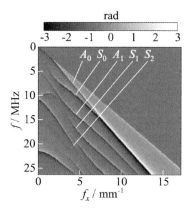

Figure 4.6 The phase of the reflectance function calculated for 0.2-millimeter brass foil; A_0, S_0, A_1, S_1, and S_2 indicate Lamb wave modes. The format is the same as that in Figure 4.5(c).

as a function of (f_x, f). A set of symmetric S and antisymmetric A modes is visualized in the image due to phase jumps at the critical angles of the leaky waves. The transfer function $H_x(f_x, f)$ of the $V(x,t)$ system is sufficiently extensive to detect the A_0, S_0, A_1, S_1, and S_2 Lamb wave modes. There is a good agreement

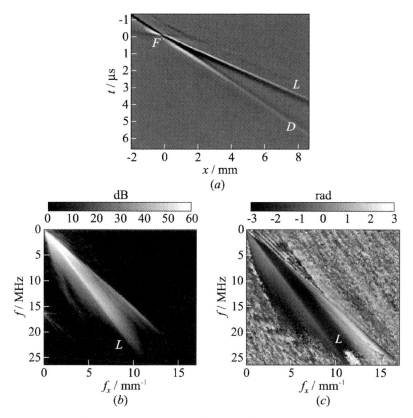

Figure 4.7 Waveform $V(x, t)$ (a), magnitude (b), and phase (c) of the spectrum $W(f_x, f)$ for polystyrene: L indicates the leaky surface-skimming wave, F corresponds to the focused position of the transducers.

between experimental and simulated data within the active (f_x, f) region of the system.

The $V(x, t)$ and $W(f_x, f)$ functions obtained for polystyrene are shown in Figure 4.7. The velocity of the bulk shear wave in polystyrene is smaller than the velocity of sound in water and a leaky Rayleigh wave is not produced in this case. The waveform L observed in $V(x, t)$ is produced by a leaky surface-skimming compressional wave. For soft materials such as polystyrene, the velocity of this leaky wave is approximately equal to the velocity of the longitudinal bulk wave [157, 158].

The $V(x, t)$ and $W(f_x, f)$ functions show some distinguishing features because of low impedance of the material. The magnitude of the reflectance function is much smaller than unity except at the longitudinal critical angle, where it reaches a maximum. The phase of the reflectance function is constant up to the critical angle and then monotonically decreases with increasing angle of incidence. As a result, the phase of the wave L in $V(x, t)$ is inverted relative to the Rayleigh wave response R. From the slope of the waveform L, the velocity of leaky skimming wave was calculated to be $c_L = 2329$ m s^{-1}. A value of $c_L = 2327$ m s^{-1} was also computed in the spectral domain from the angular positions of the experimental maxima. Both of these values are in reasonable agreement with the published values.

4.6
Temperature Stability of the $V(x)$ QSAM System

To investigate experimentally the temperature coefficient of the LSAW velocity measurement error, a set of $V(x, t)$ waveforms was recorded for a fused quartz specimen over a temperature range of 10–50 °C. The temperature dependence of the leaky wave velocity c_R, calculated in the spatio-temporal domain from the slope of the Rayleigh wave R, is presented in Figure 4.8. Assuming the linearity of this dependence, the experimental temperature coefficient of c_R was found to be $\beta_q = 0.37$ m (s °C)$^{-1}$.

The temperature coefficient of the leaky Rayleigh wave was also calculated based on temperature dependences of the bulk wave velocities and the densities of fused quartz and water. The bulk wave velocities in fused quartz obtained from the same manufacturer had been measured earlier [152] at various temperatures. The published data of the velocity of sound in water [159], the density of water, and the thermal expansion of fused quartz [160] were used in the calculations. The temperature coefficients of the longitudinal and shear wave velocities were found to be 0.70 and 0.27 m (s °C)$^{-1}$, respectively. These values are in a good agreement with the data for fused silica published in [161]. Knowing the properties of the materials at various temperatures, the LSAW velocity was calculated from the location of the pole of the reflectance function on the complex k_x plane [162], and the temperature coefficient of the leaky Rayleigh wave was found to be $\beta_t = 0.34$ m (s °C)$^{-1}$. The difference $\beta_q - \beta_t$ between the experimental and calculated coefficients can be treated as a temperature component of the measurement error, and the tempera-

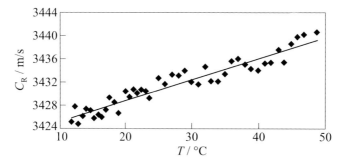

Figure 4.8 Linear-least-square curve fit for the velocity of the LSAW measured for fused quartz as a function of temperature (diamonds).

ture coefficient of the c_R measurement error, η_x, can be estimated for the $V(x,t)$ system:

$$\eta_x = \frac{\beta_q - \beta_t}{c_R} = \frac{0.03}{3430} \approx 10^{-5} \, (°C)^{-1}$$

Thus, the experimental temperature coefficient η_x is significantly lower than the theoretical estimate of $\eta_z = 10^{-3} \, (°C)^{-1}$ obtained using the $V(z)$ technique (4.1).

5
Acoustic Microscopy and Nonlinear Acoustic Effects

5.1
Nonlinear Acoustic Applications for Characterization of Material Microstructure

Interaction of an acoustic signal with matter is said to be linear if the response of matter and output signal intensity vary in linear proportion to the input signal intensity, in keeping with the Hooke law. The majority of well-known acoustic phenomena are associated with linear elasticity of materials. However, in the event of high intensity of the input signal or when materials exhibit specific properties, there occasionally may arise a variety of new "nonlinear" effects, such as wave shape distortions induced by amplitude-dependent variation of the wave propagation velocity, generation of higher harmonics, sum and difference oscillation frequencies, acoustic detection, self-focusing, etc. Certain of these phenomena have analogs in optics [163] and are being increasingly used for nondestructive examination of the microstructure of diversified materials.

5.1.1
Schematic of Experiment

A general schematic of various experimental research methods using nonlinear acoustic effects for the characterization of materials is sketched in Figure 5.1. The sketches illustrate the configurations and location of the transmitting and receiving transducers with respect to the test sample and omit the other elements of experimental setups. The diagrams of the setups for studying the acoustic nonlinearity of materials by the following methods are presented: the thermodynamic method for liquids, the acoustoelastic method for solids, the finite amplitude method, the resonance method, and the acoustic-wave interaction method.

Nonlinear properties of materials have been examined using longitudinal (focused and unfocused) and transverse bulk waves, surface acoustic waves and interface-related waves (of the Stoneley–Scholte type).

Nonlinearity of material can be defined as the dependence of the acoustic wave velocity on the applied static load. The experimental setup for studying solids and liquids is sketched in Figures 5.1(*a*) and (*b*). The picture illustrates the determina-

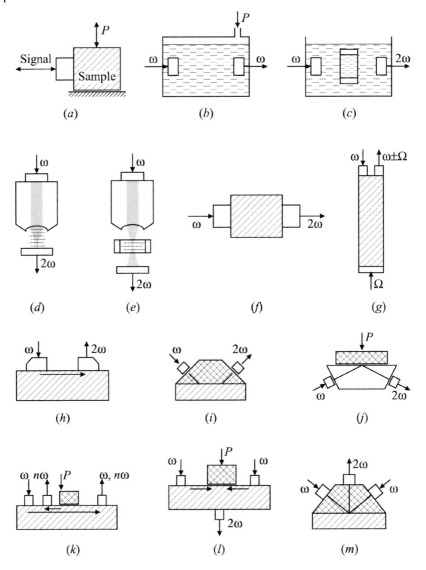

Figure 5.1 Schematic of experimental setups for studying nonlinear characteristics of materials.

tion of the acoustoelastic dependence of the ultrasound velocity on the load applied to solid materials [164]. Only one transducer is employed, and the system operates in the reflection mode. The method can also be used to visualize the stress-distribution area under load [165]. The transducer scans the sample surface in two directions; the 2D image of the stress-distribution area is formed on the basis of the velocity variation with coordinates.

The thermodynamic method of evaluating parameter B/A in liquids [166] is presented in Figure 5.1(b). Within this method, the temperature and pressure of the sample are controlled and measured. According to the following equation

$$\frac{B}{A} = 2\rho_0 c_0 \left(\frac{\partial c}{\partial p}\right)_{s,\rho=\rho_0} \equiv 2\rho_0 c_0 \left(\frac{\partial c}{\partial p}\right)_T + \frac{2c_0 T \mu}{C_p}\left(\frac{\partial c}{\partial T}\right)_p$$

where μ is the volumetric coefficient of thermal expansion, B/A can be derived by measuring the ultrasound velocity variation with temperature at constant pressure and $(\partial c/\partial p)$ under isothermal conditions.

A more refined thermodynamic procedure developed in [167] is based on the straightforward isentropic measurement of sound velocity variation with pressure rather than on the isothermal process. The key stage of both thermodynamic and acoustoelastic procedures is the measurement of the sound velocity variations.

The acoustic nonlinearity of matter can also be assessed by the finite amplitude (or spectral) method. A finite-amplitude acoustic wave generated by a transducer is distorted as it propagates through the matter and higher harmonic signals are produced in the process. The distorted acoustic wave is subsequently received either by a broad-band transducer or by a specially devised transducer at a certain frequency. When using broad-band transducers, the amplitude of each harmonic is calculated by means of fast Fourier transformation (FFT) analysis on the basis of the signal received. Once the amplitudes of the fundamental wave passed through the sample and second harmonic signals are measured, nonlinear coefficients β_2 and β_3 can be calculated from equations

$$U^0_{3\omega} = \frac{\beta_2^2}{8} k^3 U_0^4 x^2 \quad \text{for } \frac{\beta_3}{\beta_2} \gg kx \tag{5.1}$$

or

$$|\beta_2| = \frac{4 U^0_{2\omega}}{k^2 x U_0^2} \tag{5.2}$$

Yet another modification of the finite-amplitude procedure for measuring acoustic nonlinearity in liquids and biological media is demonstrated in Figure 5.1(c). Comparing the amplitude of second harmonic signals measured for the test liquid with the data for a liquid with known nonlinear parameters, it is possible to evaluate B/A. This method offers freedom from measurement of the absolute value of the acoustic wave amplitude and permits assessment of the effect of sound absorption in a sample and diffraction in a transducer on the accuracy of measurements. A similar experimental setup using a focused beam (Figures 5.1d and e) was devised in [168] for measurement of the nonlinear parameters of liquids. Transient acoustic waves were focused with an acoustic lens, and harmonic signals were detected by plane transducers after the passage of the acoustic signal through the liquid. Inasmuch as nonlinear effects emerge only in the focal area, the focused-beam technique is well suited for exploring the nonlinearity of relatively small volumes of liquids, especially with the setup sketched in Figure 5.1(e) in which a sample is placed in a container covered with a film on two sides.

Different configurations of experimental setups for studying nonlinear acoustic properties of solids in the bulk, on the surface, and at the interface can be found in the literature.

The scheme displayed in Figure 5.1(f) is most often employed to determine the acoustic nonlinearity coefficients of homogeneous media by the finite-amplitude method, based on (5.1) and (5.2) or

$$u''(x,\tau) = \frac{\beta_2 \omega u_0^2}{4\alpha c_0^2} \left[\exp(-2\alpha x) - \exp(-4\alpha x)\right] \sin 2\omega \tau \qquad (5.3)$$

A similar scheme is also appropriate for characterization of heterogeneous materials. In this case, unusually large amplitudes of higher harmonics presumably testify to the presence of certain flaws, such as disbonds [169], cracks [170], etc., in test samples. Qualitative analysis of these defects can be accomplished using the equation for disbonds:

$$S = A_1 - A_2 + A_3 - A_4 + \Lambda = \sum_{i=1}^{\infty} (-1)^{i+1} A_i \qquad (5.4)$$

or the equation for cracks:

$$\varepsilon = \frac{5\beta_{20} N}{16} \left(1 + \frac{3\alpha N_0}{8}\right)^{-2}$$

A device on the basis of the modulation technique is sketched in Figure 5.1(g). Signals of a mixed frequency are picked up by piezoelectric sound ducts or accelerometers. The length of samples is chosen so that the natural frequencies are as close to the driving frequency as possible.

Equation

$$A = A_0(1 + m \cos \Omega t)$$

where

$$m = \frac{\beta_2 \varepsilon_0 \omega_0 (\partial A/\partial \omega)_{\omega \approx \omega_0}}{A_0}$$

is applied to calculate the quantitative characteristic of the nonlinearity of materials. Other versions of the resonance method involve the detection of higher harmonics or measurement of the shift of the resonance frequency of a sample as a function of an individual amplitude of the source.

The above-presented experimental setups usually employ longitudinal waves, and it is precisely these setups that constitute the subject of the majority of the related publications.

Other types of acoustic waves, such as transverse, surface Rayleigh and Stoneley–Scholte waves, are also sometimes used for studying nonlinear properties of materials. Compared to bulk waves, SAWs are much more appealing as a tool for exploring nonlinear properties, inasmuch as the surface wave energy penetrates no deeper than through one wavelength, and therefore, the energy density of a surface wave is fairly high [171]. Owing to this fact, nonlinear effects can be revealed at lower voltage values than in experiments with bulk waves. Surface waves are commonly initiated in isotropic media using wedge-shaped transducers; finite-amplitude SAWs are most efficiently generated by interdigital transducers. Monitoring of the amplitude of second harmonics of SAWs as a function of the separation between transducers (Figure 5.1h) makes it possible to determine the contributions of dispersion and dissipation to generation of appropriate harmonics of SAWs in a versatile operation mode. Under conditions of no dispersion, the SAW-based finite-amplitude technique is suitable for calculating the third-order constants by the expression for the amplitude of second harmonics of SAW obtained in [172].

A distinction between bulk waves and SAWs is that SAW dispersion is associated with surface inhomogeneity which, in turn, is the consequence of the technical "history" of a sample – grinding, polishing, cooling, etc. Dispersion may manifest itself as oscillations of the amplitude of the second harmonic of SAW along the path of its propagation. Hence, measuring the oscillation period on the basis of equation

$$u''(x,t) \approx \frac{\beta_2 k_1 u_0^2}{c_0 \Delta k} \sin\left(\frac{\Delta k x}{2}\right) \sin\left[2\omega t - \frac{(2k_1 + k_2)x}{2}\right] \tag{5.5}$$

where $\Delta k = k_2 - 2k_1$, $k_2 = k(2\omega)$, and $2k_1 = 2k(\omega)$, one can easily reveal surface inhomogeneities in layered materials, films, etc., due to technological processing. Equation (5.5) can also be used to calculate β_2 by the finite-amplitude method in a dispersive medium. The surface of cracks and other fracture-type defects can also be identified by an increase in the second-harmonic amplitude [173] during nonlinear propagation of SAW, as is shown in Figure 5.1(h).

The analysis of the nonlinear effects of waves arising at interfaces (Stoneley–Scholte waves) carried out in [172] confirmed the possibility of using the finite-amplitude method to assess the quality of coupling of materials. The setup for studying nonlinear behavior of waves at the interface between two solids is sketched in Figure 5.1(i) [174]. In the version with a bulk wave (Figures 5.1j and k), transverse waves polarized in the vertical and horizontal directions are used [175, 176]. Considering that no transverse harmonics are generated in an isotropic medium, nonlinearity of the material can be neglected in this case (Figure 5.1j). It has been shown that at heavy loads on the contact zone, the results of measurements of the nonlinear reflection coefficient defined by expression

$$U_{2\omega} = \frac{\beta_R}{2} k U_0^2$$

are in good agreement with the nonlinear properties of the material.

The setup sketched in Figure 5.1(j) was used for qualitative and quantitative analysis of surface roughness [177]. The arrangement providing for a combination of the nonlinear effects of surface and bulk waves is illustrated in Figure 5.1(l). This configuration is based on the interaction of counter-propagating SAWs with identical frequencies. The scheme is particularly advantageous for characterization of the adhesion strength of two contact surfaces. It is also appropriate for revealing fracture-type defects on the free surface of a sample [178].

The case where fundamental bulk waves incident on the surface of a sample at some angle are used instead of SAWs is shown in Figure 5.1(m) [176, 179]. Experimental schemes presented in Figures 5.1(l) and (m) can be grouped with nonlinear tomography facilities which are also invoked to reveal surface flaws. These two configurations complement each other: at a fixed angle of incidence (including $\pi/2$ in the case of SAW attenuation), an acoustic pulse incident on the surface "reads" the nonlinear properties of the surface along a certain direction; in principle, their combination furnishes a means of reconstructing a 2D tomographic image of a nonlinear inhomogeneity.

The overwhelming majority of experimental approaches to characterization of a wide variety of solids and liquids on the basis of nonlinear acoustic effects can conceptually be classified under four basic categories of acoustic investigation techniques: (i) thermodynamic, (ii) acoustoelastic, (iii) resonance, and (iv) finite-amplitude. These four groups of methods rely on an analysis of inter-relation between various types of longitudinal, transverse, or surface waves and materials. The above-listed methods supply information about the presence of diversified structural disturbances of matter, ranging from crystal lattice deviation that is ideal to various defects and inhomogeneities. The results of measurements by the resonance and acoustoelastic methods depend on ultrasound absorption by matter to a lesser extent and, therefore, these methods are more appropriate for studying imperfect materials. In this case, nonlinear-mode reflection is a more efficient tool for revealing disbonds or fractures than the conventional methods of linear acoustics. The peculiarities of structure visualization by means of nonlinear acoustics will be covered in the next section.

5.1.2
Visualization by Nonlinear Acoustic Methods

The most obvious reason why the methods of higher harmonic generation are applied in the acoustic visualization area for the case of nonlinear media, when using systems with a focused acoustic beam, is the need for a greater resolving power. The contrast of an acoustic image, dependent on the local nonlinear properties of materials, can be enhanced dramatically in the presence of microdefects and inhomogeneities. This section covers the effect of these two factors on the performance of acoustic visualization facilities and, primarily, on the resolution of acoustic microscopes. A high local nonlinearity of microdefects provides a means for parametric acoustic modulation of the properties of test materials by an incident wave, and the signal so produced is subsequently "read" by another acoustic wave. In addition,

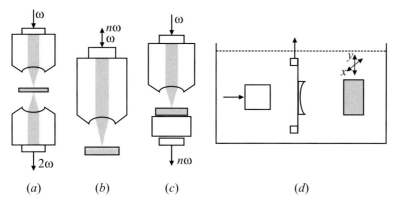

Figure 5.2 Configuration of the experimental facilities intended to improve the resolution of acoustic images obtained by the use of higher harmonics.

proper attention is given to certain new possibilities for nonlinear characterization of materials offered by the aforesaid parametric acoustic visualization.

Improvement in the acoustic microscope resolution by the use of nonlinear effects has long been known. The possibility of visualizing a structure in water with the aid of the second harmonic was first demonstrated by Kompfner and Lemons in 1976 [180] by means of a transmission acoustic microscope with a fundamental frequency of 400 MHz (Figure 5.2a). The power delivered to the transmitter was raised above the normal value typical for linear studies. The authors observed a gain in both resolution and contrast of acoustic images, compared to the images obtained within the linear mode. One most striking and impressive peculiarity of second-harmonic images was the sharp contrast at the interfaces between different structures of the sample.

Using a similar approach, Wickramasinghe and Yeack [181] obtained similar results. In both publications, attention is drawn to the fact that the second harmonic signal is formed in water between the sample and the receiving lens rather than inside the sample itself. This suggests no contribution of nonlinear effects to the resultant acoustic images.

The problem was attacked somewhat differently in [182] and [183]. The transducer and the lens were operated in the reflection mode and arranged as in Figure 5.2(b). For a transducer with an operating frequency of 4.4 GHz, the resolution was enhanced to 0.20 µm by increasing the input signal power [182]. It was found that certain details of the structure became visible, as more and more power was added to the input signal, until the point of nonlinear generation of higher harmonics was reached, following which the resolution showed no tendency to increase even with a further increase in the signal power. Using frequencies from 2.0 to 2.8 GHz and taking into account the spatial frequency response, Rugar [183] clearly demonstrated that, as the acoustic signal power was increased, the imaging sensitivity in the nonlinear mode became 1.4 times that achieved by conventional linear visualization methods.

Nonlinear effects have been studied with an acoustic microscope using immersion liquids other than water; namely, liquid nitrogen and liquid argon [183]. In this case, nonlinear effects show up more vividly owing to lower sound velocity, higher nonlinearity and lower absorption in cryogenic fluids [184, 185].

Moulthrop and co-workers [185], pursuing experiments with an acoustic microscope, operating at a frequency of 15.3 GHz with superfluid 4 He as an immersion liquid at a temperature below 0.9 K and a pressure above 30 bars, observed almost complete wave suppression at certain values of the input signal power and reconversion to the pumping-wave frequency at higher levels of the input power. When the input signal power fell within the wave-suppression range, the microscope resolution was improved. On the other hand, when the power was adjusted to the reconversion range, the microscope resolution was impaired. The authors put forward arguments to support the idea that higher harmonic generation was behind this process. This assumption was verified by Karaki and co-workers [186, 187] using a 370-megahertz microscope with 4 He at a temperature of 0.3 K as an immersion liquid. The acoustic microscope was operated under both reflection and transmission conditions, as is shown in Figure 5.2(c). For second harmonics, the resolution was improved by more than a factor of $2^{1/2}$ owing, perhaps, to distinct nonlinear properties of the immersion liquid. The power dependence of the reflected-wave effects was similar to that observed in [185].

Germain and Cheeke had explored higher harmonics using SAM at low frequencies (15 MHz) and ethanol as an immersion liquid [184]. The schematic of the experimental setup is illustrated in Figure 5.2(c). Although the frequency used was much lower than in the above-described studies, the authors managed to produce harmonics of up to 14th order. They also succeeded in isolating higher harmonics and improved the resolution in the proportion $n^{1/2}$ for $n \leq 10$ (n is the number of a harmonic). It was shown that with the 10th harmonic, the resolution improvement factor was as high as 3.

The idea that commercial diagnostic B-scanners generating signals sufficiently intense to trigger nonlinear effects would be used in clinical practice was first advanced by Muir, Carstensen, and co-workers in 1980 [188, 189]. More recently, Starrit and his colleagues [190] demonstrated nonlinear distortion of an ultrasonic wave in a human muscle tissue (*in vivo*) and bovine liver (*in vitro*). To do this, they used a conventional commercial ultrasonic setup. Ward and co-workers relatively recently explored nonlinear acoustic effects in skin tissue models with a broadband ultrasonic imaging system operating in the reflection mode [191]. Their intention was to improve the resolution of medical ultrasonic devices. A simplified scheme of the setup with a fundamental frequency of 2.25 MHz is sketched in Figure 5.2(d).

Harmonics of up to the 4th order were detected and analyzed, and two basic effects enhancing the resolution in going from the fundamental frequency to higher harmonics were revealed:

(1) reflected beam width w_n at the focal point decreased by equation $w_n/w_1 = 1/n^{0.78}$ in experiments using a skin model; and
(2) levels of the "side lobes" of reflected beams reduced by a factor ranging from 1 to 4 in accordance with the harmonic order.

For a visualization process, this implies that the lateral resolution of the system is improved, while the signal-to-noise ratio decreases.

In closing, it is worth mentioning examples making use of nonlinear effects in acoustic microscopy and biomedical studies which are available in the literature. Nonlinear effects associated with the generation of higher harmonics are seldom, if ever, used in modern diagnostic equipment; first because of the limited ability of a transducer to receive signals at frequencies of higher harmonics; and second, because a weaker signal of harmonics is impossible to enhance by increasing the acoustic signal power in view of feasible damage to the tissues studied. The former problem can be remedied either by readjusting the apparatus or by using a broad-band piezoelectric transducer, and the latter, by injecting microbubble contrast agents [192]. Intravenous injection of contrast microbubbles significantly enhances the generation of higher harmonics and subharmonics in blood and soft tissues. However, when research is performed by means of acoustic microscopy, there is no need to limit the signal power for safety reasons and therefore it is always possible to choose a signal capable of triggering the generation of harmonics. When exploring nonlinear zones, the image resolution can be improved even with a commonplace reflection-type acoustic microscope. In addition, immersion liquids with distinct nonlinear properties, e.g., cryogenic fluids, are also beneficial for enhancing nonlinear effects. Nonlinear effects in an acoustic microscope can be used as an alternative to an increase in the operating frequency, when the acoustic image quality needs to be improved. In doing so, it should be remembered that the image quality is improved by the use of the nonlinear effects in an immersion liquid or a contrast agent instead of the nonlinear properties of test objects.

5.1.3
Parametric Representation of Acoustic Nonlinearity

The majority of acoustic imaging systems form an image based on variations in reflection or transmission coefficients which, in turn, depend on certain linear parameters of test materials.

Nonlinear interaction of acoustic waves is useful for measuring the local nonlinear parameters in individual areas of a test sample. This provides a way of visualizing the nonlinear properties of matter for the purposes of biomedical engineering, nondestructive research, and acoustic microscopy.

In biology, nonlinear parametric visualization at frequencies below 5 MHz was first accomplished by Ichida and co-workers [193]; the foremost consideration underlying their research is illustrated in Figure 5.3(a). A low-frequency incident wave travels perpendicularly to a high-frequency fundamental (low-power) wave of a probing signal, which is transmitted to one side of the sample and received on

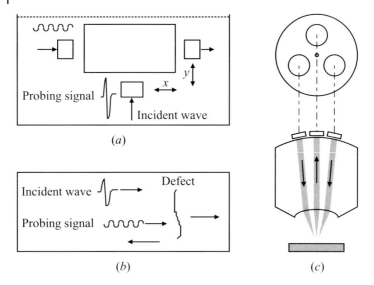

Figure 5.3 Schematic sketches of devices for parametric visualization of acoustic nonlinearity.

its opposite side. The signal frequency is varied in such a way as to generate cumulative phase shifts of the probing signal wave. Their inverse Fourier transform is proportional to the profile of the nonlinear parameter B/A along the probing signal path. The probing wave parametrically modulated by nonlinearity is subsequently detected and demodulated to derive the B/A distribution along the wave path. This method was later modified to allow the use of the pulsed form of a pumping wave, which made it possible to generate nonlinear acoustic images in the transmission mode and real time [193]. In so doing, the wave was initiated on the side of the sample opposite to the side at which the probing signal arrived, which was particularly convenient for practical use [194]. Cain deduced the theoretical expression for the phase shift of a sine-wave signal under conditions of nonlinear interaction with a pumping wave having adjusted parameters and propagating in the opposite direction [195].

The nonlinear visualization technique has been recently adapted for operation in the reflection mode and for the needs of medical diagnostics, i.e., in the case where signals can be radiated only on one side [196].

Nakagawa and co-workers [197] advanced quite a different procedure for developing nonlinear ultrasonic systems based on nonlinear interaction triggering generation of sum- and difference-frequency waves. The amplitude of the received wave with a difference frequency depends on the absorption and nonlinear parameters of the medium and, hence, tomographic images of nonlinear parameters can be reconstructed by supplementing them with the results of conventional tomography (on the basis of absorption). The quality of the thus obtained images is inadequate for clinical purposes and, in addition, the biological action of high-power pumping waves should be analyzed in more detail. However, this method has opened up fresh opportunities for parametric visualization of human tissues, and the quality of tissue visualization can be improved appreciably, compared to that achieved with the conventional B-mode back-scattering technique.

A similar method was proposed to reveal cracks in materials (Figure 5.3b [198]). A powerful incident-wave signal changes the crack opening, which causes parametric variations in the phase and amplitude of a transmitted or reflected signal. By measuring these phase or amplitude modulations of a transmitted or reflected wave, it is possible to detect and localize cracks.

It was proposed in [180, 181] that the parametric nonlinear interaction of two acoustic rays formed by two different lenses should be used in acoustic microscopy. More recently, this approach was invoked to obtain nonlinear acoustic images [199, 200]. Figure 5.3(c) illustrates a schematic of the experimental configuration of a nonlinear acoustic microscope in which two transducers with frequencies of 640 MHz (f_1) and 760 MHz (f_2) serve to initiate an ultrasonic wave, and the third transducer with frequency 520 MHz ($2f_1 - f_2$) to receive the signal [129]. The transducers were mounted on the curved surface of a quartz lens so that all were confocal with respect to one another in the aqueous medium [199]. The system parameters were chosen so that nonlinear interactions took place on the surface of the sample, and their manifestation in the immersion liquid was at a minimum. The first attempt to isolate nonlinear effects associated with the state of a sample, rather than with the immersion liquid, failed [199]. A similar device with four transducers was later described in [200]. Further investigations have disclosed that distinctions between two samples having different third-order elastic constants, i.e., differing in nonlinear properties, "invisible" as they are to a conventional acoustic microscope with linear parameters, are easily revealed in the nonlinear mode.

Acoustic contrast is most pronounced in samples containing elements responsible for nonlinearity; namely, disbonds, microcracks, etc., by virtue of contact acoustic nonlinearity (CAN). Nonlinear acoustic microscopy of CAN areas is fully considered in [201, 202]. It has been shown that an acoustic image constructed on the basis of second harmonics at a frequency of 25 MHz in the reflection mode reproduces the outlines of a "kissing bond" around a weld in steel plates. Although the inquiries into this topic are relatively few in number, each provides evidence to support the efficiency of nonlinear acoustic microscopy and new information about the properties of materials and their inhomogeneities on the microscopic level.

Reviewing the aforesaid, it is easy to outline the top-priority goals of future investigations in this field. As opposed to the case of optical waves [163], the reflection and refraction of acoustic waves, especially in solids, have long been recognized as linear processes. Experimental and theoretical results gained over the last decade indicate the need for further development of nonlinear approaches to the analysis of solids. A comprehensive inspection of the methods based on evaluation of the finite amplitude, as an ultrasonic wave propagates over nonlinear surfaces (or is reflected from them), is vital for the understanding of nonlinear approaches to studying inhomogeneous materials.

Investigation into acoustic phenomena in the nonlinear-contact area is of prime importance to a wide range of fields in materials science. Certain fresh data gained from experiments with nonlinear contacts, once scrutinized and properly interpreted, may become a building block of new efficient methods of nonlinear characterization of materials.

The results of the application of nonlinear acoustic visualization can be classified in three groups, according to frequency ranges of ultrasonic signals used in medicine and biology, namely,

(1) at frequencies above 100 MHz (high-resolution methods);
(2) at low frequencies – below 20 MHz; and
(3) at medium frequencies from 20 to 100 MHz (nonlinear effects virtually remain unexplored).

At frequencies from 20 to 100 MHz, acoustic microscopy proved to be a highly efficient tool for studying microdefects of various materials and structures, including welds [203], biological entities [204], composites, crystals, and metal alloys [81]. In addition to a reasonably high resolution, this frequency range offers freedom from purchasing expensive equipment and permits the use of relatively simple methods which are easy to introduce into industrial materials technology. On the other hand, when applied in this frequency range, nonlinear approaches yield clear and unambiguous results, especially in the case of materials with macrodefects, which significantly improves the diagnostic possibilities for detection of microcracks or microinhomogeneities which adversely affect the integral properties of matter and bring about the hazard of the disruption of materials in the aircraft industry, the nuclear power industry, microelectronics, etc.

5.2
Peculiarities of Nonlinear Acoustic Effects in the Focal Area of an Acoustic Microscope

The manifestation of nonlinear effects in acoustic microscopy owes its peculiarities to high intensities of ultrasound in both immersion liquid and test samples which, in turn, are associated with the use of focused acoustic signals. Depending on the way in which acoustic effects reveal themselves, they can be divided into three groups.

The first group of effects is associated with the action of a focused beam on test objects. There are several mechanisms for its action. Let us point out the most efficient of these and, first of all, the mechanical and thermal mechanisms essential to any object irrespective of its nature, shape, or other characteristics.

An ultrasonic beam exerts a direct mechanical action via radiation pressure forces at the boundaries of a sample or inside it as well as ultrasonic pondermotive forces in the bulk of the sample. The mechanical action on a sample can also be associated with acoustic microflows arising in an immersion liquid under the effect of ultrasonic pondermotive forces in its bulk.

The foremost nonmechanical effect of a focused ultrasonic beam on a sample is thermal, i.e., heating of the sample due to direct absorption of ultrasonic energy by the sample itself and due to heat release in an immersion liquid. Generally, in addition to the above-listed mechanisms, there may arise some special effects peculiar to one or another type of sample. Pioneering studies on the effects of focused beams on test objects in acoustic microscopes were accomplished relatively recently [205, 206]. This vital problem of acoustic microscopy persists even today.

The second group of nonlinear effects in an acoustic microscope comprises nonlinear processes of the generation of harmonics in a convergent beam and initiation of sum- and difference-frequency waves when focused beams interact with one another.

An attractive feature of an acoustic microscope, having no parallels in optical microscopy, is the feasibility of the nonlinear mode. It has been known that a highly intense acoustic wave propagating in a medium gives rise to higher harmonics. Imaging on the basis of harmonics of the fundamental frequency of radiation is possible, owing to a high concentration of the acoustic wave energy at the focal point of a lens and the nonlinear acoustic properties of test objects and immersion liquids. If the nonlinear properties of a microscopic object are more distinct than those of the environment, the harmonic-based image becomes negative with respect to its linear counterparts.

As noted above (on page 11), the nonlinear mode of SAM operation, with fundamental and output-signal frequencies of $f_0 = 400$ and 800 MHz, respectively, was achieved in [180] and other studies, e.g., in [207, 208]. The nonlinear-mode microscope was not very different from a linear SAM, apart from the fact that its receiving unit was tuned to a frequency of $2f_0$.

Nonlinear generation effects can be used in acoustic microscopy in a variety of ways. One of these was widely known even in the early stages of acoustic microscopy evolution [180, 207, 208]: as an intense convergent acoustic beam propagated in the bulk of a liquid occupied by the beam, the second harmonic formed which itself produced a convergent beam having twice the frequency of the original beam. This is a well understood phenomenon of parametric ultrasound generation covered in the literature [209, 210]. The use of a doubled frequency yields a large dividend in microscope resolution. The generation of a convergent beam at the fundamental frequency and reception at a doubled frequency constitutes the essence of the most popular nonlinear operation mode of an acoustic microscope [208]. However, under these operation conditions, an acoustic image still gives an account of the distribution of the linear acoustic properties of test objects. In order to visualize the distribution of the local nonlinear properties of a test object, one needs to receive radiation emitted during the course of nonlinear processes inside the object. These conditions of acoustic imaging can be achieved through reception of sum- and difference-frequency waves which form by nonlinear interaction of focused beams with a common focal area immediately inside the object and do not intersect outside it. Unfortunately, these undoubtedly unique operation conditions still remain to be realized.

In the third group of nonlinear processes are the effects of self-action of a convergent acoustic beam altering the beam structure and, what is most important, the size of its focal area. Among these self-action effects are self-focusing and defocusing which in the case of ultrasonic beams are associated with the heating of an immersion liquid or acoustic flows arising in it and, of course, with anisotropy of test objects, predominantly of crystals. By virtue of the aforesaid effects, a focused beam propagates in an inhomogeneous medium responsible for its defocusing or additional self-focusing, which also changes the location and structure of the focal area. This line of inquiry was developed in [209, 210] without regard to the acoustic microscopy specificity, when studying weakly converging (small-aperture) beams. To be sure, self-action phenomena will be explored in more detail in future years, partly using the data available in the literature and partly extending them to the case of wide-aperture beams, to assess their implications for acoustic microscopy.

Considering the importance of self-action effects in practical implementation of acoustic microscopy, many researchers, including my co-workers, have studied the effects of radiation pressure and heating.

In addition, one of the feasible specific effects of a focused ultrasonic beam on objects with a low resistance to shape changes was also studied under these conditions. These objects are predominantly microscopic entities such as cells and one-celled animals, gas bubbles, microparticles in dispersive systems, electron-hole drops in semiconductors, mesophases in lyotropic liquid crystals, etc. An acoustic signal in a microscope represents a pulse train and, hence, is amplitude-modulated. Modulating frequencies commonly lie in the range of tens of kilohertz, typical for the above-listed microentities [66, 211–218]. This implies that, by virtue of nonlinear forces, an acoustic signal can exert a resonance action on microparticles.

5.3
Temperature Effects in the Focal Area of an Acoustic Microscope

When studying materials with SAM, it is sometimes necessary to take into account the effects of their heating. This heating is associated both with ultrasound absorption immediately within a test object placed in the focal area of a microscope and with heat transferred to the object from the liquid in which the object is immersed. Insignificant as it usually is at low values of the input acoustic power (fractions of a milliwatt) [219], the heating can attain 20 °C or more in the case of a relatively high radiated power and strongly absorbing samples [220].

Direct measurement of the temperature distribution in a sample by means of acoustic microscopy presents problems, and therefore, it is worthwhile to resort to theoretically calculating the maximum increase in the temperature of a sample.

Heating of an immobile medium in the focal area of a spherical radiator has been analyzed both theoretically and experimentally [221–223]. The intent of these studies was, in particular, to determine the distribution of heat sources associated with ultrasound absorption in the vicinity of the focal area [223] and to explore heat transfer from the focal area [222]. The empirical equations derived from the available experimental data to relate the temperature increase to the thermal properties of test samples and parameters of an acoustic focusing system are reviewed in [221]. Unfortunately, the results of [221, 222] are unsuitable for use in calculations of the temperature of the SAM focal area where heating is dynamic in character. This is because the focal spot representing a heat source moves with respect to the test object.

This section is concerned with the estimation of the maximum temperature rise in a homogeneous sample occupying a half-space, under the assumption that the radiating lens moves uniformly over the surface of the sample. Calculations are carried out for two of the most representative cases corresponding to two of the major fields of acoustic microscopy applications.

In the first case, the thermophysical properties of test objects differ only slightly from those of water – the most popular immersion liquid. Most of these objects are biological entities [224]. It is assumed in this case that a sample and the immersion liquid together represent a medium homogeneous in its thermal properties yet having different ultrasound-absorption coefficients. The distribution of heat sources in this system depends on the ultrasound-absorption coefficient and sound-pressure distribution in the focal area of an acoustic microscope.

In the second case, ultrasound absorption in test objects is insignificant compared to absorption in the immersion liquid, and heating of samples is primarily controlled by heat diffusion from the immersion-liquid layer adjacent to the sample. Among these objects are solid-state materials, e.g., metals, semiconductors and ceramics. The thermal conductivity of these materials is usually from 2 to 3 orders of magnitude greater than that of the immersion liquid [225].

In all cases, instead of calculating the actual spatial distribution of heat sources, we will use the model approximations to the spatial distribution of the heat source density, denoted by $Q(x, y, z, t)$, such that the resultant heating of the medium is an overestimate compared to that obtained with the actual heat distribution in the focal area of an acoustic microscope. Moreover, the effect of microflows at the focal point of an acoustic lens is disregarded [226]. It is worth noting that the estimates obtained in [226] (based on the data of [227]) lead to an over-rated value of the microflow velocity for an ultrasonic beam in a duct. For an acoustic lens, this velocity must be lower by approximately the ratio of the focal-waist length to the lens radius, and is of the order of 1 cm s^{-1} at an ultrasound intensity at the focal point of the lens of about 10^3 W cm^{-2}, which is an order of magnitude lower than the scanning velocity. Heating $T(x, y, z, t)$ at a given point will be calculated in terms of the influence function [79]. In a medium homogeneous in its thermal properties, $T(x, y, z, t)$ is given by the following expression

$$T(x, y, z) = \int_{-\infty}^{t} d\tau \iiint_{\infty} \frac{Q(\xi, \vartheta, \varsigma, \tau)}{[4\pi \chi (t - \tau)]^{3/2}}$$

$$\times \exp\left\{-\frac{(x - \xi)^2 + (y - \vartheta)^2 + (z - \varsigma)^2}{4\chi (t - \tau)}\right\} d\xi \, d\vartheta \, d\varsigma \quad (5.6)$$

where χ is the thermal diffusivity of the medium.

Let us first assess the effect of point heat source $Q = \delta(x, y - vt, z)$ moving at a rate of v along the y-axis on the temperature at point $x = 0, y = 0, z = z_0$ on the z-axis. Substituting Q into (5.6) we obtain

$$T(z_0, t) = \frac{1}{\pi \chi \sqrt{z_0^2 + (vt)^2}} \exp\left\{-\frac{v}{2\chi}\left(\sqrt{z_0^2 + (vt)^2} - \frac{vt}{2}\right)\right\} \quad (5.7)$$

Multiplier $2\chi/v$ in the exponent in (5.7) conveys the meaning of a characteristic length within which the moving heat source produces a perceptible temperature change. In particular, at $v = 10 \text{ cm s}^{-1}$, $2\chi/v$ amounts to 3.2×10^{-4} cm for water, which is of the order of the focal area size of SAM with an operating frequency in the range between 500 and 1000 MHz.

Thus, the main contribution to the temperature rise in the focal area of an acoustic microscope comes from the heat sources in the vicinity of the focal waist, and therefore the heat source density $Q(x, y, z, t)$ can be approximated by quite a simple function [222] in the form of a Gaussian beam damped exponentially along the z-axis:

$$Q(x, y, z, t) = \frac{2\alpha I}{\rho C} \exp\left\{-\frac{x^2 + (y - vt)^2}{\beta} - 2\alpha z\right\} \quad (5.8)$$

where β is the half-width of the focal waist, α is the absorption coefficient, ρ is the density of the medium, C is its heat capacity, and I is the maximum intensity at the focal point of an ultrasonic lens. From this point, we will use the frame of reference with one of its axes aligned with the acoustic lens, and the acoustic lens focus is supposed to reside at the origin of the coordinates at time $t = 0$.

Certain narrowing of the heat-source distribution region outside the focal waist, stemming from Equation (5.8), leads to an imperceptible overestimation of the temperature.

Heating of biological samples is controlled both by ultrasound absorption in the samples and by heat transfer from the immersion liquid. The temperature increase on the lens axis at depth z, due to ultrasound absorption in the immersion liquid and heat transfer to the sample is derived from Equation (5.6) and is expressed by the following equation:

$$T_i(z_s) = \frac{a_s I \beta}{\rho C} \int_0^\infty \exp\left\{-\frac{(v\tau)^2}{4\chi\tau + \beta} - 2\alpha_i z_s + 4\alpha_i^2 \chi \tau\right\}$$

$$\times \operatorname{erfc}\left\{\frac{z_s}{\sqrt{4\chi\tau}} - \alpha_i\sqrt{4\chi\tau}\right\} \frac{d\tau}{\beta + 4\chi\tau} \quad (5.9)$$

where

$$\operatorname{erfc}(a) = \frac{2}{\sqrt{\pi}} \int_a^\infty \exp(-\xi^2)\, d\xi$$

5.3 Temperature Effects in the Focal Area of an Acoustic Microscope

and subscripts i and s pertain to the immersion liquid and sample, respectively. Naturally, the maximum heating due to heat transfer from the immersion liquid takes place on the surface of the sample at $z_s = 0$. Considering that in actual practice, absorption is insignificant over the characteristic thermal length, $\alpha \ll v/(2\chi)$, as well as over the focal-waist length, $\alpha \ll 1/\sqrt{\beta}$, the heating of the medium close to the surface of the sample can be represented in the form:

$$T_i(z_s) = \frac{\alpha_i I \beta}{\rho C} \exp\left(\frac{\beta v^2}{8\chi^2}\right) K_0\left(\frac{\beta v^2}{8\chi^2}\right) \tag{5.10}$$

where K_0 is the modified Bessel function. If the characteristic thermal length exceeds the focal waist half-width, $2\chi/v \ll \sqrt{\beta}$, (5.10) can be simplified to

$$T_i(0) = \frac{\alpha_i I \beta}{\chi \rho C} \ln\left(\sqrt{2} \frac{2\chi}{v\sqrt{\beta}}\right)$$

using approximation $K_0(a) \approx -\ln(a)$ [127].

For an acoustic microscope operating at a frequency of 500 MHz with water as an immersion liquid, the temperature rise was estimated at $T = 0.1\,°\text{C}$, assuming that the focal waist half-width and ultrasound intensity were $\sqrt{\beta} = 1.5\,\mu\text{m}$ and $I = 10\,\text{W cm}^{-2}$, respectively. In the following, calculations will be performed with the foregoing microscope parameters, unless otherwise specified.

At low frequencies ($f < 100$ MHz), when the focal waist half-width exceeds the thermal length, $\sqrt{\beta} \gg 2\chi/v$, the modified Bessel function can be replaced by the asymptotic expression, $e^x K_0(x) = 1.25/\sqrt{x}$ [127]. In this case, the temperature rise at the surface of the sample is governed by the equation, independent of the thermal conductivity of the sample

$$T_i(0) = \frac{1.25 \alpha_i I \sqrt{\beta}}{\sqrt{2} v \rho C}$$

For water, $T_i = 0.01\,°\text{C}$ at a frequency of 100 MHz and $\sqrt{\beta} = 7.5\,\mu\text{m}$.

It remains to be seen how heating $T_i(z_s)$ decays deep in the sample. The integrand in Equation (5.3) involves three cofactors, one of which

$$-\exp\left\{-\frac{v\tau}{4\chi\tau + \beta} - 2\alpha_i z_s + 4\alpha_i^2 \chi \tau\right\}$$

rapidly tends to zero at $\tau > v^2/(4\chi^2)$. At small τ values, $\tau < v^2/(2\chi)$, the following approximate representation

$$\text{erfc}\left\{\frac{z_s}{\sqrt{4\chi\tau}} - \alpha_i \sqrt{4\chi\tau}\right\} \approx \exp\left(-\frac{z_s^2}{4\chi\tau}\right)$$

is true, and the temperature distribution formed upon heating of the sample reduces to

$$T_{\mathrm{i}}(z_{\mathrm{s}}) \approx \frac{\alpha_{\mathrm{i}} I \beta}{\rho C} \int_0^\infty \frac{1}{\beta + 4\chi\tau} \exp\left\{-\frac{v^2\tau^2}{\beta + 4\chi\tau} - \frac{z_{\mathrm{s}}^2}{4\chi\tau}\right\} d\tau$$

$$= \frac{2I\beta\alpha_{\mathrm{i}}}{\chi\rho C} K_0\left(z_{\mathrm{s}}\frac{v}{2\chi}\right)$$

Inasmuch as at $z_{\mathrm{s}} > 2\chi/v$, the modified Bessel function decays exponentially with increasing z_{s} and the heat transferred from the immersion liquid to the sample has time to raise the temperature of just a thin layer with a characteristic size of the order of one thermal length $2\chi/v$ near the surface of the sample.

The temperature rise $T_{\mathrm{s}}(z_{\mathrm{s}})$ due to ultrasound absorption in the sample is calculated in much the same way as in the case of the immersion liquid (see (5.9)):

$$T_{\mathrm{s}}(z_{\mathrm{s}}) = \frac{\alpha_{\mathrm{s}} I \beta}{\rho C} \int_0^\infty \exp\left\{-\frac{v^2\tau^2}{4\chi\tau + \beta} - 2\alpha_{\mathrm{s}}^2 z_{\mathrm{s}} + 4\alpha_{\mathrm{s}}\chi\tau\right\}$$

$$\times \operatorname{erfc}\left\{-\frac{z_{\mathrm{s}}}{\sqrt{4\chi\tau}} + \alpha_{\mathrm{s}}\sqrt{4\chi\tau}\right\} \frac{d\tau}{4\chi\tau + \beta}$$

At $z_{\mathrm{s}} = 0$, $T(0)$ is given by an expression identical to (5.4) with α_{i} replaced by α_{s}. At $z > 2\chi/v$;

$$\operatorname{erfc}\left\{-\frac{z_{\mathrm{s}}}{\sqrt{4\chi\tau}} + \alpha_{\mathrm{s}}\sqrt{4\chi\tau}\right\} \approx 2$$

and $T(z)$ is given by the following expression:

$$T_{\mathrm{s}}(z_{\mathrm{s}}) \approx \frac{\alpha_{\mathrm{s}} I \beta}{2\chi\rho C} \exp\left(\frac{\beta v^2}{8\chi^2}\right) K_0\left(\frac{\beta v^2}{8\chi^2}\right) \exp(-2\alpha_{\mathrm{s}} z_{\mathrm{s}})$$

$$= 2T_{\mathrm{s}}(0) \exp(-2\alpha_{\mathrm{s}} z_{\mathrm{s}}) \tag{5.11}$$

As is obvious from Equation (5.11), T_{s} decreases with increasing z_{s} as $\exp(-2\alpha_{\mathrm{s}} z_{\mathrm{s}})$.

Heating $T_{\mathrm{s}}(z = 0)$ for a solid-state sample will be calculated using the method of successive approximations. By virtue of the fact that the thermal conductivity of this sample is usually 2 or 3 orders of magnitude greater than that of water, when calculating the temperature distribution in the immersion liquid, it is safe to assume that the sample is heated much less than the immersion liquid and $T_{\mathrm{i}}(z = 0) = 0$.

Under these assumptions, the temperature distribution in the immersion liquid at its interface with the sample can be determined by the reflection method [228]. We next calculate the heat flux from the immersion liquid to the sample:

$$Q_{\mathrm{s}}(x, y, z)\rho_{\mathrm{s}} C_{\mathrm{s}} = -\chi_{\mathrm{i}} \rho_{\mathrm{i}} C_{\mathrm{i}} \left.\frac{\partial T_{\mathrm{i}}}{\partial z}\right|_{z=0}$$

and the temperature at the surface of the sample. Assuming that absorption is low over the thermal length, $\alpha_i \ll v/(2\chi)$, and ultrasound is completely reflected from the surface of the sample, we obtain the following expression for the temperature gradient in the immersion liquid near the surface of the sample:

$$\left.\frac{\partial T_i}{\partial z}\right|_{z=0} = \int_{-\infty}^{t} \frac{d\tau}{[4\pi\chi(t-\tau)]^{3/2}} \iint_{\infty} d\xi\, dv$$

$$\times \int_{-\infty}^{0} \frac{\partial}{\partial z} \left\{ Q(\xi, v, \varsigma, \tau) \left[\exp\left(-\frac{(z-\varsigma)^2}{4\chi(t-\tau)}\right) - \exp\left(-\frac{(z+\varsigma)^2}{4\chi(t-\tau)}\right) \right] \right\}$$

$$\times \exp\left(-\frac{(x-\xi)^2 + (y-v)^2}{4\chi(t-\tau)}\right) d\vartheta$$

$$= -\int_{0}^{\infty} \frac{4\beta I (\rho_i C_i)^{-2}}{(\beta + 4\chi_i \tau)\sqrt{4\chi_i \tau}} \exp\left(-\frac{x^2 + (y - v(t-\tau))^2}{4\chi_i \tau + \beta}\right) d\tau \quad (5.12)$$

Obviously, the maximum temperature rise is achieved at the interface between the sample and the immersion liquid at the focal spot center. Therefore, using a conventional solution of the heat conduction equation subject to the boundary conditions, we obtain for $T_s(0)$:

$$T_s(0) = \frac{4\beta I \alpha_i \sqrt{\chi_i}}{\pi \rho_s C_s \sqrt{\chi_s}} \int_{0}^{\infty} \frac{1}{(4\chi_s t + \beta + 4\chi_i \tau)\sqrt{4t\tau}}$$

$$\times \exp\left(-\frac{(v(t-\tau))^2}{4\chi_s t + \beta + 4\chi_i \tau}\right) d\tau$$

Switching to new variables

$$R\cos\varphi = 2\sqrt{\tau\chi_s} \qquad R\sin\varphi = 2\sqrt{\tau\chi_i} \qquad u = \frac{R^2}{1-R}$$

we arrive at [229]:

$$T_s(0) = \frac{2\beta I \alpha_i}{\pi \rho_s C_s \chi_s} \int_{0}^{\pi/2} \exp\left[2\beta v^2 \left(\frac{\cos^2\varphi}{4\chi_s} + \frac{\sin^2\varphi}{4\chi_i}\right)^2\right]$$

$$\times K_0 \left[2\beta v^2 \left(\frac{\cos^2\varphi}{4\chi_s} + \frac{\sin^2\varphi}{4\chi_i}\right)^2\right] d\varphi \quad (5.13)$$

Considering that $2\beta v^2 (\cos^2\varphi/(4\chi_s) + \sin^2\varphi/(4\chi_i))^2 \ll 1$ at all or almost all φ values, the integrand can be replaced with its "upper" approximation:

$$-\ln\left[2\beta v^2 \left(\frac{\cos^2\varphi}{4\chi_s} + \frac{\sin^2\varphi}{4\chi_i}\right)^2\right]$$

In this case, Equation (5.8) reduces to

$$T_s(0) = \frac{2\beta I \alpha_i}{\rho_s C_s \chi_s} \ln\left(\frac{8\sqrt{2}\chi_i}{v\sqrt{\beta}}\right)$$

For silicon [230], $\rho_s = 2.24$ g cm^{-3}, $C_s = 0.84$ J (g · K)$^{-1}$, and $\chi_s = 0.82$ cm^2 s^{-1}. Thus, $T_s(0) = 6 \times 10^{-3}$ °C.

The heat released in the bulk of the immersion liquid and the sample and at the sample/immersion liquid interface can be augmented by the heat associated with a finite viscosity of the immersion liquid and nonadiabaticity of the processes at the immersion liquid/sample interface [13]. In order not to go into details of the specific mechanism of heat release on the surface of the sample, let us assess the temperature rise in it, assuming that the entire ultrasonic power incident on the sample is evolved on its surface. To do this, we pass to the limit in Equation (5.12), as $\alpha_s \to \infty$, and obtain the following expression for the temperature at $z_s = 0$:

$$T_\infty = \frac{I\sqrt{\pi\beta}}{4\chi\rho C} \tag{5.14}$$

In this case, $T_\infty = 4$ °C for a biological sample and $T_\infty = 0.01$ °C for silicon.

In actual practice, the heat released is proportional to $I(\gamma - 1)\sqrt{\omega\chi_i}/c_i$ or $8I\sqrt{\omega\eta_i}/(c_i\sqrt{\rho_i})$ [13] (γ is the adiabatic index, c_i is the sound velocity, and η_i is the viscosity), and hence the estimate (5.14) reduces by at least an order of magnitude.

The heating of a sample while a heat source is moving with respect to it, manifests itself as a temperature jump with a characteristic rise time of $\sqrt{\beta}/v = 10^{-5}$ s. In addition, the sample undergoes continuous heating due to its repeated scanning. This heating is primarily associated with a large amount of heat released in the immersion liquid close to the radiating lens, where the acoustic power is high: $P \approx \pi\beta I \exp(2\alpha_i R) = 2 \times 10^{-2}$ W (where R is the radius of curvature of the lens).

The order of magnitude of the continuous heating can be assessed using a simple model, assuming that acoustic lenses are planar, an immersion-liquid drop is cylindrical in shape, the lenses are temperature-controlled, and heat release depends only on coordinate z. If so, the increase in the temperature of a sample studied with a transmission-mode microscope is given by the following expression:

$$T_s = \frac{\beta I \exp(2\alpha_i R)}{2\alpha_i \xi_i \rho_i c_i A^2}$$

For a lens of radius $A = 0.015$ cm, with a radius of curvature $R = 0.03$ cm, the expected heating is $T_s = 0.4$ °C.

In the event of a reflection-mode microscope, the heating of a sample depends on absorption in the immersion liquid and heat removal from the lower boundary of the sample:

$$T_s = \frac{\beta I \exp(2\alpha_i R) L}{2\alpha_i \xi_i \rho_i c_i A^2 R}$$

For a silicon sample of thickness $L = 1$ mm, the temperature amounts to $T_s = 5 \times 10^{-3}\,°C$.

Thus, the analysis of the examples considered above has revealed that, for different relationships between the properties of a sample and an immersion liquid, the upper limit of the focal-area temperature assessed with due regard to fast-alternating and steady heating, does not exceed one degree in the case of biological samples, and $10^{-2}\,°C$ in the case of solid-state matter. Therefore, the properties of materials associated directly or indirectly with temperature variations and measured by means of acoustic microscopy can be interpreted with certainty.

5.4
Effects of Radiation Pressure on Samples Examined with an Acoustic Microscope

In the field induced by an acoustic wave during the course of its propagation there arise constant forces acting on any obstacle. For a focused acoustic beam, the radiation pressure force in the focal area of an acoustic microscope can be fairly high. In the case of biological entities, a high radiation pressure at the focal point of a microscope can alter and even cause damage [231]. Moreover, the study of radiation forces lies at the basis of the procedure for determining the integral intensity of a focused sound beam and the energy distribution in the focal area [232].

The intention of this section is to explore radiation forces at the focal point of a convergent ultrasonic beam. While the radiation pressure in a plane wave has been the subject of multiple papers (e.g., see [233]), inhomogeneous fields have not been adequately studied. The radiation pressure distribution in the field of a point spherical radiator was considered in [234], and the radiation pressure force in the near field of a planar radiator in [235]. The lenses used in acoustic microscopy have small aberrations, and the field in the focal area of an acoustic microscope is little different from that of a spherical transducer. Therefore, for simplicity, we will assess the radiation pressure formed at the focal point of a spherical transducer.

The radiation pressure force along the acoustic axis of a transducer depends on the momentum flux through the surface of a test sample [236]:

$$F_z = -\int_s \Pi_{zi} n_i\, ds \tag{5.15}$$

Here, $\Pi_{zi} = \vec{p}\delta_{zi} + \rho V_z V_i$ is the momentum flux density tensor, where $\vec{p} = \rho(\partial \Psi/\partial t)^2/(2c^2) - \rho V^2/2$ is the time-average excess pressure in a sound wave, ρ is the density of a liquid, Ψ is the vibrational velocity potential, c is the sound velocity in the liquid, $V = \mathrm{grad}\,\Psi$ is the vibrational velocity, and V_z and V_i are the vibrational velocity components; n_i is the projection of a normal to the surface of the sample in direction $\vec{i}$. For small sound wave amplitudes acoustic fields are quite adequately described by the expressions derived in the linear approximation.

To illustrate the main effects induced by the radiation pressure, consider conventional targets, namely, fixed incompressible spheres and disks. From (5.15), it follows that one has to solve the problem of focused-beam diffraction by spheres and disks before the radiation pressure on them can be calculated. Let us solve this problem for a rigid disk in the focal plane [205]. A thin disk or a slot in a screen are the limiting cases of an arbitrary spheroidal target and, therefore, the exact solution of the problem of scattering from a disk should be sought in the form of a superposition of spheroidal wave functions.

Consider a focusing spherical transducer with aperture angle α and radius of curvature f. The vibrational velocity on the transducer surface is constant and equal to V_0. For the linear approximation to be appropriate in the case of a spherical transducer, it is requisite that $\beta = kV_0 f/(c \ln(kf(1-\cos\alpha)/(2\pi))) \ll 1$ [237], where k is the wave number and β is the nonlinearity parameter. In the linear approximation, the incident field in the vicinity of the focus is represented in the form of a set of plane waves (Debye equation) [67]:

$$\Psi_i(r_0, \theta_0, \varphi_0)$$

$$= \frac{V_0 f}{2\pi} \exp(i\omega t - ikf) \int_0^\alpha \sin\theta\, d\theta \int_0^{2\pi} \exp(ikr_0 \cos\gamma)\, d\varphi \quad (5.16)$$

where $\Psi_i(r_0, \theta_0, \varphi_0)$ is the vibrational velocity potential at point $(r_0, \theta_0, \varphi_0)$, ω is the angular frequency, and γ is the angle between vectors $\mathbf{r}_0$ and $\mathbf{f}$. In a spherical coordinate system tied to the center of curvature of a focusing transducer, vectors $\mathbf{r}_0$ and $\mathbf{f}$ have coordinates $(r_0, \theta_0, \varphi_0)$ and (f, θ, φ), respectively, and $\cos\gamma = \cos\theta\cos\theta_0 + \sin\theta\sin\theta_0\cos(\varphi - \varphi_0)$.

The expression for a plane wave, $\exp(ikr_0 \cos\gamma)$, in Equation (5.16) is next expanded in terms of spheroidal wave functions [238]:

$$\exp(ikr_0 \cos\gamma) = 2 \sum_{n,m}^\infty \frac{i^n(2-\delta_{0m})}{N_{nm}} R_{nm}^{(1)}(i\,\text{sh}\,u)$$

$$\times S_{nm}(\cos v) S_{nm}(\cos\theta) \cos(m(\varphi - \varphi_0)) \quad (5.17)$$

Here, δ is the Kronecker delta, S_{nm} and $R_{nm}^{(1)}$ denote oblate angular spheroidal functions and oblate radial spheroidal functions of the first kind, respectively,

$$N_{nm} = \int_{-1}^1 S_{nm}^2(x)\, dx$$

(u, v, φ) signifies oblate spheroidal coordinates related to the Cartesian coordinates by equations

$$z = a\,\text{ch}(u)\cos(v)$$

$$x = a\,\text{ch}(u)\sin(v)\cos\varphi$$

$$y = a\,\text{ch}(u)\sin(v)\sin\varphi$$

where a is the radius of a disk placed in the $z = 0$ plane. Substitution of (5.17) into (5.16) and integration with respect to φ leaves behind only the terms with $m = 0$. Thus, the expression for the incident field can be written as

$$\Psi_i = V_0 f \exp(i\omega t - ikf) \sum_{n=0}^{\infty} g_n(\alpha) R_n^{(1)}(i \operatorname{sh} u) S_n(\cos v) \tag{5.18}$$

where

$$g_n(\alpha) = \frac{2i^n}{N_n} \int_0^{\alpha} S_n(\cos v) \sin v \, dv \tag{5.19}$$

Let us substitute S_{n0} into (5.19) in the form of an expansion in terms of Legendre polynomials $P_m(\cos v)$:

$$S_n(\cos v) = \sum_{m=0,1}^{\infty}{}' d_m^n P_m(\cos v)$$

where d_m^n are the numerical factors dependent on the ratio of the disk radius to the sound wavelength. A prime signifies that $m = 0, 2, 4 \ldots$, if n is even, and $m = 1, 3, 5 \ldots$, if n is odd. In view of the foregoing, function $g_n(\alpha)$ takes the form

$$g_n(\alpha) = \frac{2i^n}{N_n} \sum_{m=0,1}^{\infty}{}' d_m^n [P_{m-1}(\cos \alpha) - P_{m+1}(\cos \alpha)](2m+1)^{-1} \tag{5.20}$$

Equations (5.18) and (5.20) represent the desired expansion of the incident field of a focusing spherical transducer in terms of spheroidal wave functions in the vicinity of the center of curvature. The scattered wave potential sought-after in the form

$$\Psi_s = V_0 f \exp(i\omega t - ikf) \sum_{n=0}^{\infty} C_n g_n(\alpha) R_n^{(4)}(i \sin u) S_n(\cos v) \tag{5.21}$$

where $R_n^{(4)}$ denotes oblate radial spheroidal functions of the 4th kind, and unknown constants C_n are to be derived by setting the normal velocity component equal to zero on the disk surface: $C_n = 0$, if n is even, and $C_n = -R_n^{(1)'}(0)/R_n^{(4)'}(0)$, if n is odd.

The above-presented solution of the problem on diffraction by a disk will be used to calculate the radiation pressure on it in a focused beam.

To find the field scattered by a sphere, the incident field must be expanded in terms of spherical harmonics in the frame of reference tied to the center of the sphere. To do this, we proceed as proposed in [67]. First, the plane wave $\exp(ikr_0 \cos \gamma)$ from (5.16) is expanded in terms of spheroidal harmonics [239]:

$$\exp(ikr_0 \cos \gamma) = \sum_{n=0}^{\infty} i^n (2n+1) j_n(kr_0)$$

$$\times \sum_{m=0}^{\infty} \varepsilon_{nm} P_n^m (\cos \theta) P_n^m (\cos \theta_0) \cos(m(\varphi - \varphi_0)) \quad (5.22)$$

where j_n signifies the spherical Bessel functions, P_n^m stands for the associated Legendre polynomials, and $\varepsilon_{nm} = (2 - \delta_{0m})(n-m)!/(n+m)!$.

Equation (5.22) is then substituted into Equation (5.16). Integration of $\cos(m(\varphi - \varphi_0))$, appearing in the sum in (5.22), with respect to φ eliminates all except the terms with $m = 0$.

The integral of $P_n(\cos \theta) \sin \theta$ with respect to θ between the limits 0 and α equals $[P_{n-1}(\cos \alpha) - P_{n+1}(\cos \alpha)]/(2n+1)$, where P_n is the Legendre polynomial. Therefore, the incident focused wave expanded in terms of spherical harmonics takes the form [240]:

$$\Psi_i = V_0 f \exp(i\omega t - ikf) \sum_{n=0}^{\infty} B_n i^n j_n(kr_0) P_n(\cos \theta_0) \quad (5.23)$$

where $B_n = P_{n-1}(\cos \alpha) - P_{n+1}(\cos \alpha)$.

Suppose that a sphere of radius a resides at the center of curvature of a transducer. The scattered wave potential is desired in the form

$$\Psi_s = A \exp(i\omega t) \sum_{n=0}^{\infty} C_n B_n i^n h_n(kr_0) P_n(\cos \theta_0) \quad (5.24)$$

where h_n denotes the spherical Hankel function. Unknown constants C_n are to be derived from the boundary condition. Inasmuch as the sphere is incompressible and fixed, the normal component of the vibrational velocity on its surface, $\partial(\Psi_i + \Psi_s)/\partial r_0|_{r=a}$, equals zero. Substituting (5.23) and (5.24) into the boundary condition we obtain the unknown constants, $C_n = -j'_n(ka)/h'_n(ka)$. Integration over the surface of the object yields the following expression for the radiation pressure exerted by a focused beam on an incompressible sphere

$$F = 2\pi f^2 \frac{I}{c} \sum_{n=0}^{\infty} \left(1 - \frac{n(n+2)}{(ka)^2}\right) \frac{q_n q_{n+1}}{(1+n/2)(ka)^4} \quad (5.25)$$

where $I = \rho V^2 c/2$ is the sound wave intensity on the transducer surface, and $q_n = (\cos \alpha P_n(\cos \alpha) - P_{n-1}(\cos \alpha))/(j'^2_n(ka) + h'^2_n(ka))$.

When the radius of a sphere is much shorter than the sound wavelength, the sum in (5.25) can be truncated to the first two terms

$$F = \frac{W}{4c}(ka)^6 \sin^2 \alpha \left(1 + \frac{\cos \alpha (1 + \cos \alpha)}{9}\right)$$

where $W = 2\pi f^2 (1 - \cos \alpha) I$ is the total power radiated by the transducer.

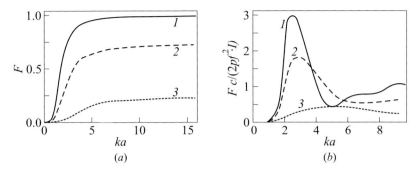

Figure 5.4 Normalized radiation force acting on a rigid sphere (a) and on a rigid disk (b): $\alpha = \pi/2$ (1), $\pi/3$ (2), and $\alpha = \pi/6$ (3).

As in a traveling plane wave, the force is proportional to $(ka)^6$ and builds up monotonically with the radius of the sphere. When a sphere is large compared to the wavelength ($ka \gg 1$), series (5.25) converges slowly. To assess the radiation pressure force when the radius of the sphere far exceeds the focal waist size one must invoke the law of conservation of momentum, according to which the pressure on the reflecting surface is equal to twice the momentum flux incident on the surface. After calculating the total momentum flux through the surface of a focusing spherical transducer along the acoustic axis, we obtain the radiation force acting on the sphere

$$F = (1 + \cos \alpha) \frac{W}{c} \qquad (5.26)$$

which depends solely on the radiator parameters. The radiation force dependence on the particle size at the focal point of an acoustic microscope is illustrated in Figure 5.4(a) for three different values of aperture angle α.

To calculate the pressure on the disk, we substitute Equations (5.18) and (5.21) into (5.15) and integrate the result over the disk surface, which yields for the radiation force:

$$F = 2\pi f^2 \frac{I}{c} \sum_{n=0}^{\infty} \sum_{m=0}^{\infty} g_{2n}(\alpha) R_{2n}^{(1)}(0) \operatorname{Re}\bigl(C_{2m+1} R_{2m+1}^{(4)}\bigr)\bigl[S_v^{nm} - S_t^{nm}(ka)^2\bigr]$$

where S_t^{nm} and S_v^{nm} are the integrals of the angular spheroidal functions

$$S_t^{nm} = \int_0^{\pi} S_{2n}(\cos v) S_{2m+1}(\cos v) \sin v \, dv$$

$$= 2 \sum_{l=0}^{\infty} \frac{d_{2l}^{2n}}{4l+1} \left[\frac{2l \, d_{2l-1}^{2m+1}}{4l-1} + \frac{(2l+1) d_{2l+1}^{2m+1}}{4l+3} \right]$$

$$S_v^{nm} = \int_0^\pi S'(\cos v) S'(\cos v) \frac{\sin^3 v}{\cos v} dv$$

$$= 4 \sum_{l=1}^\infty \frac{(-1)^{l+1} l! d_{2l}^{2n}}{\Gamma(l+1/2)} \sum_{j=0}^{l-1} \frac{(-1)^j \Gamma(j+3/2)}{J!} d_{2j+1}^{2m+1}$$

Here, Γ is the gamma function; $l, j = 0, 2, 4 \ldots$, if n is even, and $l, j = 1, 3, 5 \ldots$, if n is odd; d_n denotes the coefficients of the expansion of spheroidal angular functions in terms of Legendre polynomials

$$S_l(\cos v) = \sum_n d_n^l P_n(\cos v)$$

Contrary to the case of a sphere, the dependence for a disk exhibits well-defined oscillations (Figure 5.4b). The presence of the first maximum of the radiation pressure on the disk in a plane wave was demonstrated by Awatany [241]. In a plane wave, the maximum arises for a disk of the size corresponding to the scattered momentum maximum. In the case of a focused beam, the aforesaid cause of non-monotonic variation of the radiation force with obstacle size in a plane wave is augmented by another factor; namely, an oscillating dependence of the efficiency of diffraction by a disk on the disk size is supplemented by the inhomogeneous distribution of the momentum flux over the sound beam cross-section in the focal area. Let us consider the abovementioned factors individually. To find out how the radiation force depends on diffraction the total power W_s scattered from an obstacle must be written as a function of the disk size:

$$W_s = \frac{W}{1-\cos\alpha} \sum_{n=0}^\infty g_{2n+1}^2(\alpha) \frac{N_{2n+1}}{1 + R_{2n+1}^{(2)'}/R_{2n+1}^{(1)'}(0)}$$

This dependence was calculated numerically for a hemispherical transducer (Figure 5.5).

The momentum flux density distribution over the beam cross-section can be inferred from the expression for an acoustic field in the focal area obtained by O'Neyl [69]. The expression for the momentum flux density at an arbitrary aperture

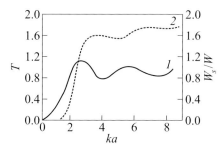

Figure 5.5 Normalized momentum flux $T = 2F_t c/W$ through the surface of a disk (1) and normalized power W_s/W scattered from the disk (2).

angle is cumbersome and therefore is omitted. For a hemispherical transducer ($\alpha = \pi/2$), the expression is much simpler:

$$\Pi_{zz} = \frac{If^2}{cr^2}\left[\sin^2 kr + J_1^2(kr) - \left(\cos kr - \frac{\sin kr}{kr}\right)^2\right]$$

where r is the distance from the observation point to the acoustic axis of the radiator, and J_n is the cylindrical Bessel function. The momentum flux density oscillates heavily and even takes negative values as the distance to the beam axis changes. This is because the excess pressure in the incident wave is also negative at these points (in other words, in antiphase).

The total momentum flux

$$F_t = 2\pi \int_0^a \Pi_{zz}(r)r\,dr$$

incident on a disk also experiences oscillations, as the disk radius changes: the radiation pressure force decreases when the disk radius falls within the region of negative Π_{zz} values. For a hemispherical radiator, the total momentum flux takes the form

$$F_t = \frac{W}{2c}\left[1 - \frac{\sin 2ka}{2ka} + \frac{\sin^2 ka}{2(ka)^2} - \frac{1}{2}\left(J_0^2(ka) + J_1^2(ka)\right)\right]$$

The normalized momentum flux T is plotted in Figure 5.5. In this case, the total momentum flux through the disk surface, as well as the power of a scattered sound wave, oscillates with increasing disk radius. As these oscillations are superimposed, the radiation pressure on a disk with a diameter close to the sound wavelength ($ka \approx 3$) increases dramatically [242]. At $ka > 3$, the oscillation amplitude does not exceed half the limiting value of the pressure and decays rapidly with increasing disk radius. In the case of very large disks ($ka > 10$), diffraction is insignificant, and the pressure force tends to limit (5.26). For a hemispherical transducer, the radiation pressure force at a maximum ($ka \approx 3$) is approximately twice the force acting on an infinitely large disk.

Estimates of the radiation pressure in the focal area of an acoustic microscope obtained at two frequencies without regard to absorption are summarized in Table 5.1.

Inasmuch as the entire energy was concentrated in the region of the principal maximum (its radius is approximately $0.6\lambda/\sin\alpha$, where λ is the sound wave-

Table 5.1 Radiation pressure force F and radiation pressure P at the focal point of an acoustic microscope ($I = 10^3$ W m^{-2}).

$\omega/(2\pi)$, MHz	f	F, N	P, N m^{-2}
100	1.5×10^{-3}	10^{-5}	1.3×10^4
200	6×10^{-3}	1.5×10^{-6}	5.3×10^4

length, and α is the lens aperture), the radiation pressure at the focal point was calculated by equation $(f/\lambda)^2 2I/c$. When absorption in an immersion liquid is taken into account, the pressure reduces by half at a frequency of 100 MHz and by three orders of magnitude at 500 MHz. However, if the wavelength is much shorter than the sphere (or disk) size ($ka \gg 1$, where k is the wave number, and a is the radius of a sphere or a disk), and a exceeds the pupil size of a transducer ($a > f \sin \alpha$), the entire momentum flux falls on the disk, and the total force, by virtue of the Borgnis theorem, remains unchanged [243].

5.5
The Theory of Modulated Focused Ultrasound Interaction with Microscopic Entities

As demonstrated in the preceding section, the radiation pressure and pondermotive forces arising in the focal area of a convergent ultrasonic beam are insignificant in acoustic microscopes. Accordingly, so is their action on fixed rigid objects. However, there is a special category of objects which perceive even these feeble forces as tangible. The case in point is the multitude of objects featuring a low resistance to shear or shape changes: biological cells and tissues; dispersed systems (namely, gels, suspended matter, and emulsions); mesophases of lyotropic crystals, etc.

It has been shown that the aforesaid objects, in addition to ultrasonic action by a conventional mechanism (deformation of samples in response to pondermotive and radiation forces), also experience the effect of a special mechanism associated with the resonant action of a focused ultrasonic beam on the structural elements of matter – biological cells, vesicles, microvessels, etc.

The bulk elastic properties of structural elements are little different from those of an ambient liquid. Therefore, their eigenmodes represent oscillations of shape rather than of volume. They depend on the elastic shear characteristics of microscopic objects, and their natural frequencies lie in the low-frequency range of the spectrum (tens or hundreds of kilohertz) because of low values of the shear modulus.

It might be useful here to point out that reflection acoustic microscopy relies on the use of relatively short probing radiofrequency pulses. This is because a single lens is employed first to excite and then to receive a probing ultrasonic signal. An acoustic-image point is a result of the accumulation of signals after processing a series of several tens or hundreds of acoustic pulses at the operating frequency of the apparatus. Therefore, an acoustic signal radiated into an immersion liquid represents an extended pulse of modulated ultrasound whose carrier frequency is dependent on the operating frequency of the microscope and the modulating frequency governed by the frequency of repetition of individual probing pulses.

As the signal hits a microscopic object, it squeezes and stretches the microparticle at regular intervals with the modulating frequency through the mediation of radiation forces acting at the microparticle/environment interface. Modulating frequencies fall into the same range as the frequencies of natural oscillations of microparticles, and therefore, a probing signal of an acoustic microscope can initiate natural oscillations. The Q factor of the oscillations is, as a rule, not high, the resonance curves are broad, and excitation of natural oscillations is possible even in the case of a considerable frequency mismatch.

Efficient excitation of shape oscillations – multipole oscillations of a microparticle – is feasible only if the condition for geometric resonance is met, i.e., if the characteristic scale of the spatial variation of nonlinear forces acting on the object is of the order of its size and, hence, nonlinear forces are at their maximum. It is this inhomogeneity of the order of the carrier wavelength in size that arises in the region of the focal waist of focused beams used in acoustic microscopes. Depending on the operating frequency of the microscope and the aperture angle of the lens, the focal area size varies from a few fractions of a micron to hundreds of microns. Therefore, nonlinear excitation of natural oscillations by a focused beam of modulated ultrasound is feasible for a variety of microscopic objects ranging from emulsion drops to biological cells and microelements of biological tissues.

The possibility of resonant excitation of natural oscillations of drops was explored by Marston and Apfel both theoretically [212] and experimentally [213, 214]. The spectra of natural oscillations of drops were analyzed by Miller and Scriven [211] and A. Prosperetti [215]. Natural oscillations of biological cells were surveyed in [66, 216–218], and the feasibility of their excitation in a standing ultrasonic wave was demonstrated by Miller [244].

In what follows, nonlinear resonant interaction of a focused beam of modulated ultrasonic radiation with biological microscopic samples (cells) is analyzed theoretically. The analysis involves formulation of the mechanical shell model of a cell, examination of cell interaction with high-frequency radiation, derivation of an expression for the radiation force acting on a cell in a high-frequency acoustic field and, finally, investigation into oscillations of microscopic objects under the action of a nonlinear force. The possible repercussions of build-up of natural oscillations of biological cells will be omitted in the present discussion because this problem deserves detailed consideration on the basis of the present-day concepts of cytology, biomechanics, and acoustics.

5.5.1
Shell Model of a Cell

Regardless of the ultrasonic field configuration, the action of nonlinear forces on a test object leads to the build-up of natural oscillations of its shape. Despite the great variability in the nature of objects capable of interacting with ultrasonic fields, the fundamental peculiarities of this interaction can be perceived within the framework of a fairly simple and yet credible model. Suppose that the test object is a sphere resistant to shape changes owing to the elastic properties and surface tension T_0 of its shell. The shell model reasonably accurately describes the mechanical properties of various biological cells, such as erythrocytes, plant cells, fat cells, etc. The perfect (spherical) shape of objects assumed in the model simplifies the description without distorting the qualitative pattern of phenomena, even in the case of intricately shaped objects.

Consider a thin spherical shell, the interior of which is filled with a liquid having density ρ^i, sound velocity c^i, bulk viscosity ζ^i, and shear viscosity η^i. The shell resides in a liquid medium with parameters ρ^0, c^0, ζ^0, and η^0. The shell itself represents a thin layer of matter of thickness h which consists of several molecular layers. The shell resists changes to its area and shape. Its resistance to area changes is characterized by compression modulus k_s, while its resistance to shear corresponds to its surface shear modulus μ. When the deformation U is small, the Hooke law for a thin shell, written in terms of spherical coordinates, takes the form

$$T_\theta = T_0 + k_s(U_{\theta\theta} + U_{\varphi\varphi}) + \mu(U_{\theta\theta} - U_{\varphi\varphi})$$

$$T_\varphi = T_0 + k_s(U_{\theta\varphi} + U_{\varphi\varphi}) + \mu(U_{\varphi\varphi} - U_{\theta\theta}) \tag{5.27}$$

where T_θ and T_φ are the tensions in the shell plane directed along unit vectors $\vec{e}_\theta$ and $\vec{e}_\varphi$, respectively; $U_{\theta\theta}$ and $U_{\varphi\varphi}$ are the normal deformations along the θ- and φ-axis. Denoting the displacements of point (r, θ, φ) by U_r and U_θ, we obtain for a sphere of radius a:

$$U_{\theta\theta} = \frac{1}{a}\left(U_r + \frac{\partial U_r}{\partial \theta}\right)$$

$$U_{\varphi\varphi} = \frac{1}{a}(U_r + U_\theta \operatorname{ctg}\theta) \tag{5.28}$$

By virtue of symmetry in φ, the equations of motion of the shell take the form

$$\frac{\partial}{\partial \theta}(a \sin\theta\, T_0) - a T_\varphi \cos\theta + a^2 \sin\theta\, F_\theta = \rho h a^2 \sin\theta \frac{\partial^2 U}{\partial t^2}$$

$$\frac{T_\theta}{R_\theta} + \frac{T_\varphi}{R_\varphi} = F_r - \rho h \frac{\partial^2 U_r}{\partial t^2} \tag{5.29}$$

where R_θ and R_φ are the principal radii of curvature, and F_θ and F_φ are the external forces directed along $\vec{e}_\theta$ and $\vec{e}_\varphi$, respectively. Inasmuch as T_θ and T_φ have a constant component T_0, one has to take into account variation of the radii of curvature of the shell in the course of its deformation. According to [245],

$$R_\theta = a + \frac{\partial^2 U_r}{\partial \theta^2} + U_r$$

$$R_\varphi = a + \frac{1}{\sin\theta}\frac{\partial}{\partial \theta}\left(\sin\theta \frac{\partial U_r}{\partial \theta}\right) \tag{5.30}$$

The frictional force F_θ acting on unit surface of the shell on the side of the ambient liquid equals:

$$F_\theta = \sigma^o_{\theta r} - \sigma^i_{\theta r}$$

Superscripts o and i signify out and into, respectively, and $\sigma_{\theta r}$ denotes the viscous stresses in the liquid given by [246]:

$$\sigma_{\theta r} = \eta\left(\frac{1}{r}\frac{\partial V_r}{\partial \theta} + \frac{\partial V_\theta}{\partial r} - \frac{V_\theta}{r}\right) \tag{5.31}$$

where V_θ and V_r are the velocity components of the medium in the spherical frame of reference.

A shell element experiences a force acting in the radial direction and equal to the difference between the radial stresses in the liquid on both sides of the shell

$$F_r = \sigma_{rr}^o - \sigma_{rr}^i \tag{5.32}$$

where σ_{rr} is the radial stress:

$$\sigma_{rr} = -p + 2\eta\frac{\partial V_r}{\partial r} + \left(\varsigma - \frac{2}{3}\eta\right)\operatorname{div}\vec{V} \tag{5.33}$$

Substituting (5.27), (5.28) and (5.30)–(5.33) into (5.29) and omitting the inertial (because h is very small) and nonlinear terms, we obtain the following equation of motion for the shell

$$\frac{k_s}{a}\frac{\partial}{\partial \theta}(2U_r + S) + \frac{\mu}{a}\left(\frac{\partial S}{\partial \theta} + 2U_\theta\right) = a\left(\sigma_{\theta r}^i - \sigma_{\theta r}^o\right)$$

$$\frac{k_s}{a^2}(2U_r + S) + \frac{2T_0}{a} - \frac{2T_0 U_r}{a^2} - \frac{T_0}{a^2 \sin\theta}\frac{\partial}{\partial \theta}\left(\sin\theta\frac{\partial U_r}{\partial \theta}\right) = \sigma_{\theta r}^o - \sigma_{\theta r}^i$$

where

$$S = \frac{\partial U_\theta}{\partial \theta} + U_\theta \operatorname{ctg}\theta = \frac{1}{\sin\theta}\frac{\partial}{\partial \theta}(\sin\theta\, U_r)$$

5.5.2
Interaction of a Cell with a High-Frequency Field within the Framework of the Shell Model. Equation for the Radiation Force

To calculate the radiation force acting on a microscopic object, consider the scattering of a high-frequency wave from the object at frequencies $\omega \sim \omega_a$. Here, ω is the carrier wave frequency, and $\omega_a = c/a$ is the geometric resonance frequency. The effects associated with the generation of a viscous wave in a liquid, as well as microflows, can be neglected because the viscous layer depth is small compared to the object size $((\eta/(\rho\omega))^{1/2} \ll a)$ [213, 214, 247, 248]. This implies that the fluids inside and outside the object can be assumed to be perfect when calculating the radiation pressure force on the cell surface at the carrier wave frequency [246]. At frequencies closely approximating the geometric resonance frequency, even the presence of the shell has no effect because the stress in the medium is many times that in the shell $(k_s/(\rho a^3) \ll \omega^2, T_0/(\rho a^3) \ll \omega^2)$.

Thus, to the first approximation, the problem reduces to the calculation of sound scattering from a liquid nonviscous sphere immersed in a perfect liquid.

The scattered wave potential Ψ_s ($V = -\mathrm{grad}\,\Psi$) is derived from the boundary conditions, one of which is the equality of the pressures at the interface

$$\rho_o \frac{\partial \Psi_o}{\partial t} = \rho_i \frac{\partial \Psi_i}{\partial t} \tag{5.34}$$

and the other is the condition for continuity of the normal velocities

$$\frac{\partial \Psi_o}{\partial r} = \frac{\partial \Psi_i}{\partial r} \tag{5.35}$$

Here, $\Psi_o = \Psi_q + \Psi_s$, where Ψ_q is the incident wave potential, and Ψ_s is the potential inside the object. The radiation force acting on the surface of the sphere is calculated subject to the boundary conditions. According to [213, 246], the normal component of the force is given by equation

$$F_{RR} = \Pi_{RR}^o - \Pi_{RR}^i$$

where Π_{RR} is the momentum flux density tensor equal to $\Pi_{ik} = \bar{p}\delta_{ik} + \rho\overline{V_i V_k}$. Here, $\bar{p} = -\rho\overline{V^2}/2 + c^2/(2\rho^*)\overline{\rho'^2}$ is the Bernoulli pressure at the modulating frequency, and ρ' is the density perturbation in an acoustic wave. With the supposition of a nonviscous liquid, the tangential component of the radiation force equals zero. Using the boundary conditions (5.34) and (5.35), we obtain for the radiation force

$$F_R = \frac{1}{2}\left\{\frac{\rho_o}{c_o^2}\left(\frac{\rho_o c_o^2}{\rho_i c_i^2} - 1\right)\left(\frac{\overline{\partial \Psi_o}}{\partial t}\right)^2 + (\rho_i - \rho_o)\left(\overline{V_R^{o2}} - \frac{\rho_i - \rho_o}{\rho_i}\overline{V_\theta^{o2}}\right)\right\} \tag{5.36}$$

Equation (5.36) specifies the force setting a microparticle into oscillation with modulating frequency Ω.

Consider a special case, typical for the majority of microscopic biological entities, where the density and sound velocity inside differ very little from their external counterparts ($\Delta\rho/\rho$, $\Delta c/c$, and $\Delta k_v/k_v \ll 1$, where k_v is the volume compression modulus). If so, one can retain just the terms linear in $\Delta\rho/\rho$, $\Delta c/c$, and $\Delta k_v/k_v$, and Equation (5.36) for the radiation force reduces to

$$F_R = \frac{1}{2}\rho_o\left\{\frac{\Delta\rho}{\rho}\overline{V_q^2} - \frac{\Delta k_v}{k_v c^2}\left(\frac{\overline{\partial \Psi_o}}{\partial t}\right)^2\right\} \tag{5.37}$$

5.5.3
Oscillations of a Microparticle under the Action of a Nonlinear Force

Consider the motion of a microscopic object under the action of the radiation force at modulating frequency Ω. To do this it is essential to relate the movements of all three components of the system (external liquid, shell, and internal liquid) to one another. The shell thickness h is small compared to the radius of the object and, therefore, the shell can be treated as an elastic surface separating the two liquids. The problem reduces to the solution of the equation of motion of a viscous liquid subject to the appropriate conditions at the surface.

The solution of the equation of motion of a viscous liquid in the spherical frame of reference is sought in the conventional form:

$$\Psi^o = \exp(-i\Omega t) \sum_{n=0}^{\infty} A_n h_n(k_l^o r) P_n(\cos\theta)$$

$$\Psi^i = \exp(-i\Omega t) \sum_{n=0}^{\infty} C_n j_n(k_l^i r) P_n(\cos\theta)$$

$$A^o = \exp(-i\Omega t) \sum_{n=0}^{\infty} B_n h_n(k_t^o r) P_n(\cos\theta)$$

$$A^i = \exp(-i\Omega t) \sum_{n=0}^{\infty} D_n j_n(k_t^i r) P_n(\cos\theta)$$

Here, Ψ is the acoustic wave potential; A is the vector potential of a viscous wave; A_n, B_n, C_n, and D_n are the unknown constants; j_n and h_n are the spherical Bessel and Hankel functions; P_n denotes the Legendre polynomials; and k_l and k_t are the wave numbers equal to ω/c and $i\omega\rho/\eta^{1/2}$, respectively. The vector and scalar potentials completely determine the vibrational velocity of a particle in a liquid:

$$\vec{V} = -\operatorname{grad}\Psi + \operatorname{rot}(\vec{r}A)$$

The relationship between the velocities (and deformations) in the internal and external liquids separated by the shell governs the joint motion of the shell and liquid. Considering that the shell is very thin, the tangential and radial components of the velocities of the liquids on both sides of the shell must be equal:

$$V_r^i = V_r^o \qquad V_\theta^i = V_\theta^o$$

By virtue of the symmetry of the problem in φ, $V_\varphi = 0$.

The stresses at the interface are related by the equations of motion of the shell, written with due regard for the radiation force. The shell displacement $\vec{U}$ is expressible in terms of the velocity of the adjacent liquid layer:

$$\vec{V} = \frac{\partial}{\partial t}\vec{U}$$

Substituting the expansions of the potentials into a series in terms of spherical harmonics into the boundary conditions and equations of motion of the shell, we arrive at an inhomogeneous set of linear equations for every harmonic n: $\hat{x}\hat{b} = \hat{f}$, where

$$x_{11} = n+1 \qquad x_{12} = n \qquad x_{13} = n(n+1) \qquad x_{14} = -n(n+1)$$

$$x_{21} = 1 \qquad x_{22} = 1 \qquad x_{23} = \xi_o h'_n(\xi_o) + h_n(\xi_o)$$

$$x_{24} = -\xi_i j'_n(\xi_i) + j_n(\xi_i) \qquad x_{31} = 0 \qquad x_{33} = 0$$

$$x_{32} = \frac{\rho_i}{\rho_o} + \frac{2\omega_k(n-1)}{(n+1)\omega^2} - \frac{4\omega_T^2}{(n+1)\omega^2} - n(n-1)q$$

$$x_{34} = \left(\frac{2\omega_k^2}{\omega^2} - n(n+1)q\right)\frac{j'_n(\xi_i)\xi_i}{j_n(\xi_i)} - \left(\frac{2\omega_k^2 + \omega_T^2}{\omega^2} - n(n+1)q\right)$$

$$x_{41} = 0 \qquad x_{42} = \frac{\omega_k^2(n-1)}{(n+1)\omega^2} - \frac{\omega_\mu^2}{\omega^2} - (n-1)q \qquad x_{43} = 0$$

$$x_{44} = \left(\frac{\omega_k^2 + \omega_\mu^2}{\omega^2} + \frac{\rho_i}{\rho_o} - (n^2+n-1)q\right) - \left(\frac{\omega_k^2 + \omega_\mu^2}{\omega^2} - q\right)\frac{j'_n(\xi_i)\xi_i}{j_n(\xi_i)}$$

$$b_1 = A_n h_n(z_o) \qquad b_2 = B_n j_n(z_i) \qquad b_3 = C_n h_n(\xi_o) \qquad b_4 = D_n j_n(\xi_i)$$

Here,

$$F_n = \frac{2n+1}{2}\int_{-1}^{1} F_R(x) P_n(x)\, dx \tag{5.38}$$

Quantities

$$\omega_k^2 = \frac{n(n+1)k_s}{\rho_* a^3}$$

$$\omega_\mu^2 = \frac{(n-1)(n+1)\mu}{\rho_* a^3}$$

$$\omega_T^2 = \frac{(n-1)n(n+1)(n+2)T_0}{\rho_* a^3}$$

are the frequencies defining the restoring (elastic) forces in the shell. They depend on moduli k and m, as well as on constant tension T_0. Quantity $\rho_* = n\rho_o + (n+1)\rho_i$ is the reduced density dependent on the spherical harmonic number n. Parameter $q = 2(\xi_i^2 \rho_i/\rho_* + \xi_o^2 \rho_o/\rho_*)$ characterizes the process of viscous relaxation.

At the frequencies of interest, ω_μ and ω_T, the cell shell can be regarded as inelastic and, therefore, it is possible to retain only those terms of the solution which contain ω_k. In this case, the shell displacement under the action of radiation pressure forces at the modulating frequency takes the form:

$$U_R(n) = \frac{n(n+1)F_n(\omega)}{a\rho_* \Omega^2 d_C^2(\Omega)} \tag{5.39}$$

where

$$d_c^n = 1 - \frac{\Omega_n^2}{\Omega^2} - \frac{\rho_o}{\rho_*} H_n(n+2)^2 - \frac{\rho_i}{\rho_*} J_n(n-1)^2 - 2q(n-1)(n-2)$$

$$\Omega_n^2 = \omega_T^2 + \frac{4\omega_k^2 \omega_\mu^2}{\omega_k^2 + \omega_\mu^2} \qquad H_n = \frac{h_n(\xi_o)}{h_{n-1}(\xi_o)\xi_o} \qquad J_n = \frac{j_n(\xi_i)}{j_{n-1}(\xi_i)\xi_i}$$

Displacement (5.39) depends on the parameters of eigenmodes at modulating frequency Ω and driving force $F_n(\omega)$. At given carrier frequency ω, the cell displacement is specified by the determinant in Equation (5.39). The roots of this determinant are the frequencies of natural oscillations of a biological sample. If the depth of viscous-wave penetration at the resonance frequency is much shorter than the sample itself, the cell response is distinctly resonant in nature. For the majority of biological cells, the roots of the dispersion equation are imaginary and, as such, natural oscillations are damped.

Consider the excitation of natural oscillations of a cell residing at the focal point. Prior to calculating $F_n(\omega)$, the field in the neighborhood of the focus is expanded in terms of spherical harmonics (5.23):

$$\Psi = A \exp(i\omega t) \sum_{n=0}^{\infty} B_n i^n j_n(kr_0) P_n(\cos\theta_0)$$

Here, r_0 is the distance from the observation point to the focus, and θ is the angle between $\vec{r}$ and the acoustic axis of a transducer. Let us denote the sums by

$$\varphi(\vec{r}) = \sum_{n=0}^{\infty} B_n i^n j_n(kr) P_n(\cos\theta)$$

$$\varphi^*(\vec{r}) = \sum_{n=0}^{\infty} B_n i^{-n} j_n(kr) P_n(\cos\theta)$$

If so, a wave with frequency ω propagating in the positive direction can be expressed as

$$\Psi_\omega = \frac{1}{2} A\varphi \exp(i\omega t) + \text{c.c.} \qquad (5.40)$$

where c.c. stands for the complex conjugate function.

A wave with frequency $\omega + \Omega$ also traveling in the positive direction is given by equation

$$\Psi_\Omega = \frac{1}{2} A_\Omega \varphi \exp(i(\omega + \Omega)t) + \text{c.c.} \qquad (5.41)$$

with an accuracy of Ω/ω, where $A_\Omega = V_0 f \exp[i(\omega + \Omega)f/c]$. Substituting (5.40) and (5.41) first into (5.37) and then into (5.38) yields for $F_l(\omega)$:

$$F_l(\omega) = \frac{2n+1}{4}\frac{I}{c}(kf)^2 \exp(-2ikf - i\Omega t)$$

$$\times \left\{ \frac{\Delta\rho}{\rho(ka)^2}\left(S_R^l(z_o j_n'(z_o)) + S_\theta^l(j_n(z_o))\right) - \frac{\Delta k}{k_v} S_R^l(j_n(z_o)) \right\}$$

Functions S_R^l and S_θ^l result from integration of F_R with respect to θ in (5.38):

$$S_R^l(y_n) = \frac{1}{2}\sum_{n=0}^{\infty} B_n^2 y_n^2 I_1(n,n,l) + \sum_{n=0}^{\infty}\sum_{m=n+1}^{\infty} i^{m-n} B_n B_m y_n y_m I_1(n,m,l) \quad (5.42)$$

if l is even and $S_R^l(y_n) = 0$, if l is odd;

$$S_\theta^l(y_n) = \frac{1}{2}\sum_{n=0}^{\infty} B_n^2 y_n^2 I_2(n,n,l) + \sum_{n=0}^{\infty}\sum_{m=n+1}^{\infty} i^{m-n} B_n B_m y_n y_m I_2(n,m,l) \quad (5.43)$$

if l is even and $S_\theta^l(y_n) = 0$, if l is odd.

As is customary, the triple integrals of Legendre polynomials in Equations (5.42) and (5.43) are denoted by

$$I_1(n,m,l) = \int_{-1}^{1} P_n(x) P_m(x) P_l(x)\, dx$$

$$I_2(n,m,l) = \int_{-1}^{1} P_n'(x) P_m'(x) P_l(x)\, dx$$

These integrals are expressible in terms of $3j$ symbols [249, 250]:

$$I_1(n,m,l) = 2\begin{pmatrix} n & m & l \\ 0 & 0 & 0 \end{pmatrix}^2$$

$$I_2(n,m,l) = -2\sqrt{n(n+1)m(m+1)}\begin{pmatrix} n & m & l \\ 0 & -1 & 0 \end{pmatrix}\begin{pmatrix} n & m & l \\ 0 & 0 & 0 \end{pmatrix}$$

Explicit expressions for the $3j$ symbols are available in [249, 250].

The equation for the nonlinear force is cumbersome; however, it is obvious that when a biological sample is placed at the center of curvature, modulated sound excites only its even modes. Considering that the modulating frequency Ω is much lower than the geometric resonance frequency ω_a, the liquid can be treated as incompressible, and the zero mode can be neglected. The frequency dependences of the quadrupole component $F_2(\omega)$ and the tetrapole component $F_4(\omega)$ are displayed in Figure 5.6. By adjusting the carrier wave frequency, it is possible to achieve efficient build-up of the quadrupole mode (at $ka \approx 2$, $F_2/F_4 \sim 10$).

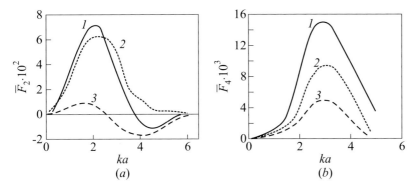

Figure 5.6 Frequency dependences of the quadrupole (a) component of the radiation force $\bar{F}_2 = F_2 4c/(I(kf)^2(2n+1))$ and tetrapole (b) component of the radiation force $\bar{F}_4 = F_4 4c/(I(kf)^2(2n+1))$: $1 - \Delta c/c = 1$, $\Delta k/k_v = 1$, $2 - \Delta c/c = 0$, $\Delta k/k_v = 1$, and $3 - \Delta c/c = 1$, $\Delta k/k_v = 0$ ($\alpha = \pi/2$).

Consider the excitation of multipole oscillations of a cell by the radiation pressure force. At given carrier wave frequency ω, the amplitude of cell oscillations at the modulating frequency Ω is controlled by the frequency dependence of the determinant $d_c^n(\Omega)$ (5.39).

This section deals solely with low-frequency oscillations with modulating frequency $\Omega \ll \Omega_T, \omega_\mu$. Oscillations are assumed to be relaxational in nature, $\omega_R > \omega_T, \omega_\mu$. For relaxation oscillations, the determinant $d_c^n(\Omega)$ has the form

$$d_c^n(\Omega) = \left(i\Omega\left(\frac{\rho_i}{\rho_*}N_i\omega_{Ri} + \frac{\rho_o}{\rho_*}N_o\omega_{Ro} + N_s\omega_{Rs}\right) - \Omega_n^2\right)/\Omega^2 \quad (5.44)$$

where

$$N_i = (n-1)(2n^2 + 5n + 5)$$

$$N_o = (n+2)(2n^2 - n + 2)$$

$$N_s = 4(n-1)(n+2)$$

Equation (5.44) was derived using the expansions of H_n and J_n at small values of their argument ξ ($\xi \ll 1$):

$$H_n(\xi) = \frac{2n-1}{\xi^2} - \frac{1}{2n-3}$$

$$J_n(\xi) = \frac{2n+1}{\xi^2} - \frac{1}{2n+5}$$

It is evident from (5.44) that the cell response at the modulating frequency depends on the ratio between the modulating frequency Ω and the shell relaxation frequency ω_n^p:

$$\omega_n^p = -i\Omega_n^2 / \left(\frac{\rho_i}{\rho_*} N_i \omega_{Ri} + \frac{\rho_o}{\rho_*} N_o \omega_{Ro} + N_s \omega_{Rs} \right)$$

Assuming that the liquid densities inside and outside the cell are equal, the shell relaxation frequency takes the form

$$\omega_2^p = -i \frac{\mu + 1.5 T_0}{\eta \mu + 1.7 \eta_i a + 2 \eta_o a}$$

for the quadrupole mode.

At frequencies $\Omega \ll |\omega_n^p|$, the displacement $U_R(n)$ is independent of the modulating frequency Ω:

$$U_R(n) = -\frac{n(n+1) F_n(\omega)}{a \rho_* \Omega_n^2} \tag{5.45}$$

At frequencies $\Omega \gg |\omega_n^p|$, the displacement $U_R(n)$ decreases with increasing frequency as $1/\Omega$:

$$U_R(n) = -i \frac{n(n+1) F_n(\omega)}{a \rho_* \Omega (N_i \omega_{Ri} + N_o \omega_{Ro} + N_s \omega_{Rs})}$$

Let us assess the oscillation amplitude in the focal area of an acoustic microscope, taking as an example swollen erythrocytes ($T_0 = 2.5 \times 10^{-2}$ N m^{-1}, $\mu = 10^{-5}$ N m^{-1}, $k_s = 0.5$ N m^{-1}, $a = 5 \times 10^{-4}$ cm, and $h = 10^{-2}$ μm). From (5.39), it follows that $\Omega_2 = 10^6$ s^{-1} for swollen erythrocytes, and at modulating frequencies $\Omega \ll \Omega_2$, the amplitude of quadrupole oscillations of a swollen shell is given by (5.45), where $\rho_* \approx 1$ g cm^{-3}. The quadrupole force component F_2 was estimated at $F_2 \approx 10^{-4}$ by considering that $\Delta\rho/\rho \sim \Delta k/k \sim 10^{-2}$ for erythrocytes. At the focal point of an acoustic microscope with an operating frequency of 500 MHz and a focal length of $f = 500$ μm, the energy flux density I was estimated at 0.1 W cm^{-2}, $(kf)^2 \approx 1.5 \times 10^4$, and hence,

$$F_2 = \overline{F_2} \frac{5I}{4c} (kf)^2 \approx 75 \left[\frac{g}{cm \cdot s^2} \right]$$

The relative deformation amplitude U/a is thus seen to be

$$\frac{U_R}{a} \sim \frac{23 F_2}{a \rho_* \Omega_2^2} \approx 0.2 \times 10^{-2}$$

This deformation of swollen erythrocytes can acquire considerable values, depending on the erythrocyte membrane tension.

6
Investigation of the Local Properties and Microstructure of Model Systems and Composites by the Acoustic Microscopy Methods

6.1
Study of the Viscoelastic Properties of Model Collagen Systems by the Acousto-Microscopic Methods. Experimental Setup

Studies of acoustic properties of composite materials are very important in understanding the processes that govern the imaging of real objects and also to reveal the factors that control the contrast of acoustic images.

A desk-top transmission acoustic microscope was developed at the Institute of Chemical Physics and employed to explore model systems such as thin gelatin and collagen films. The transmission acoustic microscopy system developed at the Institute operated at a 450-megahertz frequency; the system provided a resolution of about 3 µm in water at a focal zone length of about 12 µm. The main microscope unit consists of a confocal lens system comprising two acoustic waveguides made of sapphire single crystals of z-cut with acoustic lenses at their ends. The acoustic lenses are fabricated as spherical recesses with a radius in the range from 50 to 300 µm; the aperture half-angles of the top and bottom lenses are $\theta_m = 60°$ and 45°, respectively. The distance between the lenses is about 500 µm which permits objects of a thickness ranging from 5 to 400 µm to be examined, ruling out the possibility of their mechanical damage by lenses. However, when choosing the thickness of an object to be studied, one has to take into account acoustic wave attenuation in the sample at the microscope operation frequency.

Adjustment of the system is implemented by moving one lens with respect to the other via three coordinate axes and is controlled by the signal level.

The object studied is fixed in the holder of the scanning system and is placed in a drop of immersion liquid between the two lenses. The object is scanned over two coordinates in the object plane, namely, fast (raster) scanning in the X-direction and slow (framing) scanning in the Y-direction. The system with a sample on a rod allows the angle between the X-scanning axis and sample plane to be adjusted; a similar adjusting device is provided for Y-scanning. The feasibility of changing the object position in two directions allows the object plane to be adjusted parallel to the scanning plane. A fast scanning device provides plane-parallel sample motion with a maximum amplitude of 2 mm and a sampling frequency of 25 Hz. Slow

Acoustic Microscopy. Roman Gr. Maev
Copyright © 2008 WILEY-VCH Verlag GmbH & Co. KGaA, Weinheim
ISBN: 978-3-527-40744-6

scanning is permitted by means of a hydraulic device providing uniform motion of the object. The scanning system makes it possible to obtain an acoustic image of the object cross-section within a 2 × 5 mm field (time per frame is 8 and 10 s, respectively) and to change the object position between the lenses within a 0.5-millimeter range at 1-micron steps.

Data about the coordinate in the sample scanned and amplitude at the acoustic microscope output are fed into the memory of a recording computer where they are saved as a three-dimensional (3D) data file and then, after appropriate processing, are exposed on the display screen. Depending on a processing procedure used, one obtains either a gray-scale acoustic image (C-scanning), one-dimensional acoustic profiles (A-scanning), or bar charts characterizing the distribution of ultrasound transmission over the scanning plane (B-scanning).

At the first investigation stage, we studied the acoustic properties of gelatin films formed by collagen in the denatured state. Edible gelatin was used in experiments. A gelatin sample, swollen in water for 1 h, was heated to 70 °C until it completely dissolved. The transparent homogeneous solution was poured into a Petri dish in the form of a thin layer. Ten-micron films were obtained after drying the solution layer over 24 h. Observed variations of the film thickness within ±1 µm are the main source of scatter of the measured speeds of sound. Gelatin films were studied by the technique of $A(z)$-curves in which water or mercury were used as immersion liquids.

The disadvantage of water as an immersion liquid in studies of collagen systems is that the film swells in it. Fortunately, as films swell quite slowly, successive measurements showed quite good reproducibility of the results. The $A(z)$ curve for a sonicated gelatin film in water exhibits a single maximum. The data on the velocity and attenuation of sound waves in gelatin films are listed in Table 6.1. The observed variations of the measured speeds of sound in various film areas amount to about 3% and are attributed to thickness variations. As the water content in the film increased during the course of tests, the shift in the $A(z)$ maximum with respect to the appropriate $A(z)$ maximum for water reduces, which is indicative of the acoustic velocity decreasing in a swelling film.

When mercury was used as an immersion liquid in studies of dried films, the $A(z)$ curve showed two maximums pertaining to transmission of longitudinal and shear waves through the sample. The data on velocities of longitudinal and shear acoustic waves in a dried 10-micron film are listed in Table 6.1.

A microscope with mercury as an immersion liquid was also used to study gelatin films containing an amount of water (up to 10% of the total film weight). The water content in the film was assessed through comparison of its weight with that of the completely dried film (for which repeated weighing showed no changes in the film weight). The film thickness was 8.5 ± 1.0 µm. As the acoustic lenses came closer to each other, the $A(z)$ curve exhibited several equally spaced maximums with a decaying amplitude. Generally speaking, rays belonging to the beam that are incident on the plate at different angles can excite various oscillation modes. However, when the film thickness is small and an aperture of the incident acoustic beam is limited, only a single wave oscillation mode is excited in the film.

Table 6.1 Density and acoustic characteristics of gelatin and collagen films.

Material	Thickness μm	Density g cm^{-3}	Speed of sound km s^{-1}	Attenuation coefficient α cm^{-1}	α/f^2 10^{18} s^2 cm^{-1}
Gelatin film (measurements in water)	14	1.2	1.9	405 ± 40	20 ± 2
Dried gelatin film (measurements in mercury)	10	1.3	$c_L = 2.62$, $c_T = 1.09$	590 ± 60	29 ± 3
Collagen film	12	13	2.4	1320 ± 130	65 ± 6
Dried collagen fibers	10	1.35	2.8	890 ± 90	44 ± 4

It is re-emitted in the immersion liquid when spreading along the plate and produces the maximums recorded as the lenses come together. The wave guide mode decays during the course of its propagation along the plate; in addition, as the waves are re-emitted in the liquid, the distance to the merging point of waves at the axis of the lens system increases and, hence, so does damping of the re-emitted waves in the immersion liquid. Therefore, the amplitude of the maximums diminishes as $|z|$ increases. It follows from a simple inference of the ray model that the neighboring maximums are formed by rays which have additionally passed twice through the plate, the distance between two neighboring maximums given by the formula:

$$\Delta z = 2d \frac{\sqrt{1 - \sin^2 \theta}}{\sqrt{(c_0/c_L)^2 - \sin^2 \theta}} \tag{6.1}$$

Thus, knowing the plate thickness d, the speed of sound in a sample can be assessed from a measured distance Δz between the maximums by the formula

$$c_L = \frac{c_0}{\sqrt{\cos^2 \theta + (\Delta z/2d)^2 \sin^2 \theta}} \frac{\Delta z}{2d} \tag{6.2}$$

The $A(z)$ has a set of subordinate maximums when the acoustic impedance values of the sample studied and immersion liquid significantly differ. The transmission coefficient T behavior is of a pronounced resonant nature.

We consider the angular dependence of ultrasound wave transmission through a layer ($c = 1.6$ km s^{-1} and $\rho = 1.1$ g cm^{-3}) submerged in mercury ($c_0 = 1.45$ km s^{-1} and $\rho_0 = 13.546$ g cm^{-3}) when the parameter $fd = 4.5 \times 10^3$ MHz μm (f is the ultrasound frequency and d is the layer thickness), which corresponds to a plate thickness equivalent to about three wavelengths. Only one angle of full transmission falls into the angular aperture (θ_m is about 40°) because of the small plate thickness; hence, only a single fundamental oscillation mode is excited. Rays aligned with this direction produce, as a result of multiple reflections, a regular set of maximums of the $A(z)$ curve. As the plate thickness increases, so does the number of full transmission angles falling into the angular lens aperture; therefore, several fundamental oscillation modes are excited in the plate that interfere. As a result, the regular system of maximums is violated. The onset of an irregular system of maximums was observed in our experiments.

The $A(z)$-curve technique was also employed to study layers made of molecular collagen solutions. Molecules in the films are positioned so that their long axes lie in the layer plane. Collagen films were fabricated by drying a molecular solution of a pig skin collagen over 48 h at 4 °C. We believe that no collagen denaturing occurred in the course of film fabrication. Ultrasound in collagen films attenuates more intensely than it does in gelatin films, which is tentatively ascribed to the scatter caused by inhomogeneities of the film structure. The data on the velocity and attenuation of sound in the collagen film are summarized in Table 6.1.

It is noteworthy that the high-frequency collagen properties depend on the ordering degree of its structure. A higher speed of longitudinal sound and lower attenuation is observed in collagen fibers whose structure is more ordered than that of films.

An acoustic microscope can be used not only to measure acoustic parameters of solid and elastic samples, but also to study the acoustic properties of liquids, structured solutions, and so on. We will consider ultrasound attenuation in gelatin gels. Ultrasound attenuation in structured gelatin solutions at low frequencies was investigated in [251, 252]. Mikhailov and Tarutina [251] report results of measurements of ultrasound attenuation at frequencies of 8.2 and 10.4 MHz in gelatin solutions of a concentration ranging between 1% and 7%. A comparison of the data has shown that the quadratic frequency dependence of the attenuation coefficient holds quite well in this frequency range. The attenuation coefficient was found to be independent of the gelatin concentration, starting with its value of about 4% and attaining a constant level of $\alpha = 0.09$ cm^{-1} ($\alpha/f^2 = 8.5 \times 10^{18}$ s^2 cm^{-1}). Data on the viscoelastic properties of gelatin solutions at high frequencies are virtually absent in the literature.

We explored the attenuation coefficient dependence on the gelatin concentration P in water in a transmission acoustic microscope at a 450-megahertz frequency. Sound wave attenuation was measured employing the $A(z)$ technique in which the maximum $A(z)$ signal amplitude was measured in gelatin at various distances between the lenses. First, we plotted the $A(z)$-curve for water and measured the maximum signal amplitude A_0. The coefficient of sound attenuation in gelatin α_2 was calculated using the known value of the attenuation coefficient for water α_1 by the formula:

$$\alpha_2 = \alpha_1 + \frac{1}{x} \ln \frac{A_0}{A_g} \tag{6.3}$$

where x is the length of the acoustic cell (distance between the lenses).

No shift of the signal amplitude maximum was observed at the gelatin concentrations studied. Thus, the acoustic velocity in dilute gelatin solutions is believed to be

Table 6.2 Temperature of gelatin transition from the liquid state to gel.

Concentration P %	Temperature T °C
3.33	19.0 ± 0.2
4.05	19.8 ± 0.2
5.12	20.5 ± 0.2
6.67	22.5 ± 0.2
10.07	23.5 ± 0.2
14.29	24.5 ± 0.2
16.70	26.7 ± 0.2

Table 6.3 Results of measurements of acoustic signal attenuation in gelatin solution.

Concentration P %	Attenuation coefficient α cm^{-1}	α/f^2 10^{18} s^2 cm^{-1}
3.33	67 ± 2	3.3 ± 0.1
4.05	71 ± 2	3.5 ± 0.1
5.12	74 ± 2	3.6 ± 0.1
6.67	81 ± 3	4.0 ± 0.2
10.07	94 ± 3	4.6 ± 0.2
14.29	114 ± 4	5.6 ± 0.3

virtually identical to its value in water. In assessing the attenuation coefficient, the difference in reflection at the lens–water and lens–gelatin interfaces was ignored (the difference in the reflection coefficients at normal incidence of waves amounts to about 1% in terms of amplitude).

A conventional edible gelatin was used in the experiments (as in [251]). A gelatin sample was held in distilled water for 1 h and then heated to 70 °C until it completely dissolved. A liquid transparent solution was poured between the acoustic microscope lenses with the aid of a pipette and gradually cooled down to room temperature $T = 18\,°C$, converting into jelly. All the solutions studied were in the jelly state at the aforesaid temperature. First, the temperature of the liquid–gel transition was ascertained for every solution sample (Table 6.2).

The maximum gelatin concentration at which the acoustic signal induced by a wave which has traversed the acoustic cell filled with gelatin is still detectable is about 15%. The acoustic wave attenuation in gelatin was measured after the solution had gelatinized. No less than five measurements were performed at each concentration and the data were averaged. Samples with six various gelatin concentrations were examined. The data are listed in Table 6.3.

As can be seen, unlike measurements at low frequencies, our experiments indicate that the linear attenuation behavior holds up to the 15% gelatin concentration. The discrepancy in the concentration dependence at low and high frequencies is attributed to the nature of relaxation processes in structured solutions.

Table 6.4 Results of measurements of acoustic signal attenuation in gelatin solution performed at various frequencies.

f MHz	P %	α cm^{-1} (experiment)	α cm^{-1} (linear extrapolation)	α/f^2 10^{18} s^2 cm^{-1}
8.25	3	0.052 ± 0.006	154.7	7.7 ± 0.8
	5	0.060 ± 0.007	178.5	8.8 ± 1.0
10.4	1.5	0.048 ± 0.003	90	4.5 ± 0.3
	3	0.082 ± 0.003	153.8	7.6 ± 0.3
	5	0.093 ± 0.012	174.4	8.6 ± 1.1
	7	0.090 ± 0.010	168.8	8.4 ± 0.9

The major contribution to the attenuation of acoustic waves at low frequencies comes from large segments in the gelatin molecular network. As the solution concentration increases, a spreading acoustic wave is no longer capable of inducing motion in such segments due to sterical hindrances. No energy is spent to move the segments; therefore, a further increase in the solution concentration does not cause attenuation of sound waves. Attenuation of acoustic waves at high frequencies is mostly controlled by the motion of small molecular segments (side branches of chains). The motion of such small segments is virtually unhindered because there is always some free space for it. Therefore, attenuation of acoustic waves at high frequencies increases as the solution gets thicker (more side chains appear that contribute to the acoustic wave attenuation).

Table 6.4 lists the results of measurements of sound attenuation, as borrowed from [251]; it also lists the attenuation coefficient values at 450 MHz we assessed using linear extrapolation from the low-frequency range.

Comparison of the α/f^2 parameter calculated for, e.g., a 5% gelatin concentration at frequencies of 450 and 10 MHz, shows that the quadratic frequency dependence of the ultrasound attenuation coefficient does not hold over such a wide frequency range.

6.2
Microstructure Investigations of Multilayer Photographic Film Structures Using Scanning Acoustic Microscopy Methods

The use of ultrasound high-frequency waves in studying microstructure and viscoelastic properties of photographic materials seems to be very promising in the first place because it offers the opportunity of studying processed and nonprocessed samples not exposed to light.

Photographic film is a material consisting of two layers, namely, celluloid support (50–80 μm) and emulsion layer (20–30 μm). The system as a whole is fairly thick; therefore, reflection acoustic microscopy, imposing no restrictions on the thickness of samples examined, is best suited for studying properties and microstructure of commercial photographic materials.

The microstructure of films was additionally investigated with the aid of an ELSAM (Leitz Ltd.) reflection acoustic microscope [253, 254]. The ELSAM reflection microscope comprises a conventional optical microscope equipped with replaceable objectives and an acoustic microscope with exchangeable lenses. The two microscopes can very precisely (to within ±5 µm) be installed above the same site of the sample. This permits comparative investigations of the optical and acoustic object images to be performed. The acoustic microscope operates in the frequency range of 0.1–2 GHz. The high-frequency range is covered with two acoustic objectives operating in the 0.8–1.3 and 1.3–2-gigahertz ranges, respectively. The remainder of the frequency range of the acoustic microscope is covered by lenses operating at 100, 200, and 400 MHz.

The maximum resolution of the ELSAM system is 0.6 µm at a frequency of 2 GHz. Short-focus lenses are used to attain high resolution, which entails a reduction in the distances between the lens and object studied. This distance amounts to about 30 µm at frequencies ranging from 0.6 to 2 GHz.

Quantitative data can be obtained when the reflected signal amplitude is recorded with linear scanning of the object or by using automated recording at 0.1-micron steps of the output signal of microscope V as a function of the coordinate Z of the focused beam point in the sample.

All the data on operation microscope parameters (frequency, focus magnification, temperature) are displayed on the screen.

To reduce sound wave attenuation in water at frequencies ranging from 1.5 to 2 GHz, both the lens and immersion liquid are preheated to 60 °C.

The emulsion layer of a photographic film is a composite material consisting of a gelatin matrix and silver halogenide particles (silver bromide) or reduced silver. The concentration of particles is quite high (30%–50%). The size varies depending on the film assignment. The coarsest grains 3–6 µm in size are inherent in X-ray films, grains in dosimetric films are 1–3 µm in size, while in fine-grained films (MZ-3L, MICRAD-200) they range between 0.5 and 1 µm. There are also films where grains are less than 0.5 µm in size. To resolve individual particles in acoustic images, one has to work at high frequencies. The optimal operation frequency of a microscope is 1.3–1.4 GHz. Resolution at this frequency in water amounts to 0.85 µm at an angular aperture of the lens equal to $\theta_m = 50°$. The acoustic properties of the gelatin matrix and inclusions differ: the speed of sound in gelatin $c_L = 1.7$–1.8 km s^{-1}, its density $\rho = 1.2$ g cm^{-3}, the acoustic impedance $\rho c = 2.04 \times 10^5$ g (cm^2 s)$^{-1}$; the speed of longitudinal sound in silver $c_L = 3.6$ km s^{-1}, the density $\rho = 10.5$ g cm^{-3}, the acoustic impedance $\rho c = 37.8 \times 10^5$ g (cm^2 s)$^{-1}$; the speed of sound in silver bromide $c_L = 2.9$ km s^{-1}, the density $\rho = 6.5$ g cm^{-3}, the acoustic impedance $\rho c = 18.85 \times 10^5$ g (cm^2 s)$^{-1}$.

The acoustic properties of the matrix differ only a little from those of the immersion liquid (water), whereas the properties of particles differ greatly. Silver particles in the acoustic images of photographic films taken at a 1.3-gigahertz frequency (Figure 6.1) are seen as bright luminous spots against the dark background of

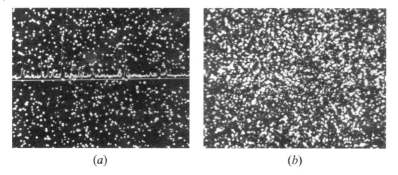

(a) *(b)*

Figure 6.1 Acoustic images of photographic films taken in the reflection regime, $f = 1.3$ GHz: (a) dosimetric film; the scanning field length is 200 μm, and (b) fine-grained MZ-3L film; the scanning field length is 100 μm.

gelatin, which poorly reflects acoustic radiation. Thus, the microphotograph displays the distribution of silver particles in the emulsion layer. Although the particle concentration is fairly high, only few particles are seen in the image (Figure 6.1a) because of a very short resolution depth at a great magnification level (the scanning field is about 62.5 μm in size). The boundaries of particles falling into the focal zone are very sharp. The light stains seen in the microphotographs represent particles positioned in deeper layers. As the lens is displaced toward the sample (the focus shifts deeper inwards into the emulsion), the images of some particles are gradually blurred because they leave the focal zone, while other particles with sharp boundaries appear. Silver crystals are oriented chaotically with respect to the film plane; therefore, signals reflected from different particles are dissimilar.

When a photographic film is examined under an optical microscope, it seems that the number of particles seen in the same field of vision is much greater. This is accounted for by the fact that the resolution depth of an optical microscope is much longer. The focal zone length of the acoustic microscope at $\theta_m = 50°$ and frequency $f = 1.3$ GHz, is $d = 6.44$ μm, while at $f = 1.4$ GHz, $d = 5.99$ μm. Thus, only one or two particles find themselves in the focal zone when coarse-grained films are studied at high frequencies. If the microscope resolution is less than the particle size, silver grains look like light spots with smeared boundaries.

Acoustic image photographs of a film taken when the lens is focused on its surface show interference fringes around particles in the emulsion layer depth (Figure 6.2). Only signals commensurate in their amplitude can produce interference fringes. In the above considered case, interference arises due to reflection from the poorly reflecting gelatin surface (the reflection coefficient R is about 0.15 in terms of amplitude), which is located in the focal plane, and also from a highly reflective silver particle located deeper (R is about 0.9). No theory of $V(z)$ dependencies for composite materials containing particles, the dimensions of which are of the order of the acoustic wavelength, has been developed as yet. As the lens is dis-

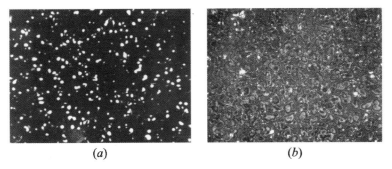

Figure 6.2 Acoustic images of an X-ray film taken in the transmission regime, $f = 1.4$ GHz, the scanned field is 62 μm in size: (a) the lens is focused on particles lying underneath the film surface, and (b) the lens is focused on the film surface.

placed toward the object, domains of the materials possessing different properties (either grain or matrix) alternatively find themselves in the focal zone, all of them contributing to the output microscope signal.

6.3
Investigation of the Microstructure Peculiarities of High-temperature Superconducting Materials by Scanning Acoustic Microscopy Methods

The acoustic properties of superconducting materials (HTSC-materials) are traditionally studied by employing integral techniques [255]; therefore, they characterize the material volume as a whole. We have demonstrated that acoustic microscopy is an effective tool for studying both local acoustic properties of individual crystallites and various defects in HTSC materials. It is worth noting that acoustic microscopy allows one: (1) to observe defects deep in an opaque sample as distinct from optical microscopy, (2) to ascertain the distribution of Rayleigh wave velocities over a small (of order 10 μm^2) surface area (e.g., within an individual crystallite), and (3) to draw conclusions about the presence of defects in a near-surface layer based on measurements of acoustic wave attenuation.

In interpreting acoustic image formation when the beam is focused beneath the sample surface, one can assume that only rays emanating from the lens focus are efficiently received by the detector. For example, (Figure 6.3), ray B contributes to the signal much more than does ray A. Therefore, when the beam is focused on the sample surface, the surface defects would be out of focus and appear in the image as blurred spots or miss completely. These defects would find themselves in the focus as the lens approaches the object.

Of importance for practical applications is the assessment of the defect depth in a sample. We consider a case when the defect image is produced by rays reflected once from it (Figure 6.3b), i.e., when rays re-reflected between the surface and defect can be disregarded. This case takes place, first, if the object is small enough

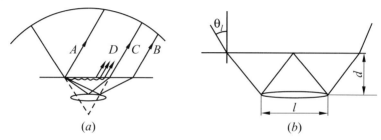

Figure 6.3 Geometric interpretation of output signal formation in SAM at $z < 0$. (a) A is the mirror reflected ray, B is the longitudinal wave ray transformed in the sample and reflected from a surface defect, C is the ray transformed from a transverse wave, and D is a leaky Rayleigh wave. (b) Re-reflected rays do not enter the object because of a small value of $l < 2d(c_T/c_L)(1 - c_T^2/c_L^2)^{1/2}$, where c_T and c_L are velocities of the longitudinal and transverse acoustic waves in the sample.

for the re-reflected beams to hit it solely when they spread at a small angle with respect to the acoustic lens axis z, and secondly, if the sound wave attenuation in the sample is significant:

$$d = \delta z \frac{\text{tg}\,\theta}{\text{tg}\,\theta'} \tag{6.4}$$

where θ is the angle between the incident beam and z-axis and θ' is the refraction angle in the sample. At $\theta \ll 1$, terms of the order of θ^3 and higher can be neglected and sine can be replaced by tangent (paraxial approximation). We arrive at the well-known formula in which δz is independent of θ [111]:

$$d_{\text{par}} = \delta z \frac{c_1}{c} \tag{6.5}$$

where c_1 is the speed of sound in the immersion liquid (water) and c is the acoustic velocity in the sample. However, apertures used in the majority of acoustic microscopes are wide; therefore, acoustic signals experience significant aberrations after the wave traverses the layers. That is, the convergence points of beams reflected at various angles arrive at different points of the acoustic axis.

To localize defects more precisely, one has to consider more thoroughly the formation of the output signal in the microscope when a layer is present on an ideally reflecting substrate. A commonly accepted procedure in analyzing formation of the output signal of an acoustic microscope is the angular-spectrum approach [59]. We introduce a frame of reference centered at the lens focus. The distance z in this frame of reference corresponds to the lens displacement with respect to the sample surface. When the signal is focused on the sample surface, z is assumed to be equal to zero. Then, the output signal can be written as a superposition of signals generated in the transducer by the planar wave front component:

$$V(z) = \int_0^{\theta_m} R(\theta) \exp(2ik_1 z \cos\theta) \sin\theta \, d\theta \tag{6.6}$$

where $R(\theta)$ is the coefficient of reflection from the object, θ_m is the aperture of lens, and k_1 is the wave vector in the immersion liquid. Furthermore, because its form affects to some extent the position of the principal maximum of the $V(z)$ curve, the lens aperture function is plotted for angle θ_m.

Now, we can assess the focus shift due to the presence of a defect possessing high reflectivity and located at depth d beneath the surface. These calculations are based on our previous studies [256]. The defect is assumed to be plane, parallel to the surface, and much larger than the wavelength. Excitation of transverse waves in the sample is also neglected. The expression for the reflection coefficient $R(\theta)$ between the liquid layer and gas in this case reads:

$$R(\theta) = \frac{(1-\tau)\exp[-2i\phi - (1+\tau)]}{(1+\tau)\exp[-2i\phi + (1-\tau)]} \tag{6.7}$$

where $\tau = \rho_1 c_1 \cos\theta'/(\rho c \cos\theta)$, $\phi = kd\cos\theta'$, k is the wave vector in the sample, ρ_1 and ρ are the densities of the immersion liquid and sample, respectively [48]. If the acoustic properties of the sample are assumed to be similar to those of the immersion liquid, substitution of $\tau \sim 1$ yields, with accuracy corresponding to condition

$$\frac{(\tau-1)^2}{2\tau} \ll 1 \tag{6.8}$$

the following expression for the reflection coefficient:

$$R(\theta) \sim \exp(2ikd\cos\theta') \tag{6.9}$$

where k is the wave vector in the sample. Hence,

$$V(z) \sim \int_0^\infty \exp(i\Phi)\sin\theta\, d\theta \tag{6.10}$$

where $\Phi(\theta, z, d) = 2k_1 z\cos\theta + 2kd\cos\theta'$. Extending the expression for Φ to a situation when small parameter $x = 1 - \cos\theta \ll 1$ yields

$$\Phi(x, z, d) = 2k_1\left[z + \frac{dc_1}{c} - x\left(z + \frac{dc}{c_1}\right) + \frac{dcx^2}{2c_1}\left(1 - \frac{c^2}{c_1^2}\right)\right] \tag{6.11}$$

Retaining the linear and quadratic conditions in terms of x in Equation (6.10), we can express $V(z)$ via a complex Fresnel integral

$$V(z) \cong \int_{S_1}^{S_2} \exp\left(\frac{i\pi s^2}{2}\right) ds \tag{6.12}$$

where

$$s = 2\sqrt{\frac{k_1 g}{\pi}}\left(x + \frac{e}{2g}\right) \qquad e = -z - \frac{dc}{c_1} \qquad g = \frac{d(1 - c^2/c_1^2)c}{2c_1}$$

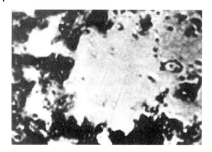

Figure 6.4 Low-density $YBa_2Cu_3O_{7-x}$ ceramics. Frequency is 1.7 GHz.

The behavior of $V(z)$ dependence can conveniently be analyzed in terms of a Cornu spiral. In this case, the $V(z)$ maximum is attained when $s_1 = s_2$. Based on these conditions, the z_0 point of the $V(z)$ curve maximum is found from the expression

$$z_0 = \frac{dc}{c_1}\left[1 - \left(1 - \frac{c^2}{c_1^2}\right)\frac{1-\cos\theta_m}{2}\right]$$

Then

$$d = \frac{z_0 c_1/c}{1 - (1 - c^2/c_1^2)(1-\cos\theta_m)/2} \tag{6.13}$$

At small apertures, this expression yields the same result as does Equation (6.5).

The physical meaning of the term $(1 - c^2/c_1^2)(1-\cos\theta_m)/2$ is that the focus in the sample is formed by rays which are intermediate between the paraxial and extreme rays. The greater effect of the lateral rays is accounted for by the fact that the relative contribution of rays in the output signal, specified by (6.4), is proportional to the space angle value, i.e., to $\sin\theta$. In deriving the formula, we assumed that the acoustic sample properties differed insignificantly from those of the immersion liquid (Equation (6.8)). The subsurface image of perofskite ceramics with acoustic velocities $c_L = 5$ km s^{-1} and $c_T = 3$ km s^{-1} is shown [257] to be produced at $z < 0$ exclusively by transverse waves. Then, if the defect size meets the following condition

$$l < \frac{2dc_T}{c_L}\sqrt{1 - \frac{c_T^2}{c_L^2}} \tag{6.14}$$

where l is the defect size in the x–y plane, Equation (6.13) still holds because, due to its small size, re-reflected beams do not affect the defect image (Figure 6.4). For perofskite ceramics, we obtain $l < 3d/2$.

A subsurface sample image of perofskite ceramics is mainly created by transverse waves. Indeed, let us isolate the defect contribution R_{def} from the reflection coefficient:

$$R = R_{\text{def}} + R' \tag{6.15}$$

The R_{def} magnitude is specified by rays B and C. Assuming that the penetration depth of a Rayleigh wave is less than d, we neglect the defect influence on its propagation. Therefore, along with ray A, reflected directly from the surface, the Rayleigh wave contributes to R'. Inasmuch as both longitudinal and transverse acoustic waves spread in the sample, which is solid, the SAM signal generated by the defect equals

$$V_{\text{def}} = \int_0^{\theta_L} T_L T_L^* \sin\theta \, d\theta + \int_{\theta_L}^{\theta_T} T_T T_T^* \sin\theta \, d\theta \qquad (6.16)$$

where T_L and T_L^* are the transformation coefficients for the longitudinal wave in the sample spreading inwards into the immersion liquid and backwards, respectively likewise, T_T and T_T^* are the appropriate coefficients for transformation of the wave spreading from the immersion into a transverse wave in the sample and *vice versa*.

We assume that the influence of the following phenomena can be disregarded: (1) excitation of transverse waves in the $0 < \theta < \theta_L$, $\theta_T < \theta < \theta_m$ angular ranges (θ_L and θ_T are the cutoff angles, $\sin\theta_L = c_1/c_L$ and $\sin\theta_T = c_1/c_T$), and (2) nonuniform longitudinal wave at angles in the $\theta_L < \theta < \theta_m$ range. The ratio of the signals due to the contributions of longitudinal and transverse waves reads:

$$r = \frac{\int_0^{\theta_L} T_L T_L^* \sin\theta \, d\theta}{\int_{\theta_L}^{\theta_T} T_T T_T^* \sin\theta \, d\theta} \qquad (6.17)$$

The transmission coefficients in Equation (6.8) are specified by the appropriate expressions for transmission coefficients for a plane wave incident normally to the interface between liquid half-spaces [48]:

$$T_{L,T} = \frac{2}{\tau + 1} \qquad (6.18)$$

This indicates that for $YBa_2Cu_3O_{7-x}$ ceramics with the acoustic velocity values $c_L = 5 \text{ km s}^{-1}$ and $c_T = 3 \text{ km s}^{-1}$ [223], longitudinal and transverse waves are efficiently excited in the 0°–17° ($\theta_L = 17°$) and 17°–30° ($\theta_T = 30°$) intervals, respectively, so that $r \sim 3$. Hence, transverse waves make the major contribution to the formation of the subsurface defect image.

The output SAM signal depends heavily on distance z (Equation (6.5)); therefore, only the topography of uneven samples can be studied by this technique. Variation of the function $R'(x, y)$ significantly exceeds that of $R_{\text{def}}(x, y)$ (see Equation (6.15)), where x and y are the coordinates in the scanning plane; hence, the acoustic contrast of images specified by the values of $\partial V/\partial x$ and $\partial V/\partial y$ in uneven samples depends, in the first place, on the topography of their surface.

As mentioned above, SAM furnishes quantitative information about a sample studied; namely, it permits the local velocity of the Rayleigh and re-emitted Rayleigh waves to be assessed, based on the $V(z)$ dependence. It can be shown that the fol-

Figure 6.5 Hot-pressed YBaCuO ceramics, the image size is 300×200 μm. The frequency is 1.7 GHz.

lowing expression holds for the Rayleigh wave velocity in a homogeneous isotropic object [113]:

$$V_R = \frac{c_1}{\sqrt{1 - (1 - \lambda_1/(2\Delta z))^2}} \qquad (6.19)$$

where Δz is the period of $V(z)$ oscillations and λ_1 is the wavelength in the immersion. Homogeneity of the images obtained with focusing on the surface and in the object bulk serves as a criterion of smoothness and in-depth homogeneity of the object. That is, variations δV in signal V within the zone where the speed of sound is measured must meet condition $\delta V \ll V$.

We investigated ceramics of the 123 $A\mathrm{Ba_2Cu_3O_{7-x}}$ (A = Y, Yb, Ho) and $\mathrm{Bi_2Si_2CaCu_2O_{8-x}}$ type, both hot-pressed to 70 kbar, and also ordinary ones, with the aid of an ELSAM microscope (see Section 6.2 for a description). Water preheated to 333 K or methanol at room temperature were used as an immersion liquid. Figure 6.4 displays a surface fragment of coarse ceramics 123 prepared using a conventional procedure. Crystallite domains are seen as light spots with sharply outlined boundaries, dark areas are pores. Focusing on the bottom of pores using the maximum of an acoustic wave reflected from them as an indicator allows the pore depth to be measured with an accuracy of 0.05 μm. Normally, the depth value ranges between 1 and 3 μm.

The number of pores and crystallites was counted automatically with the use of an advanced image processing technique especially developed by our group including new hardware and new software "ELSAM," which were implemented into the ELSAM system. For example, pores occupy 5.5% of the total area in the image of hot-pressed ceramics 123 shown in Figure 6.5. A small fraction of the pore surface area is inherent in high-density ceramics (cf. Figures 6.4 and 6.5).

The Rayleigh velocity value V_R was assessed from the $V(z)$ dependence at a frequency of 1.7 GHz with z being varied between 0 and 20 μm; hence, the maximum distance traveled by the Rayleigh wave amounted to $h = z \tan\theta \sim 10$ μm. Clearly, V_R is a Rayleigh wave velocity averaged over all directions in the sample surface. V_R was measured in fairly large crystallites so that their boundaries did not affect the Rayleigh wave propagation. Ten or fifteen crystallites were chosen on a surface area of 2 mm to measure the $V(z)$-dependencies. The curves thus plotted were compared by subtracting one from another on a computer, the majority of curves (70%–80% for individual samples) were identical. This is evidence of the lack of ac-

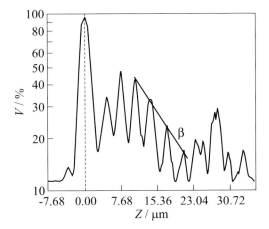

Figure 6.6 $V(z)$ curve of a $Bi_2Sr_2CaCu_2O_{8-x}$ crystal. β is the slope of line enveloping auxiliary maximums. The lens radius is 40 μm; therefore, the $V(z)$ curve is distorted by waves multiply reflected between the lens and object surface at $z < -20$ μm.

cidental factors affecting the output microscope signal. Such factors would include crystallite boundaries, subsurface defects, and small crystallite thickness (the latter of which would result in reflection of the acoustic wave from the back crystallite surface, which, in turn, would inevitably lead to accidental distortion of the $V(z)$ curve because of accidental variations in the crystallite thickness).

Thus, the velocity v is identical in various crystallites of the same sample. It is well known that the speed of sound along the c-axis in perofskite structures studied, differs significantly from the acoustic velocity in the (a, b) plane ($V[100] \sim V[010] < V[001]$) [258], while crystallites in ceramics are plates whose dimensions in the [100] and [010] directions are much greater than their thickness in the [001] direction. This suggests that axis c in the crystallite examined is perpendicular to the ceramics surface. The following values of the Rayleigh velocity were obtained for ceramics: 2.51 km s^{-1} for $Bi_2Sr_2CaCu_2O_{8-x}$, 2.80 km s^{-1} for $YBa_2Cu_3O_{7-x}$, and 2.86 km s^{-1} for $HoBa_2Cu_3O_{7-x}$ and $YBa_2Cu_3O_{7-x}$. With an accuracy of 0.5 km s^{-1}, they amount to 0.8–0.9 of c_T in the [001] plane for the appropriate materials.

The value of the Rayleigh wave absorption α_R is specified by the slope of the line tangent to the auxiliary $V(z)$ curve maximums (angle β in Figure 6.6) [259], which allows α_R to be evaluated for every crystallite. Thus, the lack of side maximums caused by the Rayleigh wave at $z < 0$ is the evidence of significant wave damping on the surface of crystallites studied. For example, full absorption $\alpha_R > 0.1k_R$, $k_R = 2\pi v/V_R$, can be estimated by the lack of a maximum on the $V(z)$ curve at $z < -15$ μm, assuming the signal to noise ratio for ELSAM to be equal to 20 dB. A great number of microcracks in the sample can be responsible for this wave attenuation even if the crack opening size (e.g., 0.01 μm) is far beyond the microscope resolution (0.4 μm). This is because water does not penetrate in microcracks and the hypersound of the frequencies applied is damped in air almost completely within the wavelength. Such an estimate of α_R is an integral value of the Rayleigh wave attenuation within a segment 10 μm in diameter; it is fairly sensitive to the density of crack distributions in crystallite.

Such large absorption values were observed in crystallites of many ceramics, particularly high-density ones. No cracks are seen in Figure 6.5 because they have been "polished out" (nonetched sample); moreover, some cracks are not seen because they are very small. As the electron microscope photograph of the same (but etched) sample shows, the representative crack opening value amounts to 100 Å. It is the presence of such cracks which is responsible for abnormally high Rayleigh wave attenuation. It should be pointed out that the α_R value is assessed by an order of magnitude because the value of auxiliary maximums that determine the angle β depends heavily on the recording instrument parameters, particularly on the delay in receiving the reflected pulse after the transmitted pulse (the so-called gate); furthermore, of practical importance is α_S, the portion of total absorption α_R associated with acoustic damping and structure scattering, rather than with the total value of the Rayleigh wave attenuation (which depends mostly on how rapidly the wave leaks in low-defective and low-porosity samples). The attenuation $\alpha = \alpha_R - \alpha_S$ pertaining to wave leaking proper, can be assessed using the c_L and c_T values indicated above, with water as an immersion liquid. From the expression for attenuation α derived in [260], we obtain

$$\alpha \sim \frac{\rho_1 c_1}{\rho c_L} \frac{k_R}{2\pi} \qquad (6.20)$$

and, hence, $\alpha_R \sim 0.01 k_R$. Thus, the presence of a large number of microcracks significantly increases the sound wave attenuation.

Observation of individual crystallite sections reveals domains which are distinguished by their brightness. These can be attributed to a locally changed phase or structural composition of the material. Different ceramic phases exhibit dissimilar acoustic properties and, hence, different reflection coefficients. This effect can be assessed in Equation (6.4) by using a simple expression for the reflection coefficient R derived for the reflection of a longitudinal wave incident normally on the sample surface from the liquid half-space [48]. This yields longitudinal wave velocities of 4.9 and 4.4 km s^{-1}, for the $YBa_2Cu_3O_{6.5}$ and $YBaCu_3O_7$ phases, respectively, [257], and a difference in the reflection coefficient values of as low as 2%. This suggests that the difference in reflection coefficients for high-reflection objects is small; therefore, to visualize different phases, one should employ a so-called quasi-3D imaging method in which the oscilloscope beam, along with variation in its brightness, deviates from its scanning line on the screen in proportion to the output SAM signal value. Figure 6.7 illustrates a quasi-3D image of a high-density Y-ceramic in which phase inclusions are exposed as spikes.

We also observed a twin structure in some crystallites, which, generally speaking, could not be observed with the aid of an axially symmetric spherical lens. This observation should presumably be attributed to the origination of strained zones

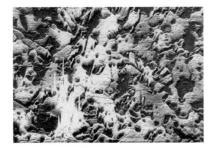

Figure 6.7 Quasi-3D image of the surface of a hot-pressed $YBa_2Cu_3O_{7-x}$ ceramics, the image size is 200×130 μm. The $YBa_2Cu_3O_6$ phase with a greater speed of sound is distinguished, the frequency is 1.7 GHz.

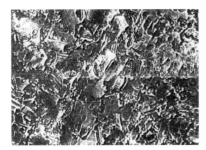

Figure 6.8 The same site as in Figure 6.7. The depth of focusing $z = -5.5$ μm. Twins are seen.

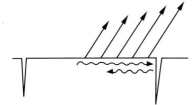

Figure 6.9 Mechanism of twin visualization with a Rayleigh wave.

around the twinning boundaries, which is supported by the fact that no twin structure is seen at $z = 0$ (Figure 6.5), appearing only at $z < 0$ (Figure 6.8) when the lens starts receiving the leaky Rayleigh wave.

Figure 6.9 illustrates a plausible mechanism of visualization of twin boundaries. The Rayleigh wave spreading over the single crystallite surface is reflected from the twinning boundary because it is inhomogeneous, thereby enhancing wave-leaking near the boundary. As a result, the twinning boundary is contoured with a light line and shows up in the image even if the transverse boundary dimension is much less than the wavelength. A similar contouring effect is observed for any boundaries at $z < 0$. To improve the contrast of a twin structure image, one should use an asymmetric lens, e.g., either shifted or equipped with an elliptic converter. This would allow twin structure components of different orientation to be distinguished by the difference in reflection coefficient values in a manner similar to pictures taken in polarized light [261].

$YBa_2Cu_3O_{7-x}$ films on various supports were also studied. Figure 6.10 illustrates the image of a $YBa_2Cu_3O_{7-x}$ film applied to a ZrO_2 support by evaporating the target with the aid of a pulse-periodical laser at 1200 K. The image shows weak variations of the acoustic signal pertaining to a dome-like relief of the film 123

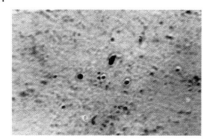

Figure 6.10 YBa$_2$Cu$_3$O$_{7-x}$ film on a ZrO$_2$ support. Image size is 300 × 200 μm. Frequency is 1.8 GHz.

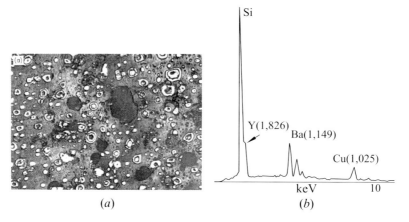

(a) (b)

Figure 6.11 YBaCuO films on a Si support. (a) Dark spots belong to the "green" BaCuO$_5$ phase and light sites are separations. (b) Electron microscopy spectrum in the "green" phase zone; numbers in parentheses indicate exact keV values pertaining to each inclusion.

surface. Individual microscopic droplets 1–2 μm in diameter are also clearly discernible; they arise due to inhomogeneous material atomization (so-called "sputtering" associated with the unsteady evaporation regime).

Films prepared by subsequent thermal depositions of Y, Ba, and Cu on a silicon support at 720 K in an oxygen atmosphere were studied. These multilayer films fabricated were then annealed at 1120 K for 4 min to produce a superconducting phase. The above film fabrication technique is depicted in detail in [262, 263]. Investigations have demonstrated that the films were multiphase (Figure 6.11). There are Y$_2$BaCuO$_5$ ("green" phase) spots, CuO, Y$_2$O$_3$, and BaCuO$_2$ inclusions along with the YBa$_2$Cu$_3$O$_{7-x}$ phase. The 123 and 211 phases are unambiguously identified by the shape of the $V(z)$ dependencies. Light spots with interference fringes are also seen in Figure 6.11(a), belonging to separations caused by insufficient adhesion and a difference in the temperature expansion coefficients during the course of thermal treatment.

As well as the ceramic samples studied crystalline YBa$_2$Cu$_3$O$_7$ and BiSrCaCuO samples were also investigated. Both surface defects (Figure 6.12) and structure defects (Figure 6.13) were revealed in the samples. Twinning with a period of

Figure 6.12 $Bi_2CaSr_2Cu_2O_{8-x}$ crystal surface, image size is 60×40 μm.

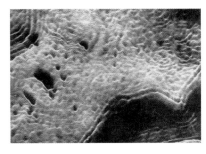

Figure 6.13 The same site as in Figure 6.12, focusing depth is $z = -6.6$ μm.

about 1 μm was observed in $YBa_2Cu_3O_7$ samples. The V_R values measured in the (a, b) single-crystal planes were close to V_R measured in unstrained crystallites of the appropriate ceramics.

Thus, it is shown that acoustic microscopy offers broad opportunities for studying and controlling the quality of HTSC materials and also for assessing local acoustic properties and revealing various defects in crystallites, crystals, and films.

6.4
Application of Acoustic Microscopy to the Study of Multilayer Reinforced Fiber–Glass Graphite Composites

This section discusses the advantages offered by application of the acoustic microscopy methods in revealing microcracks and other structure defects in composite materials based on a polymer matrix, reinforced by glass-graphite fibers. Two samples of the material were investigated: these were 2-millimeter thick plates 41×99 mm in size, fabricated at the University of Bath, England. Each sample was subjected to mechanical impact to induce an internal defect.

The investigation was implemented by using a desk-top broad-field short-pulse SAM designed and manufactured at Tessonics Inc. (Canada) and operating at ultrasonic frequencies in the 10–100-megahertz range [264].

A sample immersed in a cell filled with water was fixed on a movable table.

The acoustic system consisted of a 50-megahertz spherical lens made of sapphire. All reflected signals in the reflection operation regime were successively received by the same lens, alternatively transmitting and receiving the waves. Fast scanning was performed along the y-axis when forming columns in the image

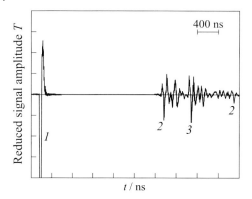

Figure 6.14 Oscilloscopic trace of a reflected acoustic signal: 1 – sound pulse, 2 – signal reflected from the surface, and 3 – signal reflected from an inhomogeneity in the sample bulk.

raster; slow scanning was implemented along the x-axis. The scanning step was 50 µm along both axes. The maximum scanning surface area was 12 × 12 mm and the amplification coefficient was 20–60 dB. A narrow-aperture acoustic lens with a 11-degree aperture and focal distance in water of 16 mm was used in this study. The acoustic lens was connected to the high-frequency microscope with a coaxial cable. The high-frequency microscope system was comprised of a unit generating the master pulse, commuting unit, and an analog-to-digital converter operating as a stroboscope to increase the effective digitizing frequency. The software, developed by Borland Delphi, allowed the A-, B-, and C-scans to be obtained. The possibility of image saving in one of the graphical formats and of importing the image to other programs for subsequent processing was also provided.

Samples were studied using a short-pulse regime of ultrasound probing; the pulse duration did not exceed 2–3 periods. Most of the ultrasound pulse energy went to the reflected signal and the acoustic structure image was arranged based on the measured reflected signal intensity at various points of the sample surface.

The velocity of longitudinal sound propagation in the sample, as assessed by dividing the doubled sample thickness by the delay time of the ultrasound signal arrival, was $2d/\Delta t = 2800 \pm 200$ m s^{-1}. Given the velocity of longitudinal sound propagation in the sample material, the depth of defect localization was assessed by the time of the ultrasound pulse reflection from the structures of deformed zones (Figure 6.14).

It was found that the defect was located in the middle of the sample. The optimal ultrasound signal frequency was 15 MHz. A further increase in the frequency of the ultrasound signal brought about significant absorption, while at lower frequencies the image resolution was fairly poor.

A 3D image of an internal defect was obtained by mechanically scanning the sample surface. Focusing of the ultrasound signal appreciably improved the image resolution and power characteristics of the system. The reflected signal intensity at every individual point was presented by an appropriate gradation of the gray scale.

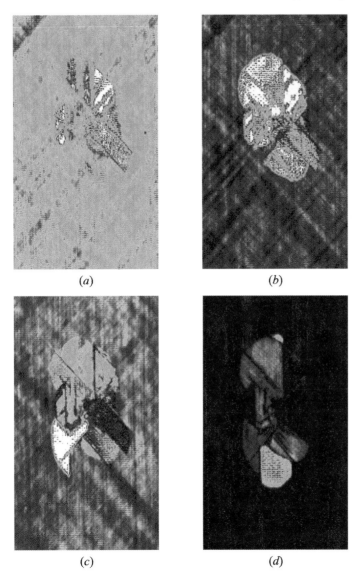

Figure 6.15 C-scans of a defective area in sample 2JS(1) at various depths: (a) 0.6 mm, (b) 0.9, (c) 1.2, and (d) 1.5 mm. The scanned area size is 30 × 50 mm.

The precise shape of the defected zone can be ascertained by C-scanning implemented subsequently at various depths from the sample surface. Figure 6.15 displays a set of C-scans implemented at various depths in the sample. The depth resolution was 120 ns. Contours of images taken at various depths differ from each other. This stems from the fact that deformations at various depths affect different structure components.

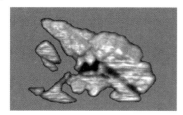 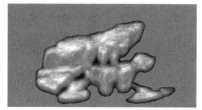

Figure 6.16 Three-dimensional image of a defect in the sample. Two different views are shown.

The volume reconstruction of the 3D damaged zone was implemented using the method developed by Alex Denisov from our group. The 3D image in Figure 6.16 furnishes complete information about the defect shape and size.

The results of the investigations performed convincingly demonstrate that the use of various acoustic microscopy techniques allows one to rapidly localize a defect and to determine with a high accuracy the shape, size, and localization of strained zones in materials and constructions.

7
Scanning Acoustic Microscopy of Polymer Composite Materials

Investigation of permolecular polymer structure and its relation to the basic physicochemical characteristics and destruction processes is a traditional topic of physical chemistry and material science of polymer materials [265–269]. Because of the extensive use of polymer mixtures and composites in modern technologies, the aforesaid problem is an urgent issue for multiphase polymer materials. As a rule, up to the present time, these investigations were mostly performed by employing optical and electron microscopy techniques that exhibit, along with obvious advantages, significant drawbacks. Optical methods are unsuitable for exploring opaque and turbid media or inhomogeneous transparent media with noappreciable difference in their optical properties. Electron microscopy methods necessitate special pretreatment of samples, namely, contrasting or fabrication of replicas; therefore, as a matter of fact, they turn out to be indirect techniques [270, 271]. This is why recent attention of investigators was focused on acoustic methods for studying the structure and properties of polymer materials based on the use of acoustic microscopes.

Although acoustic methods were extensively used previously in studies of polymers, their application basically concentrated on exploring the internal relaxation processes and dynamics of molecular motion [30, 272–276]. They were based on measurements of the frequency dependence of the speed and absorption of sound in polymers which, in turn, characterized the bulk properties of a sample as a whole. Acoustic phenomena also underlie the acoustic microscopy methods that have provided basic data on molecular dynamics in polymers of various sorts [30, 272, 276]. New opportunities are offered by the acoustic microscopy techniques in which ultrasound is used to visualize the microstructure of polymer samples and to gain quantitative information about their morphologic peculiarities [1, 4, 277–280].

The present chapter is aimed at developing methods for obtaining and interpreting acoustic images of polymer composite materials, methods for quantitative analysis of changes in local viscoelastic properties of composite polymer objects, and methods for mathematical processing acoustic images to obtain quantitative characteristics of composite polymer systems. It also aims at assessing structural peculiarities of polymer mixtures prepared using various technological techniques,

7.1
Acoustic Methods for the Investigation of Polymers

A fact specific to acoustic methods is that their use entails no alteration or modifications of the polymer material structure and no damage to the samples studied. Acoustic techniques permit one to quickly and precisely measure the most important parameters characterizing the physical and mechanical properties of polymers, such as dynamic elasticity modules and acoustic loss coefficients.

There exists a definite correlation between the acoustic parameters of polymer materials and their chemical structure [276]. The speed of sound, reflection index, acoustic absorption factor, and some other parameters, depend on the permolecular arrangement of macromolecular materials [274, 281, 282].

However, this correlation is extremely difficult to trace using the values of these parameters proper. It is the frequency dependence of the speed of sound and absorption factor which is much more sensitive to the composition and structure of polymers. This dependence is dictated by relaxation processes. The energy of directed motion of material particles in an acoustic wave is transferred to internal degrees of freedom due to the recovery of thermodynamic equilibrium. As a result, the energy of a wave reduces as it spreads in the material. The speed of sound depends also on dissipative processes in the material. The energy exchange between vibrational and rotational degrees of freedom and the translational energy of directed motion, the excitation of motion of side macromolecule chains, and of their individual segments of various lengths, are among the basic representative mechanisms of acoustic relaxation in polymers. A relaxation process is usually characterized by a relaxation time which is the time interval it takes for a parameter of the material which has experienced deviation from its thermodynamic equilibrium value to be equilibrated. The effect of relaxation processes on the propagation of an acoustic wave depends on the ratio of the wave period T and relaxation time τ. The lower the sound frequency the more completely is the perturbed equilibrium recovered and the more energy of the ordered motion is lost. This gives rise to an appreciable dependence of the internal friction (viscosity) on the sound frequency:

$$\eta = \eta_0 + \delta_\eta \frac{1}{1 + \omega^2 \tau^2} \qquad (7.1)$$

where the first term is the contribution of very fast relaxation processes, the second term specifies the contribution of the relaxation process considered, and δ_η is the contribution of this relaxation process to the viscosity at frequencies lower than $1/\tau$.

Relaxation processes contribute also to the elastic material characteristics, thereby bringing about a frequency dependence of the modulus of elasticity G:

$$G = G_\omega + \Delta G \frac{\omega^2 \tau^2}{1 + \omega^2 \tau^2} \qquad (7.2)$$

where G_ω is the value of the elasticity modulus G at a low-frequency $\omega < \gamma/\tau$.

As the ultrasound frequency increases so does the modulus value, and at $\omega \gg 1/\tau$, $G = G_\omega + \Delta G$. Thus, dependence of the elasticity modulus and viscosity coefficients on the external impact frequency or, as they say, the frequency dispersion of G_ω and η_ω, is inherent in polymer materials. Peculiarities of the frequency dispersion of elastic and viscous material properties are governed by the nature of the molecular motions in the material, as well as its molecular structure and properties. Slow relaxation processes are lacking in many materials, i.e., relaxation times are commensurate with intermolecular collision times ($\tau \approx 10^{-12}$–10^{-13} s). As a result, such materials as water, a great many crystals, and others, exhibit virtually no frequency dispersion. Polymer structure peculiarities (long chain molecules weakly interacting with each other) give rise to a continuous relaxation time spectrum within a very wide range, from some fractions of an hour, to minutes, to nano- and femtoseconds [25, 276]. The contribution of relaxation processes with a representative relaxation time τ is specified by the so-called relaxation time function that, naturally, depends on both external conditions (temperature, pressure, aggregate state) and on the type of material. As the acoustic field frequency varies, some of the relaxation processes freeze (as the frequency increases) or unfreeze (as the frequency decreases), bringing about appreciable changes in the viscoelastic material properties. Thus, a steep rise in the shear modulus is observed in elastomers at high frequencies (10^7–10^9 MHz) which is attributed to freezing the motion of long segments of polymer molecules and to a transition of the material from a high-elasticity state to the conventional solid state [283].

A great many papers published during 1950–1980 dealt with studies of the frequency dependences of absorption and the speed of acoustic waves, i.e., with the so-called acoustic spectroscopy. The major purpose of these works was to find out how the characteristic peculiarities of the frequency dependences are related to particular types of molecular motion and to study, based on this correlation, the dynamics of molecular motion. Unfortunately, the researchers failed to find a stable and reliable correlation between the frequency dependences of the absorption and the velocity of acoustic waves and particular physical processes in the materials. It is associated, to a great extent, with a smooth behavior of the frequency dependences, which, in turn, stems from the very wide spectrum of relaxation times inherent in any particular relaxation type.

It should be emphasized that acoustic properties of polymers are not confined to their so-called physicomechanical properties, which term is extensively used in the physical chemistry of polymers (see Table 7.1). The fact is that normally, by the physical and mechanical properties, one means static or low-frequency values of the elasticity modulus G and viscosity η, whereas the dynamic values of these parameters $G(\omega)$ and $\eta(\omega)$ essential for acoustics are

$$G(\omega) = \sqrt{\frac{G_\omega}{\rho}} \tag{7.3}$$

$$\eta(\omega) = \frac{\eta_\omega \omega^2}{2\rho G_\omega^3} \tag{7.4}$$

Static properties are measured, as a rule, in samples of significant size that are subjected to destruction; therefore, the measurements require a great amount of the polymer material. Moreover, studies of the polymer material properties of natural samples by measuring stretching strain, failure stress, fatigue, etc., are associated with changes in the polymer material structure. Acoustic microscopy techniques require small-size polymer samples and measurements are performed applying oscillations of a very low amplitude; hence, both the polymer structure and properties remain invariable during the course of investigation. Furthermore, a sample can repeatedly be used by varying the temperature range.

7.2
Methods for Studying and Visualizing the Dispersed Phase in Polymer Blends

A number of techniques have been developed in ultrasonic investigations. Pulse techniques are most popular in measurements of the velocity, and absorption of ultrasonic waves in polymers. Any pulse technique intended to measure ultrasonic absorption is based on a comparison of the amplitude of pulses which have traveled certain distances in the sample studied, and on the calculation of the absorption coefficient γ, by the formula [147]

$$\gamma = \frac{1}{x_2 - x_1} \ln \frac{u_1}{u_2} \tag{7.5}$$

where x_1 and x_2 are the coordinates of the points, and u_1 and u_2 are the oscillation amplitudes at points x_1 and x_2, respectively.

The velocity of ultrasound waves is evaluated as the ratio between the acoustic path length L and the time of travel of the acoustic pulse, τ_0

$$c = \frac{L}{\tau_0} \tag{7.6}$$

Pulse-phase methods of measuring the ultrasound speed employ a correlation between changes in the acoustic signal phase, arising as a result of sample traversing by the pulse and the speed of sound [284]

$$c = \frac{\omega L}{\Delta \zeta} \tag{7.7}$$

where ω is the circular frequency of ultrasonic oscillations, L is the acoustic path length, and $\Delta\zeta$ is the change in the acoustic signal phase.

Normally, researchers observe changes in the phase shift as a function of frequency and acoustic path length. Basic methods for measuring ultrasound wave velocity in solids, and specifically in polymers, were discussed in a number of papers and monographs [272, 284–291]. The Williams–Lamb method [292, 293] and pulse superposition technique [294] are used very rarely to measure the speed of ultrasound in polymers because of the much higher absorption of acoustic waves in them compared to metals.

The phase comparison method [295] that permits the speed of sound to be measured in small samples is extensively employed in studies of polymers. When using this method, one has to take into account the effect of a contact layer through which the sample studied communicates with a piezoelectric transducer or buffer rod. The immersion media and buffer rod techniques [281, 284, 287, 294, 296] are used most frequently in measurements of the speed and absorption of sound. Defectoscopy instruments operate at frequencies in the 20 kHz–25 MHz range. There are two basic methods of ultrasound examination, namely, shadowgraphic (through transmission) and echographic (reflection) methods; there are also a few versions of immersion and contact ultrasound methods. In addition, a natural oscillation (sound) method [297] based on the use of the natural frequencies (usually of the first harmonics) of the sample studied is also applied. The natural oscillations are detected with a resonance instrument but with high amplification of the excited signal. The torsional vibration method, which is a version of the latter method, is based on impact excitation of a controlled sample and on an analysis of its natural oscillations [298].

The impedance (reaction) method is based on the assessment of the mechanical resistance of a sample studied at the site where it contacts the transducer exciting the acoustic waves [283].

Among the problems of primary importance in acoustic microscopy are: interpretation of acoustic signatures, elucidation of the nature and mechanisms of acoustic contrast for various types of objects, the development of quantitative methods for studying local viscoelastic properties of samples [60, 63, 110, 201, 299, 300], and also for depicting their integral characteristics [301]. A great many acoustic microscopy techniques were developed for the investigation of biological objects [252, 302–305]. However, acoustic microscopy techniques intended to study polymer materials are very scarce. Those techniques must provide qualitative images of the microstructure of composite materials and films of blends, which implies selection of optimal media and development of appropriate techniques for preparing the samples to be examined.

7.3
Objects of Investigation

Polymer blends are objects that are of importance both for practical applications and for fundamental research. Their application, physical and mechanical properties, and structure are closely inter-related. This chapter discusses investigations of commercial polymers of the polyolefin group chosen as illustrative objects. This selection is dictated by the fact that polyolefins make up the largest group of commercial thermoplastic polymers. Some of the most wide-spread polyolefin blends are those based on low-density polyethylene (LDPE). Articles made of LDPE and polyvinylchloride (PVC) blends are widely used to manufacture synthetic paper and nonwoven materials; LDPE, polystyrene (PS), and polymethylacrylate (PMMA) blends are used to fabricate films, coatings, and fibers; mixtures of LDPE with graft ethylene copolymers with styrene or vinyl acetate are used to manufacture articles by casting under pressure and extrusion. Taking into account the importance of these blends, many researchers have studied their structural characteristics and how these are related to the physical and mechanical material properties.

This section deals with LDPE-based polymer blends because LDPE is not only a commercial material that is most widely used in practice but is also the most thoroughly studied polymer. Polymethylacrylate, PS, or PVC serves as the second blend component.

The materials used in the work are:

- granulated LDPE with the following characteristics: $\rho = 0.91$ g cm^{-3}, melt flow index (MFI) $= 2.0$ g/10 min, melting point $T_m = 110\,°C$, molecular weight (viscosity-mean) $\approx 45\,000$;
- PS (bulk polymer), MFI $= 3.2$ g/10 min, molecular weight $\approx 290\,000$, softening temperature $80\,°C$, $\rho = 0.90$ g cm^{-3};
- PMMA, MFI $= 1.10$ g/10 min, softening temperature $110\,°C$, $\rho = 0.94$ g cm^{-3};
- PVC, softening temperature $80\,°C$, $K = 73$, $\rho = 1.39$ g cm^{-3}.

The ratio between blend components (wt.%) was varied as follows:

LDPE	Second component
90	10
80	20
50	50
20	80

7.4
Basic Requirements Imposed on Polymer Mixtures and Methods for their Study by Acoustic Microscopy

The great importance of acoustic microscopy methods in the study of polymer blends makes the development of sample fabrication techniques one of the major problems to be solved. Two approaches have been proposed to do this: investigation of individual particles and of films of polymer blends.

Individual polymer blend particles were studied with the aid of Mylar films stretched on a holder. The speed of sound in Mylar $c = 2.54$ km s^{-1}, $\rho = 1.18$ g cm^{-3}, and the absorption of sound within it (at a frequency of 450 MHz) $\gamma = 0.0186$ μm^{-1} [306]. The amplitude of a signal after it passes through a Mylar film 2 μm thick is $0.964 A_0$ (A_0 is the incident wave amplitude). The Mylar film absorbed sound only negligibly. The image had greatest contrast when the difference between the levels of the signals from two neighboring zones with different ρ, c, and γ values was the greatest. The object position at which the boundaries separating various zones were sharpest were found by displacing the object between the lenses along a vertical direction.

Polymer composite films were manufactured on a 15-ton hydraulic press. A polymer blend sample was placed on a polyamide support and covered with a polyamide film. The press was preheated to 160 °C (for LDPE-based blends). The mold was held in the press for 3 mins without pressurizing. Then the pressure was raised to 100 atm under which the sample was held for 1 min. This procedure of film manufacturing is the best one for preserving the blend structure because it virtually rules out aggregation of heterogeneous zones.

Films retaining the structure of the material were made by cutting a thin layer on a Leitz (Germany) microtome. To make microscopic sections of a good quality which retain the initial structure inherent in the blend type, one has to simultaneously cool the knife and table. The microscopic sections were glued to the support with a resin, so that the surface studied remained clean.

A microtome section was cut from a pellet fabricated with the aid of a manual press at $T = 160$ °C in which the sample was held for 5 mins under a constant pressure. The blend partially melted inside the pellet and fastened, its structure being retained completely; therefore, the section cut through the pellet center furnished a pattern of the distribution of heterogeneous phase particles in the blend.

7.5
Investigation into the Mechanisms of Acoustic Contrast in Polymers

Three basic physical phenomena play a significant role in the formation of the acoustic contrast, they are:

(1) phase aberrations due to sound beam refraction in the sample;
(2) ultrasound attenuation in the sample; and
(3) partial reflection of the incident beam at the front and rear sample boundaries.

If the formation of transverse waves in the sample is neglected and propagation through the sample of longitudinal waves only is taken into account, the transmission coefficient T reads [126, 307]:

$$T(\theta) = \frac{4\xi e^{i\varsigma} e^{-\gamma_L d/\cos\theta'}}{(1+\xi)^2} \frac{1}{1 - ((1-\xi)/(1+\xi))^2 e^{2i\varsigma} e^{-2\gamma_j d/\cos\theta'}} \tag{7.8}$$

where $\xi = [(\rho c)/(\rho_L c_L)](\cos\theta'/\cos\theta)$ is the ratio between the acoustic impedance values of the immersion liquid (ρ is the density of the liquid and c is the speed of sound within it) and of the sample material (ρ_L is the density of the sample material and c_L is the velocity of longitudinal waves within it), $\varsigma = (\omega d/c)\cos\theta'$ is the change in the phase of a longitudinal wave spreading at velocity c_L after it passes through a plate of thickness d, and ω is the ultrasound frequency.

The first multiplier in Equation (7.8) specifies changes in the amplitude and phase of the wave during its journey through the plate in the form of a longitudinal wave; the second term sums up contributions of the waves arising beyond the plate as a result of secondary reflections. The ratio of the acoustic impedance values $\rho c/(\rho_L c_L)$ for the majority of plastics falls in the range:

$$\frac{1}{2} \leq \frac{\rho c}{\rho_L c_L} \leq 2 \tag{7.9}$$

Inequality (7.9) indicates that the reverberation effects are small in polymers. Hence, the second term in (7.8) can be set equal to unity:

$$T(\theta) \approx T_1(\theta) = \frac{4\xi e^{i\varsigma} e^{-\gamma_L d/\cos\theta'}}{(1+\xi)^2} \tag{7.10}$$

At a small plate value, $\gamma d < 1$, sound attenuation in a sample affects the transmission coefficient only slightly; therefore, $T(\theta)$ can be written as

$$T(\theta) \approx T_2(\theta) = \frac{4\rho c/(\rho_L c_L)}{(1+\rho c/(\rho_L c_L))^2} e^{i\varsigma} \tag{7.11}$$

All three mechanisms of acoustic contrast formation compete with each other and different mechanisms dominate in samples of various thickness.

We studied the contrast formation mechanism in an LDPE–PS system. The basic acoustic characteristics of materials, such as density ρ, longitudinal speed of sound c_L, the ratio between the velocities of longitudinal waves in the plastic and water, the angle of total internal reflection θ_L, the longitudinal acoustic impedance ρc_L, and the ratio between the impedance values in the plastic and water, are listed in Table 7.1. The data are borrowed from [201, 289, 290]. Sound attenuation in both plastics is estimated using a linear approximation of the frequency dependence of the ultrasound attenuation coefficient γ:

$$\gamma = mf + b \tag{7.12}$$

where m and b are the empirical constants, and f is the frequency.

Table 7.1 Basic acoustic parameters of LDPE and PS, water is taken as an immersion agent for comparison ($c = 1.49$ km s^{-1} and $\rho = 1.0$ g cm^{-3}).

Parameters	LDPE	PS
Density ρ_1, g cm^{-3}	0.920	1.050
Velocity of longitudinal acoustic waves c_L, km s^{-1}	1.950	2.400
Ratio between the velocities of longitudinal waves c_L/c	1.304	1.605
Angle of total internal reflection $\theta_L = \arcsin(c/c_L)$ with respect to water	50	38.50
Longitudinal acoustic impedance $\rho_1 c_L$	1.794	2.52
Ratio between impedance values $(\rho_1 c_L)/(\rho c)$ with respect to water	1.204	1.691
Reciprocal ratio of the impedance values $(\rho c)/(\rho_1 c_L)$	0.831	0.591

Table 7.2 Parameters of acoustic wave attenuation in LDPE and PS.

Parameters	LDPE	PS
m, dB (cm MHz)$^{-1}$	5.25	2.16
b, dB cm^{-1}	−1.72	−0.267

The linear approximation is inherent in ultrasound attenuation in plastics at megahertz frequencies due to the continuous relaxation time spectra of these materials. We used this linear approximation in the range of higher frequencies which is justified by the fact that our experimental data, on the whole, support the attenuation values used to estimate γ. To assess the coefficient γ of ultrasound attenuation in LDPE and PS, we borrowed the parameters m and b reported in [289] (Table 7.2). The data indicate that acoustic properties of the blend components studied, differ significantly. Polystyrene is more rigid acoustically, its longitudinal acoustic velocity is fairly high (about 30% as high as the appropriate speed of sound in polyethylene) and appreciably differs from the speed of sound in water: $n_{st} = c_L/c = 1.6$. Therefore, one should expect significant phase aberrations in PS and reflections at the interfaces. Ultrasound absorption in PS is much less (about 2.5-fold) than it is in polyethylene. However, its value at the microscope operation frequency $f = 450$ MHz $\gamma_{st} \approx 230$ cm^{-1} is high enough for the attenuation effects to start dominating in the formation of the output signal of the acoustic microscope for PS films thicker than 45–50 μm. Polyethylene is softer acoustically, the difference of the velocity of longitudinal acoustic waves in it from the speed of sound in water is not so significant, $n_e = c_L/c = 1.3$. Phase aberrations and reflections at the boundaries of films made of this material are unimportant when the film thickness exceeds 15–20 μm because of the intense attenuation of acoustic waves in LDPE ($\gamma_e \approx 550$ cm^{-1}).

To correctly interpret images of LDPE–PS blend films obtained by transmission acoustic microscopy and to understand the principles of acoustic contrast forma-

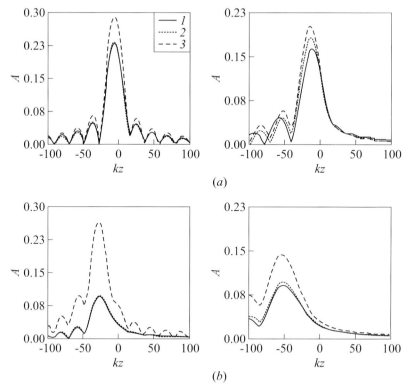

Figure 7.1 Output signal of an acoustic microscope A vs coordinate of the receiving lens z for LDPE (left-hand column) and PS (right-hand column) films calculated under various assumptions concerning the character of incident beam propagation in the sample: (a) $kd = 10$, $d = 5.3$ µm; (b) $kd = 50$, $d = 26$ µm; $1 - A(kz)$ dependence is calculated with an allowance for all phenomena arising as the beam passes through the plate; $2 - A_1(kz)$ dependence is calculated neglecting the reverberation processes, and $3 - A_2(kz)$ dependence is calculated ignoring both the reverberation processes and sound attenuation in the sample.

tion in images of these systems, we calculated the $A(z)$ dependences for LDPE and PS films of various thicknesses (Figures 7.1–7.5): $d = 0$, 5.3, 26, 42, 79, 158, 237, and 315 µm. The values of the parameter $kd = 2\pi d/\lambda$ (λ is the wavelength of sound used in water; in the case at issue, $\lambda = 3.3$ µm) corresponding to the above-listed film thickness values are: 0, 10, 50, 80, 150, 300, 450, and 600, respectively.

The calculated dependences are displayed in Figure 7.5 and will be discussed below. Tentatively, to bring to light the contribution of one physical factor or another to the formation of the output signal of an acoustic microscope, we calculated three versions of the $A(z)$ dependences for LDPE and PS at every value of $kd = 10$, 50, 80, and 150. In the first version ($A(kz)$ curves in Figures 7.1–7.4), the following physical phenomena, which arise when a focused acoustic beam passes through a plane-parallel plate, were taken into account (except for the partial transformation of the incident beam into transverse waves):

7.5 Investigation into the Mechanisms of Acoustic Contrast in Polymers

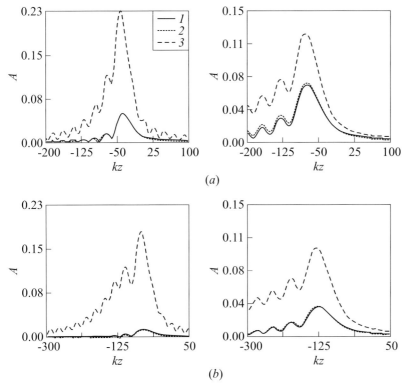

Figure 7.2 Output signal of an acoustic microscope A vs coordinate of the receiving lens z for LDPE (left-hand column) and PS (right-hand column) films calculated under various assumptions concerning the character of incident beam propagation in the sample: (a) $kd = 80$, $d = 42$ μm; (b) $kd = 150$, $d = 79$ μm; $1 - A(kz)$ dependence is calculated with an allowance for all phenomena arising as the beam passes through the plate; $2 - A_1(kz)$ dependence is calculated neglecting the reverberation processes, and $3 - A_2(kz)$ dependence is calculated ignoring both the reverberation processes and sound attenuation in the sample.

(1) the partial reflection of incident waves at the front and rear sample boundaries;
(2) the formation in the plate of a system of re-reflected waves and their contribution to the beam escaping the plate (effects of reverberation or repeated echo signals);
(3) the attenuation of the direct transmitted signal and of repeated echo signals; the attenuation values are different for various angular beam components because the distances traveled by the wave in the sample are diverse in the beam components; and
(4) phase variations in transmitted waves because of the diverse path lengths in the plate for different angular beam components (phase aberrations).

To ascertain the contribution of reverberation effects to the formation of the output signal, we disregarded the contribution of repeatedly reflected waves to the

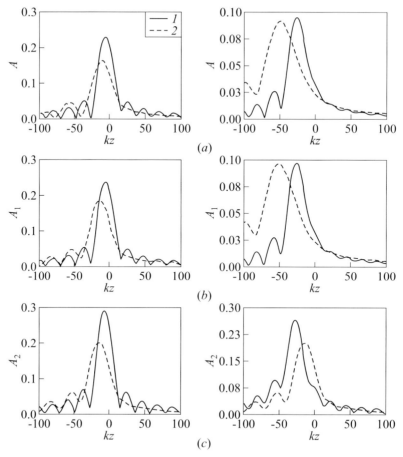

Figure 7.3 Comparison of the $A_e(kz)$ dependences for LDPE (1) and $A_{st}(kz)$ for PS (2) calculated under various assumptions concerning the nature of acoustic beam propagation through the sample: left-hand column: $kd = 10$, $d = 5.3$ μm; right-hand column: $kd = 50$, $d = 26$ μm. (a) $A(kz)$ dependence is calculated with an allowance for all phenomena arising as the beam passes through the plate; (b) $A_1(kz)$ dependence is calculated neglecting the reverberation processes, and (c) $A_2(kz)$ dependence is calculated ignoring both the reverberation processes and sound attenuation in the sample.

acoustic beam field beyond the plate in the second variant of the calculations of $A(z)$ dependences ($A_1(kz)$ curves in Figures 7.1–7.4). To this end, we used an expression for the transmission coefficient $T(\theta)$ in the form of Equation (7.10). Finally, when calculating the third set of $A(z)$ dependences ($A_3(kz)$ dependences in Figures 7.1–7.4), we neglected both the reverberation effects and attenuation of the incident wave in the course of its propagation through the plate and used Equation (7.11) for the transmission coefficient $T(\theta)$. The discrepancy between the $A(z)$ dependences for LDPE and PS in the latter case is dictated predominantly by the phase aberration level in these materials. A small contribution to the contrast is

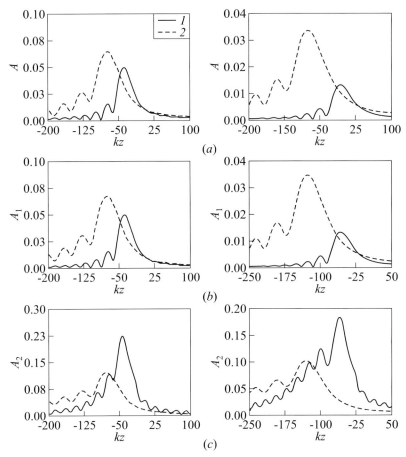

Figure 7.4 Comparison of the $A_e(kz)$ dependences for LDPE (*1*) and $A_{st}(kz)$ for PS (*2*) calculated under various assumptions concerning the nature of acoustic beam propagation through the sample: left-hand column: $kd = 80$, $d = 42$ μm; right-hand column: $kd = 150$, $d = 79$ μm. (a) $A(kz)$ dependence is calculated with an allowance for all phenomena arising as the beam passes through the plate; (b) $A_1(kz)$ dependence is calculated neglecting the reverberation processes, and (c) $A_2(kz)$ dependence is calculated ignoring both the reverberation processes and sound attenuation in the sample.

also made by the difference between the reflection effects at the immersion liquid–sample interface. The results of calculations are displayed in Figures 7.1 and 7.2. The figures show the $A(z)$ dependences calculated under the above-indicated assumptions (three variants) separately for polyethylene (left-hand columns) and PS (right-hand columns).

Calculations demonstrate that the effects of repeated reflections in the sample, and of reverberation, only insignificantly affect the $A(z)$ dependences. Discrepancy between the $A(z)$ curves calculated with and without reverberation effects taken into account (curves *1* and *2* in Figure 7.1, right-hand column) is discernible only

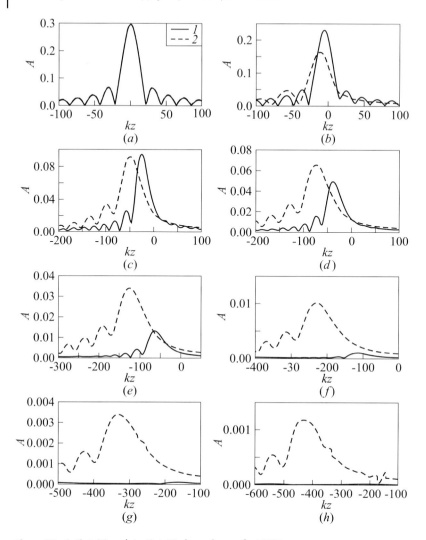

Figure 7.5 $A_e(kz)$ (1) and $A_{st}(kz)$ (2) dependences for LDPE and PS samples of various thickness: (a) $kd = 0$, $d = 0$ μm; (b) $kd = 10$, $d = 5.3$ μm: (c) $kd = 50$, $d = 26$ μm; (d) $kd = 80$, $d = 42$ μm; (e) $kd = 150$, $d = 79$ μm; (f) $kd = 300$, $d = 158$ μm; (g) $kd = 450$, $d = 237$ μm; and (h) $kd = 600$, $d = 315$ μm.

for PS, the elastic properties of which significantly differ from the elastic properties of water, in thin films whose thickness is commensurate with the wavelength of probing ultrasound. This notable contribution of the reverberation phenomena stems from the fact that a discrepancy between the acoustic impedance values of water and PS is great ($\rho_L c_L/(\rho c) \approx 1.7$); therefore, the effects of reflections at the interfaces are quite prominent. At the same time, the plate thickness is insufficient

($d \approx \lambda$) for the reverberation processes to be indiscernible against the background of phase aberration and attenuation. Attenuation of an incident ultrasound wave is appreciable in films of virtually arbitrary thickness. Curves 3, calculated with no account taken of this attenuation, differ notably from the true $A(kz)$ dependences (see Figure 7.1a), even at $kd = 10$ ($d \approx 1.5\lambda$). The attenuation effects show up in the main peak of the $A(z)$ dependence and in the overall signal level (see Figures 7.1 and 7.2).

Phase aberrations specify the shape of the $A(z)$ curves. They are responsible for the existence of the main maximum in the confocal position of the lens system for oscillations arising when the kz value (i.e., the distance between the lenses) varies. As the ratio of the speeds of the sound c_L/c increases, so does the contribution of the phase aberrations. As the c_L/c ratio increases, the central maximum of the $A(z)$ curve decreases and its shift relative to its position at $z = 0$ (no object) increases. The values of main maximums and of their shifts control the acoustic contrast nature in a transmission microscope.

Figures 7.3 and 7.4 compare the $A(kz)$ curves for LDPE and PS, calculated assuming that the major mechanisms of acoustic contrast are identical.

Figures 7.3(a) and 7.4(a) compare the $A_e(kz)$ and $A_{st}(kz)$ curves calculated with an allowance for all the above-listed factors governing the acoustic contrast. $A_{1e}(kz)$ and $A_{1st}(kz)$ curves calculated disregarding the reverberation effects are compared in Figures 7.3(b) and 7.4(b). As mentioned above, reverberation processes affect the value and behavior of $A(kz)$ functions only insignificantly. Therefore, the graphs shown in Figures 7.3(a) and 7.4(a) differ only slightly from the graphs in Figures 7.3(b) and 7.4(b). Finally, Figures 7.3(c) and 7.4(c) display the $A_{2e}(kz)$ and $A_{2st}(kz)$ curves calculated with no account taken of the attenuation of acoustic waves in the two plastics.

It can be seen that attenuation effects in thin samples are not prominent (see Figure 7.3, left-hand column, at $kd = 10$). The central peak value is governed by reflection at the plate boundaries and most notably by phase aberrations. Inasmuch as the difference between the speeds of sound in water and PS is much greater than the appropriate difference for LDPE, the phase aberrations in PS are more intense. Therefore, the central peak in the $A_{st}(kz)$, $A_{1st}(kz)$, and $A_{2st}(kz)$ curves is higher than the central peak in the appropriate LDPE curves. Attenuation effects are incapable of changing this relation because the sound pathway length in the plastic is too short. In addition, it is worth noting that the $A(kz)$ curves in PS lose their symmetric shape because of large phase aberrations.

As the thickness of the film studied (kd parameter) increases, phase aberrations remain significant because they control the $A(kz)$ curve shape. The $A(z)$ curves in PS also lose their symmetric shape because of large phase aberration values. As follows from Figures 7.3(c) and 7.4(c) when attenuation is neglected, phase aberrations are responsible for the lower signal values in PS, as compared to those in LDPE samples of any thickness. The attenuation effects change the pattern drastically. Attenuation of acoustic waves in LDPE is much more intense; therefore, as the film thickness increases, the signal in LDPE drops much more steeply than it does in PS. As a result, the signals from films of moderate thickness are commensurate for both plastics (see Figure 7.3a, right-hand column, and Figure 7.4a, left-hand column). On the other hand, at larger film thickness values the signal from PS samples significantly exceeds the signal from LPDE samples (see Figure 7.4a, right-hand column).

The analysis performed allows the formation of acoustic image contrast of polymer mixture films to be traced (see Figure 7.5). The attenuation effects can be neglected only for very thin samples with $kd \leq 20$–30. This corresponds to film thickness $d \leq 10$–16 μm at the microscope operation frequency $f = 450$ MHz. The maximum output signal value for such thin samples is controlled by reflection at the sample boundaries and by phase aberrations, i.e., in the final run by the longitudinal speed of sound c_L in the sample. Inasmuch as the speed of sound in LPDE is lower ($n_{st} = 1.6$, $n_e = 1.3$), the central peak value for its samples is greater (see Figure 7.5b). The shift of the central peak is insignificant because of a small sample thickness; therefore, the main $A(z)$ curve maximums for LPDE and PS samples are superimposed on each other (see Figure 7.5b). Accordingly, the acoustic contrast of images of thin LDPE–PS blend films is determined by the distribution of the longitudinal acoustic velocity over the sample studied; LDPE-filled domains transmit sound better than do domains containing mostly PS. Domains in which LDPE dominates appear as light areas in the acoustic images, while domains in which PS dominates are dark.

As the sample thickness increases, the contrast pattern in acoustic images changes because the attenuation effects come into play. Sound is attenuated in LDPE more intensely than it is in PS; therefore, the attenuation effects in LDPE show up at smaller sample thickness values. For this reason, the central peak values in the $A(z)$ curves calculated for LPDE and PS samples at $kd \approx 40$–120 (at film thickness ≈ 20–50 μm at operation frequency $f = 450$ MHz) are of the same order. Because of the drastically differing speeds of sound in these two materials, the peaks are separated in space (see Figures 7.5c and 7.5d). At a lens position more distant from the initial confocal position $z = 0$, the signal at the receiving lens is much larger for domains in which PS dominates; at smaller shifts of the receiving lens with respect to the $z = 0$ position, the acoustic beam passes more easily through LDPE-abundant domains. Thus, a high-quality acoustic image of a two-phase LDPE–PS film of an intermediate thickness can be obtained at two different positions of the receiving lens. As the lens is shifted from the $z = 0$ position, first a picture arises in which LDPE domains are light and PS domains are dark. After further displacement of the receiving lens away from the $z = 0$ position the image contrast reverses, LDPE-containing domains become dark while PS-containing domains become light.

As the sample thickness increases further, acoustic wave attenuation in PS comes into prominence. Absorption in LDPE spots under these conditions is so high that the receiving lens actually records only the central peak belonging to PS-containing domains. The set of Figures 7.5(e)–(h) demonstrates how the peaks in the $A(kz)$ curves for LDPE and PS at $kd = 150$, 300, 450, and 600 (film thickness values ranging from 75 to 300 μm, respectively, at $f = 450$ MHz) separate while the overall level of the peaks diminish due to aberrations and attenuation. The cen-

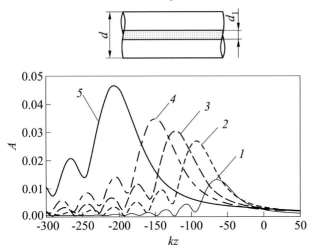

Figure 7.6 Output signal of an acoustic microscope A vs the shift of the receiving lens z for a laminated polyethylene–PS–polyethylene system at various PS inclusion thickness values. Sample thickness $kd = 150$ ($d \approx 79$ μm in a microscope operating at frequency $f = 450$ MHz). The $A(kz)$ function is independent of the inclusion position in the polyethylene matrix: $1 - kd_1 = 0$, neat polyethylene; $2 - kd_1 = 20$ ($d_1 = 10$ μm); $3 - kd_1 = 40$ ($d_1 = 20$ μm); $4 - kd_1 = 60$ ($d_1 = 30$ μm); and $5 - kd_1 = 100$ ($d_1 = 50$ μm) [240].

tral peak in the $A_e(kz)$ for LDPE diminishes much faster, so that at $kd \geq 150$ ($d \geq 80$–10 μm), the whole $A_e(kz)$ curve lies beneath the $A_{st}(kz)$ curve. As far as acoustic image formation is concerned, this signifies that LDPE-containing domains show up in the acoustic signature as dark spots at any position of the receiving lens.

Thus, variation of both the elastic and viscous properties of the material studied can be displayed in acoustic signatures. Variation with respect to which properties show up in the image depends on the sample thickness. In thin samples ($kd \leq 100$–120), there are elastic properties that play a significant role in acoustic imaging. Reversal of the acoustic contrast can be observed in an intermediate thickness range ($kd \approx 50$–100) in which the elastic mechanisms of acoustic imaging give way to the viscosity-controlled ones. In a high-frequency acoustic microscope (in our case, $f = 450$ MHz), the intermediate thickness range corresponds to very thin polymer blend films (not thicker than 50 μm). Distribution of the components in such thin films must predominantly be lateral, i.e., along the film. The reversal effect in these films would be observed just as it is.

Acoustic contrast for thicker films ($kd \geq 150$) is entirely controlled by attenuation effects. This is true both for LDPE- and PS-containing domains and for domains containing the two components in which a PS layer is normally incorporated in a polyethylene matrix. The position of the central $A(z)$ curve peak for such laminated domains depends on the PS inclusion thickness (Figure 7.6). Polystyrene is more rigid and absorbs acoustic radiation less than does LDPE; as the inclusion thickness increases, so does the peak in the $A(kz)$ curve with the peak position moving toward greater z values (the distance between the lenses diminishes). This implies that the brightness of a spot in the acoustic image is unambiguously related to the inclusion thickness; hence the brighter the spot the thicker the inclusion. A relative change in the signal value allows one to quantitatively assess the inclusion thickness. A simple version of such an assessment procedure in which the signal value is assumed to exponentially depend on the thickness is discussed in Section 6.6.

Two important comments should be made. The first is associated with the fact that all the above-mentioned dependences are of a universal nature within the assumption of a linear frequency dependence of ultrasound attenuation, provided dimensionless kd and kz variables are used (i.e., distance is measured in wavelength units). Inasmuch as $\gamma d = (\gamma/k)kd \approx (m/(2\pi))ckd$, all the curves plotted in Figures 7.1–7.6 are independent of the frequency used in the acoustic microscope. Hence, the conclusions drawn from an analysis of these dependences are of a universal nature. The results obtained can be used in a wide range of operation frequencies. The only thing to be done is to convert sample thickness d and lens shift values z from wavelength units to dimensional length values.

The second comment concerns the dependence of the results of theoretical analysis on the angular aperture of the lens system in a microscope. Calculations were performed at a fairly large angular aperture $2\theta_m = 90°$. As the aperture decreases, so does the contribution of phase aberrations; in particular, all the $A(kz)$ curves widen. Accordingly, the sample thickness at which one can observe the contrast of acoustic images caused by the distribution of elastic material properties or the contrast reversal phenomenon reduces.

The conclusions drawn from the above theoretical analysis that concern acoustic imaging of LDPE–PS blend films can successfully be applied to other polymer blends such as LDPE–PMMA and LDPE–PVC. In the latter case, one has to keep in mind that ultrasound attenuation in PVC is ten times higher than that in PS; therefore, attenuation effects will be a dominating mechanism of acoustic imaging at any sample thickness.

7.6
Acoustic Imaging of the Spatial Phase Distribution in Polymer Mixtures

The available acoustic microscopy methods furnish information about the thin structure of polymer blends [65, 301].

The transmission scanning microscopy method allows one to obtain images of the phase distribution throughout the sample volume in the plane scanned. Data on the output acoustic signal amplitude are directed to a computer where they are stored in the form of a 3D file and displayed after preliminary processing on a screen with a 256×256 pixel resolution and 16 brightness gradations. Depending on the type of processing, the result of imaging is either halftone acoustic images,

7.7 Investigation of Structure and Homogeneity of Mixture Components Distribution

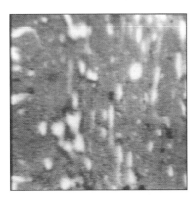

Figure 7.7 Acoustic signature of a fragment of a LDPE:PS = 90:10 blend film. The image size is 500 μm in length [63, 66].

or acoustic profiles, or, alternatively, bar charts of the ultrasound transmittance in the scanning plane.

Figure 7.7 displays the acoustic signature. As can be seen, the acoustic contrast is good, with the acoustic image of the polymer composite clearly showing phase distribution. The film thickness is about 100 μm; therefore, the major mechanism of the acoustic image contrast formation is the sound absorption in LDPE. In accordance with the results obtained above, darker areas in the acoustic signature belong to LDPE-rich zones. This is the way in which the distribution of components in polymer blend samples can be visualized. Acoustic images furnish information about the homogeneity of the distribution of one phase in the other, the extent of agglomeration of inclusions, and also their shape and size.

The signal amplitude on the receiving transducer for inclusion B of thickness l_b located in the focal zone can be written as [65]:

$$A = 4A_0 \frac{\rho_b c_b \times \rho_a c_a}{(\rho_b c_b + \rho_a c_a)^2} e^{(\gamma_a - \gamma_b) l_b} \tag{7.13}$$

where ρ_a, c_a, and γ_a are the density, velocity, and absorption of ultrasound in material A; ρ_b, c_b, and γ_b are the density, velocity and absorption of ultrasound in material B, A_0 is the signal amplitude at the receiving transducer when no inclusion is present, and A is the signal amplitude when an inclusion is in the focus. Given the elastic and viscous properties of the two polymers, the inclusion thickness l_b is evaluated based on the measured A/A_0 ratio.

7.7
Investigation of the Structure and Homogeneity of the Mixture Components Distribution within each other. Measure of Homogeneity

Assessment of both the size of each individual inclusion and the number of inclusions in the sample bulk, as well as the averaged characteristics of the distribution of one phase in the other, are of special interest and importance for polymer blends. Various quantitative methods [308–313] offer the ability to characterize the size of dispersed phase particles, their shape, specific surface area, and other structural parameters. Depending on the concentrations of components and mixing conditions, the following structure types are observed in two-component polymer composite materials:

(1) individual grains of one polymer are distributed in the matrix of another polymer;
(2) agglomerates of individual grains of one polymer are distributed in the matrix of another polymer;
(3) individual grains and agglomerates of one polymer are distributed in the matrix of another polymer;
(4) zones of one polymer of a complicated shape irregularly alternate with grains of another polymer; and
(5) complex structures are found consisting of structures 2, 3, and 4.

By grains we mean an inclusion of material B with clear-cut boundaries contained in the matrix of material A.

It should be emphasized that acoustic microscopy furnishes exhaustive information about the arrangement of inclusions, their sizes, and degree of agglomeration throughout the sample bulk rather than at its surface [313]. Such parameters as strength and deformation properties of polymer composites depend on the size distribution of dispersed phase particles in the bulk matrix, as well as the presence of agglomerated particles and the homogeneity of their distribution [311, 314–318].

A quantitative measure characterizing the homogeneity of the distribution of grained material particles in a matrix of another material was developed for studying polymer blends. The following characteristics are of interest in analyzing images of a distribution of one phase in the other:

- the number of grains in the field of vision of the microscope, coordinates of their centers of mass, and ratio between the total grain surface area and the sample surface area;
- the surface area, perimeters and sizes of grains (maximum separation of points belonging to a grain), and also an average value and variance (mean-square deviation) of the above-mentioned quantities; and
- the homogeneity of the material distribution in the microscope field of vision. If grains are agglomerated, of particular interest are all the aforesaid characteristics, as well as which grains are incorporated in each agglomerate, the average number of grains in agglomerates, and their variation.

We assumed that the samples studied are films, and that the possible superposition of grain images on each other can be neglected, and also that grains have clear-cut boundaries.

Under these assumptions, points of the acoustic image belonging to different material phases can be distinguished quite correctly with the aid of threshold processing. Let $q(x, y)$ be a function corresponding to the brightness of the

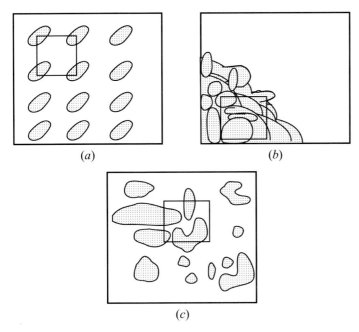

Figure 7.8 Models of arrangement of material B grains in the matrix of material A: (a) highest homogeneity, (b) highest inhomogeneity, and (c) random distribution of grains [240, 279, 313].

acoustic signature at point (x, y) which is bounded within the interval $[0, 1]$; its zero and unity values specify the lowest and highest brightness values, respectively. The result of threshold processing of a given image is expressed as follows:

$$b(x, y) = \begin{cases} 0 & \text{if } q(x, y) < q_0 \\ 1 & \text{if } q(x, y) > q_0 \end{cases}$$

where q_0 is a certain threshold value chosen experimentally.

As a result of such a transformation, we obtain a binary image $b(x, y)$ of white grains against a dark background, or *vice versa*, which reflects the grain structure of the object studied.

The most ordered arrangement of identical grains (similarly oriented grains are arranged so that their centers of mass are positioned at nodes of a square grid (Figure 7.8a)) is assumed to be a highly homogeneous structure. The case when all grains form a continuous agglomerate (Figure 7.8b) is naturally assumed to be the least homogeneous structure.

To measure a real structure which differs from these two extreme cases, one has to construct a continuous function based on binary images $b(x, y)$, the value of which changes monotonically as the distribution of the grain material in the matrix becomes more ordered.

Consider a square image fragment of a surface area

$$S_q = \frac{\bar{S}_{qr}}{\Delta} \tag{7.14}$$

where $\bar{S}_{qr}$ is the average grain surface area and Δ is the fraction of grains in the field of vision (Figure 7.8a).

The ratio between the total surface area of grains that find themselves in the square is designated by R. When grains are uniformly distributed in the field of vision, the quantity R is constant and independent of the square fragment position in the field of vision. If the grains are dissimilar and positioned irregularly (Figure 7.8b), quantity R would exhibit some scatter depending on the position of the square. The spread of R values is greater the more nonuniform the grain arrangement in the field of vision and it tends to unity when the grain distribution in the field of vision is most inhomogeneous. The variance (standard deviation averaged over the position of square fragment S_q) of R can be used as the measure of its spread

$$\delta(R) = |R - \bar{R}| \tag{7.15}$$

The value of $\delta(R)$ depends not only on the uncertainty of the grain material distribution but on the average ratio between phases in the field of vision of the microscope as well. Therefore, quantity $\delta(R)$ reduced by its maximum value δ_m at a given ratio of the total grain surface area to the image surface area can serve as a homogeneity of degree G:

$$G = \frac{\delta(R)}{\delta_m} \tag{7.16}$$

Obviously, if identical grains are regularly distributed over the field of vision (maximum homogeneity), $G = 0$, because $\delta(R) = 0$ in this case, but when the inhomogeneity of the grain distribution in the field of vision is at a maximum, i.e., when grains form a single continuous agglomerate, δ tends to unity. The δ_m value is assessed as follows. We consider a limiting case in which a great number of grains are incorporated in a single continuous agglomerate so that the distance between the grain surfaces can be neglected (Figure 7.8b). The square surface area used in calculations of $\delta(R)$ is much less than the agglomerate surface area. We calculate an approximate value $\bar{R}$ in this situation. If the positions of a square, in which both grain and matrix materials fall into it are disregarded, assuming thereby that quantity R is either 1 inside the agglomerate or 0 beyond it, $\bar{R}$ approximately equals:

$$\bar{R} \approx \frac{1 \times S_{\text{agl}} + 0 \times S_{\text{matr}}}{S_{\text{agl}} + S_{\text{matr}}} = \frac{S_{\text{agl}}}{S_{\text{mic}}} = \Delta \tag{7.17}$$

where S_{agl} is the agglomerate surface area, S_{matr} is the matrix material surface area, and S_{mic} is the surface area of the microscope field of vision.

The agglomerate surface area in the case at issue equals the sum of the surface areas of all grains; therefore, $\bar{R}$ is approximately equal to the ratio of the sum of grain surface areas to the surface area of the field of vision. Now, we estimate an approximate $\delta(R)$ value. The situations when square S_q finds itself at the agglomerate boundary being again disregarded, the expression for $\delta(R) = \delta_m$ can be written as follows:

$$\delta_m = \sqrt{\frac{(R_1 - \bar{R})^2 S_{agl} + (R_2 - \bar{R})^2 S_{matr}}{S_{agl} + S_{matr}}} \qquad (7.18)$$

where R_1 is the value of R within the agglomerate and R_2 is its value beyond the agglomerate.

Within the agglomerate, $R - \bar{R} = 1 - \Delta$, and $R - \bar{R} = 0 - \Delta$ beyond it, because R equals 1 and 0, respectively. Also, $S_{agl} + S_{matr} = S_{mic}$. Substituting these quantities in (7.18) yields

$$\delta_m = \sqrt{\frac{(1-\Delta)^2 S_{agl} + \Delta^2 S_{matr}}{S_{mic}}} \qquad (7.19)$$

In the case considered, $S_{matr} = (1 - \Delta) S_{mic}$ and $S_{agl} = S_{mic}\Delta$; therefore, (7.19) can be rewritten as

$$\delta_m = \sqrt{(1-\Delta)^2 \Delta + \Delta^2(1-\Delta)} = \sqrt{\Delta(1-\Delta)} \qquad (7.20)$$

Thus, the homogeneity degree reads:

$$G = \frac{\delta(R)}{\sqrt{\Delta(1-\Delta)}} \qquad (7.21)$$

Thus derived, the homogeneity degree is independent of the ratio of the total grain surface area to the acoustic image surface area.

7.8 Numerical Processing of Acoustic Images of Granulated Structures

The development of methods for quantitative characterization of objects containing grain agglomerates calls for an adequate formalization of the notion of an agglomerate. In analyzing images of composite materials, it is reasonable to assume that an agglomerate is a group of grains which exhibits any phenomena relevant to the interactions between them.

The authors of [311, 315, 319, 320] also place among the agglomerates groups of grains in which the distance between neighboring grains is less than their radius. The stress fields around grains are expected to significantly superimpose at these distances; hence, such groups exhibit properties that differ qualitatively from the properties of groups far removed from each other. This criterion is at minimum acceptable at low filling degrees (10%–15%) because at high filling degrees, all or the majority of grains are positioned closer to each other.

At high filling values (50% of one phase and 50% of the other phase), we formalize the algorithm notion as follows: $L(a_i, a_j)$ is the separation of grains a_i and a_j, defined as the shortest distance between points belonging to their boundaries. Grains a_i and a_j for which condition $L(a_i, a_j) \leq \max[L(a_i, a_r), L(a_j, a_r)]$ holds, are assumed to be the nearest neighbors. Here, a_r is any grain in the field of vision of the microscope other than a_i or a_j. Also, two grains for which there is no third grain positioned closer to them than they are separated from each other are said to be closest neighbors. If all the points of the image boundary are assumed to belong to grains, from consideration of the $L(a_i, a_r)$ dependence on the mutual disposition of grains and of their location in the field of vision, one can easily show that, in the presence of one or several agglomerates, the closest neighbors belonging to one and the same agglomerate are characterized by $L(a_i, a_j)$ values that are much less than

$$L_{\max} = \max[L(a_i, a_j)] \tag{7.22}$$

We assign a_i to a certain agglomerate if the closest neighbor a_j belongs to the same agglomerate and if

$$L(a_i, a_j) < \beta L_{\max} \tag{7.23}$$

where β is an empirical constant in the 0.3–0.5 interval.

The quantitative characteristics of agglomerates and grains are identical, among these are: their number in the field of vision of the microscope, the surface area, and so on.

Thus, using an acoustic signature, one can assess the following quantitative characteristics of polymer blend samples:

(1) the number of grains in the field of vision;
(2) the total surface area of the grains;
(3) the mean grain size;
(4) the variation in the grain size values;
(5) the mean surface area of a grain;
(6) the variation in the grain surface area;
(7) the degree of grain homogeneity;
(8) the number of agglomerates in the field of vision;
(9) the total surface area of the agglomerates;
(10) the mean surface area of an agglomerate;
(11) the variation in the agglomerate size;
(12) the mean agglomerate surface area;
(13) the variation in the agglomerate surface area; and
(14) the degree of agglomerate homogeneity.

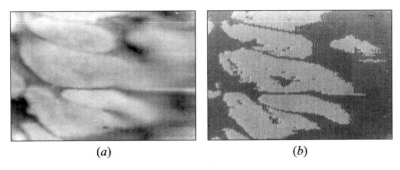

(a) (b)

Figure 7.9 The acoustic signature (a) and the result of threshold processing (b) of a fragment of an LDPE:PS = 50:50 film image. The image size is 500 μm in length [240, 279, 313].

A set of computer programs has been developed for calculating the aforesaid characteristics of acoustic signatures. Figure 7.9(a) displays an acoustic signature resulting from discretization of the output microscope signal on a 128×128 raster with a 256 tone scale at each point (the photo is from the graphic display screen). The result of threshold processing of the image is shown in Figure 7.9(b).

7.9
Exploring the Microstructure of Polymer Blends in an Acoustic Microscope and Comparison with other Techniques

7.9.1
Studies of the Microstructure of Individual Particles in a Blend

In studying the processes that occur during the course of polymer mixing, either in melt or by grinding under elastic-strain conditions, visualization of the internal structure or of an individual extrudate or powder particle is of great interest. Individual powder particles were examined in alcohol with an anticoagulant added to avoid their sticking to each other. Figure 7.10 demonstrates optical (on the left) and acoustic (on the right) photos of a single powder particle (Figure 7.10a) and of an extrudate particle (Figure 7.10b).

The optical photo of a sample pretreated with tetrahydrofurane (subjected to extraction) furnishes no information about distribution of the LDPE and PS phases within an individual particle.

The acoustic signatures of particles of the same mixture LDPE:PS (80:20) provide information about the distribution of phases within extrudate particles and in powder particles. As seen from Figure 7.10(b), the components in extrudate particles are poorly mixed; LDPE and PS are distributed in the form of zones that virtually do not overlap. Images of mixtures prepared by grinding under elastic-strain conditions clearly show that the individual fine PS particles are distributed in the LDPE phase. Processing of the images yields: a PS particle size of about 23 μm, a variance

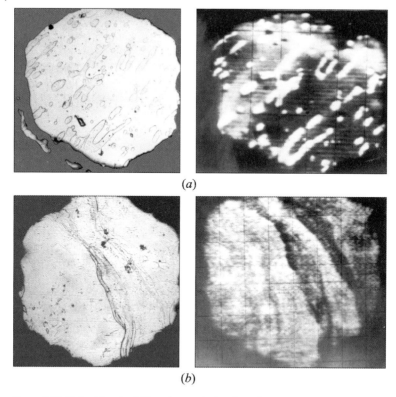

Figure 7.10 Optical (on the left) and acoustic (on the right) images of LDPE:PS = 80:20 blend particles. The particles are prepared by grinding under elastic-strain conditions (a) (the image size is 200 μm in length), and by mixing in melt (b) (the image size is 500 μm in length [65, 313].

of the grain size of ±15%, and a mean surface area of the PS phase grains of about 575 μm.

Based on the results of PS extraction from the mixture with the aid of tetrahydrofurane, we can state that the PS and LDPE phases are fully separated. The acoustic histogram shows LDPE alone (for a sample washed in tetrahydrofurane). This observation is supported by infrared (IR) spectroscopy measurements in extracted samples which show peaks inherent in LDPE, namely, 1471 cm^{-1}, 1369, 1354, 1304, 729, and 720 cm^{-1}. No peaks belonging to PS, namely, 1601, 1493, 760, and 798 cm^{-1} showed up in the IR spectra. This allows us to assert that occlusion of one phase by the other is ruled out in LDPE:PS blends in samples prepared both by conventional extrusion and by grinding under elastic-strain conditions. Investigations of 80:20, 50:50, and 20:80 LDPE:PMMA and LDPE:PVC mixtures yielded similar results.

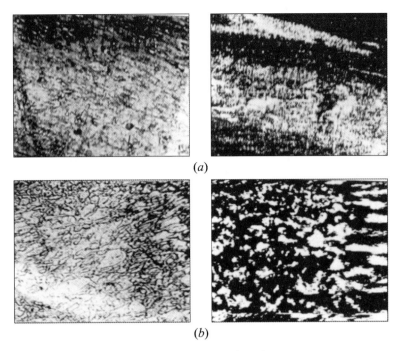

Figure 7.11 Optical (on the left) and acoustic (on the right) photos of films made of an LDPE:PS = 80:20 blend mixed (a) in a melt (the film thickness is 150 μm and the image length is 600 μm) and by grinding under elastic-strain conditions (the film thickness is 150 μm) [65, 313].

7.9.2
Studies of Film Structure and the Homogeneity of Phase Distribution in Polymer Blend Films

Exploring blend films prepared by mixing in melt and by grinding under elastic-strain conditions furnishes very interesting results. Figure 7.11 shows optical (on the left) and acoustic (on the right) images of films fabricated from LDPE:PS = 80:20 blends mixed in a melt (Figure 7.11a) and by grinding under elastic-strain conditions (Figure 7.11b).

Optical photos furnish no information about the structure of the films studied.

On the other hand, sharply outlined dark zones filled with LDPE and light zones filled with PS are clearly distinguished in the acoustic signatures. Acoustic signatures provide unequivocal evidence of a more homogeneous phase distribution in samples mixed by grinding under elastic-strain conditions.

The homogeneity degree calculated by Equation (7.21) for a blend mixed in melt equals 0.934, while for a film made of a blend mixed by grinding under elastic-strain conditions, it is 0.427. Hence, films made of blends mixed by grinding under elastic-strain conditions are much more homogeneous than are films made of blends mixed in melt.

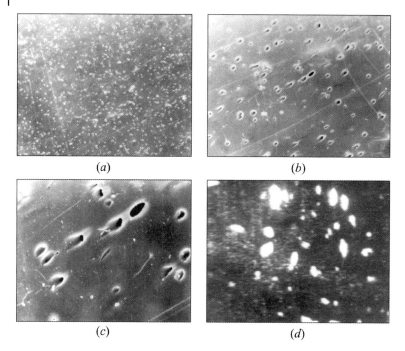

Figure 7.12 Electron microscopy photos of an LDPE:PS = 90:10 blend film after PS extraction from it; magnification is ×60 (a), ×200 (b), and ×600 (c), and the acoustic image of the same film before extraction (d), the scan length is 120 μm.

Figures 7.12(a) to (c) show electron microscopy photos of LDPE:PMMA = 90:10 films taken at various resolutions after PMMA extraction with tetrahydrofurane. The acoustic image of the same film taken before preparing it for examination by electron microscopy is shown in Figure 7.12(d). As can be seen, identification of phases by acoustic microscopy requires no destructive pretreatment.

Electron microphotographs of low-temperature chips of a PDPE:PS = 50:50 blend before and after PS extraction with tetrahydrofurane are shown in Figures 7.13(a) and (b). Figure 7.13(c) displays an acoustic microphotograph of the same sample without surface pretreatment. Obviously, identification of phases by electron microscopy is impeded, whereas acoustic microscopy illustrates perfectly the phase topography in a sample.

7.9.3
Assessment of the Component Distribution in Polymer Blends at Various Sizes of the Mixture Particle Fractions

To characterize polymer blends prepared by various methods, it is important to know how the blend homogeneity depends on concentrations of the blend compo-

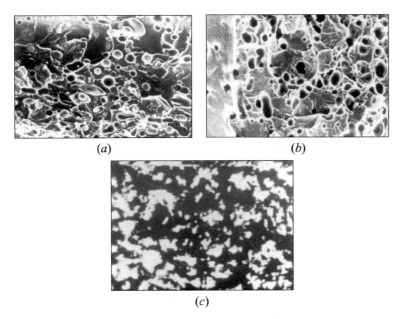

Figure 7.13 Electron microscopy image of LDPE:PS = 50:50 blend films taken before (*a*) and after (*b*) PS extraction; acoustic signature of the same film before extraction (*c*). The image length size is 400 μm.

nents. Table 7.3 lists the homogeneity values assessed by mathematical processing of acoustic microphotographs on a computer. The table clearly demonstrates that the homogeneity degree depends appreciably on the concentrations of the basic blend components and on the technique used to mix them. The greater the LDPE concentration in a blend, the higher the homogeneity of the resulting composition for a given treatment technique – which is natural because it is a dividing component. The homogeneity of samples mixed by grinding under elastic-strain conditions is much better than the homogeneity of blends of the same composition mixed in melt.

The content of the components as a function of the size of the powder fractions in a blend was assessed with the aid of acoustic microscopy. The mixture particles were sieved into fractions: > 315 μm, 315–200, 200–100, 100–50, and < 50 μm. Films were fabricated from each fraction. As seen from the acoustic signatures of films shown in Figure 7.14, the percentage content of the components in films made of fractions of a different size is different at identical initial concentrations.

The acoustic signatures of films were obtained after extraction of various fractions from mixtures containing the components in different allowed concentrations, with the aid of a suitable mathematical processing. The percentage content of LPDE and of the second component is to be evaluated as a function of the size

Table 7.3 Mean grain size and homogeneity degree as a function of the mixing technique and component concentrations in a polymer mixture [65].

Mixture composition	Percentage ratio of the components	Mean grain size μm	Variation in grain size μm	Homogeneity measure
		Mixing in melt		
LDPE:PVC	80:20	76	79	0.983
	50:50	68	73	0.969
	20:80	83	86	0.977
LDPE:PMMA	80:20	81	87	0.975
	50:50	74	105	0.928
	20:80	79	82	0.944
LDPE:PS	80:20	73	82	0.961
	50:50	70	78	0.933
	20:80	75	79	0.969
		Mixing by grinding under elastic-strain conditions		
LDPE:PVC	80:20	7	6	0.347
	50:50	25	29	0.596
	20:80	33	42	0.783
LDPE:PMMA	80:20	8	8	0.384
	50:50	27	30	0.608
	20:80	29	35	0.732
LDPE:PS	80:20	8	9	0.335
	50:50	28	32	0.621
	20:80	32	39	0.724

of a powder fraction prepared by grinding under elastic-strain conditions. The data obtained are summarized in Table 7.4.

As can be seen from Table 7.4, the 100–200-micrometer fraction provides the best compliance with the initial percentage ratio between the components. Blends with coarser fractions are enriched with LDPE, while finer fractions yield blends enriched with the second component (PS, PVC, or PMMA). Similar measurements were performed using the conventional labor-consuming balance technique. Here the second component was extracted from the mixture with tetrahydrofurane, then the samples were dried and weighed and the percentage contents of the second component and LDPE within a fraction of a given size were calculated. The two methods yielded consistent results.

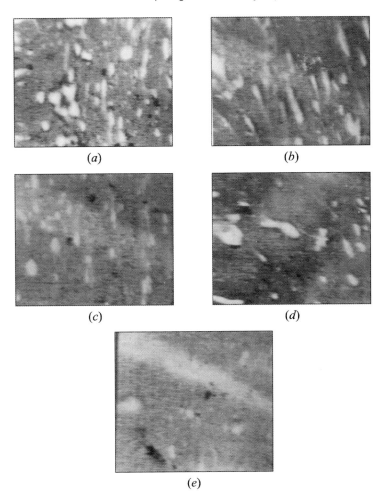

Figure 7.14 Acoustic signatures of films made of LDPE:PS = 90:10 mixture fractions mixed by grinding under elastic-strain conditions. Fraction size: (a) < 50 μm, (b) 50–100, (c) 100–200, (d) 200–315, and (e) > 315 μm. The film thickness is 110 μm and the scan length is 720 μm.

7.9.4
Investigation of the Distribution Homogeneity and the Physical and Mechanical Polymer Blend Properties

Physical and mechanical properties pertain to the most important characteristics of polymer systems. As shown above, structures of polymer blends, mixed by two different techniques (grinding under elastic-strain conditions and mixing in melt), are quite different.

Table 7.4 Percentage contents of LDPE and second component in blends as a function of the fraction size of a powder mixed by grinding under elastic-strain conditions [65].

Fraction size μm	LDPE:PMMA			LDPE:PS			LDPE:PVC		
	80:20	50:50	20:80	80:20	50:50	20:80	80:20	50:50	20:80
> 315	98:2	87:13	36:64	97:3	84:16	31:69	98:2	83:17	30:70
315–200	89:11	64:36	26:74	88:12	66:34	28:72	87:13	65:35	29:71
200–100	85:15	52:48	19:81	86:14	51:49	18:82	85:15	50:50	20:80
100–50	69:31	63:37	13:87	67:33	37:63	14:86	67:33	39:61	13:87
< 50	59:41	10:90	6:94	62:38	15:85	9:91	63:37	13:87	8:92

An analysis of how the physical and mechanical properties of polymer systems depend on the distribution of components in blends seems to be of great interest and practical importance.

The following parameters are chosen as basic structure characteristics of blends: degree of homogeneity, mean grain size, and variation in the grain size. Statistical data were collected 15 times for each blend sort with a given percentage ratio between the components. The basic measured data are listed in Table 7.5. This table suggests that materials made of polymer blends mixed by grinding under elastic-strain conditions exhibit higher elasticity. The results of tests show a fairly good repeatability from sample to sample, namely, 2% or 5%, depending on the mixture components and their concentrations; whereas materials made of granules mixed in melt show a greater spread of the data in the range 13%–28%. Materials mixed in melt demonstrate a still greater spread of the data of strength tests. Thus, the scatter of ultimate strength values of such LDPE:PS materials amounts to 25% while their scatter for blends mixed by grinding under elastic-strain conditions does not exceed 3%.

The spread of elastic and strength properties of polymer blends correlates remarkably with the degree of homogeneity derived from acoustic characteristics; the more homogeneous the material, the more stable are its elastic and strength properties. Thus, the basic inference from an analysis of the correlation between the structure of a sample and its acoustic parameters is as follows: as the sample homogeneity increases and, hence, the grain size reduces at a fixed ratio of the mixture components, the elasticity modulus of a system increases and its stability is enhanced drastically. The mean-square spread of the basic mechanical properties of a composite (Young's modulus, yield stress, ultimate strength, and stretching strain limit) increases several-fold (from 1.5 to 10 times) when going from a more homogeneous blend mixed by grinding under elastic-strain conditions, to a material with a large mean grain size and low homogeneity degree, fabricated by mixing in melt. This result is of practical importance because examination of the initial material by acoustic methods allows the stability and macroscopic homogeneity of the mechanical properties of articles from which it is made to be assessed.

Table 7.5 Correlation between physical and mechanical properties of polymer composites and their topological structure assessed by acoustomicroscopic methods.

Ratio of components %	Young's modulus E kgf mm^{-2}	Standard deviation of E %	Yield strength σ_T kgf mm^{-2}	Standard deviation of σ_T %	Ultimate stress σ_p kgf mm^{-2}	Standard deviation of σ_p %	Percent elongation ε_p %	Standard deviation of ε_p %	Homogeneity degree	Mean grain size	Variation in grain size
					Mixing in melt						
LDPE:PVC											
80:20	25.80 ± 0.87	3.4	0.76 ± 0.01	1.3	0.77 ± 0.01	1.5	56.7 ± 6.3	11.1	0.347	7	6
50:50	35.51 ± 0.92	2.0	0.47 ± 0.01	1.5	0.48 ± 0.01	1.5	11.4 ± 1.4	12.1	0.596	25	29
LDPE:PS											
80:20	29.11 ± 1.07	3.6	0.79 ± 0.01	1.6	0.76 ± 0.01	1.1	54.9 ± 5.7	10.6	0.335	8	9
50:50	45.65 ± 1.32	2.9	0.55 ± 0.02	3.2	0.56 ± 0.02	3.1	33.8 ± 3.4	12.1	0.624	28	32
LDPE:PMMA											
80:20	28.96 ± 1.02	4.3	0.75 ± 0.01	1.3	0.79 ± 0.01	1.1	36.1 ± 2.5	7.2	0.384	8	8
50:50	50.69 ± 2.60	5.1	0.80 ± 0.01	1.2	0.44 ± 0.02	4.5	12.0 ± 0.4	3.9	0.608	27	30
					Elastic-deformation mixing						
LDPE:PVC											
80:20	19.36 ± 3.29	17.0	0.83 ± 0.03	4.7	0.83 ± 0.07	8.1	73.7 ± 58.1	73.7	0.983	76	79
50:50	27.87 ± 5.50	27.8	0.60 ± 0.05	8.3	0.52 ± 0.08	17.3	25.5 ± 12.9	51.3	0.969	68	73
LDPE:PS											
80:20	20.69 ± 2.98	14.1	0.64 ± 0.16	25.0	0.63 ± 0.16	26.2	39.1 ± 13.4	34.4	0.961	73	82
50:50	18.16 ± 4.41	24.0	0.35 ± 0.12	23.2	0.52 ± 0.13	25.4	32.9 ± 15.5	46.7	0.933	70	78
LDPE:PMMA											
80:20	21.95 ± 3.11	13.8	0.69 ± 0.63	9.2	0.78 ± 0.08	5.3	41.3 ± 16.5	39.8	0.975	81	87
50:50	16.65 ± 5.13	16.5	0.63 ± 0.09	15.1	0.58 ± 0.05	9.4	24.5 ± 9.9	40.1	0.928	74	105

7.9.5
Examination of the Polymer Film Structure via Surface Defects

Acoustic microscopy makes it possible to observe the bulk structure of polymer blend films even, in many cases, when the film surface has defects, e.g., when dirt spots, scratches, grazes, and so on, are present. This possibility depends on the nature of the defect. Fine roughness that is much smaller in size than the wavelength of an acoustic wave (in the immersion at the microscope operation frequency) only slightly affects the penetration of an ultrasonic beam into the sample bulk. In our case of a transmission microscope operating at a frequency of 450 MHz, the maximum size of these rough edges is estimated to be 1.0–1.5 μm. The contribution of coarser roughness depends on the ratio between its representative size and the size of the ultrasonic beam spot on the sample surface of the focused incident beam. The fact is that, when imaging a sample bulk structure, the beam is focused somewhere near the film midplane; therefore, the size of an ultrasound spot on the film surface is much larger than the wavelength of the ultrasound used. A defect on the film surface becomes a serious obstacle on the beam path if its area within the ultrasound spot is commensurate with the spot area. In other cases, the defect can be distinguished in acoustic images. Such a critical defect size must, in principle, be commensurate with the spot thickness. To this end, an aperture angle of the focusing lens is recommended to be about 60° or greater, a condition which is normally met in transmission microscopes. The effect of surface roughness on the quality of the acoustic signatures of bulk structures can additionally be reduced by proper matching of the acoustic properties of water (immersion liquid) and polymer film to minimize reflection contributions and to make the refraction angles less significant.

The feasibility of observing the internal structure of a film through large spots on its surface depends appreciably on the spot material and thickness. This is be-

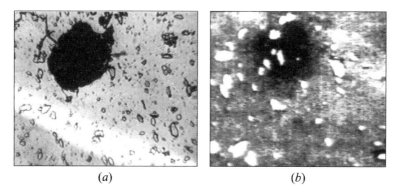

Figure 7.15 Optical (*a*) and acoustic (*b*) images of LDPE:PS = 90:10 film manufactured by grinding under elastic-strain conditions with a dirt spot on its surface. Film thickness is approximately 110 μm. The dirt spot is transparent to ultrasound.

cause a sufficient amount of acoustic energy must be transmitted through the spot; also, the structure of the converging beam must not be perturbed significantly by refraction aberrations. The bulk film structure is clearly seen through a thin metal coating (less than 1 or 2 microns thick, in our case); as for other spot materials, their effect depends on the spot thickness and the acoustic properties (impedance and absorption). Optical and acoustic photos of a composite polymer film with a dirt spot on its surface are displayed in Figure 7.15 for illustration. The optical photograph (Figure 7.15a) is incapable of distinguishing even the surface film structure under the spot, whereas the acoustic signature (Figures 7.15b) perfectly demonstrates both the dirt spot (its blurred edges provide evidence of an uneven thickness distribution over the spot surface) and the bulk film structure under the spot.

7.10
Application of Acoustic Microscopy Techniques for Investigation of the Multi-layered Polymer System Structure

Measurements of individual internal layers in materials during the course of their service are of both practical and theoretical interest [321, 322]. The fact that an ultrasound pulse is multiply reflected from the layer boundaries in layered materials presents the greatest difficulty in ultrasound studies. One of the approaches to overcoming this difficulty is to choose the probing signal frequency to be nearly equal to the resonance frequency of the layer material studied [322]. In the case of polymer structures, the problem becomes more complicated because of high sound absorption and weak reflection from the internal layers. On the other hand, weak reflection simplifies the solution of the problem associated with distinguishing the signal from a layer studied against the background of signals multiply reflected in the multilayer system.

The objective of this section is to explore the possibility of the application of short-pulse acoustic microscopy to investigations of the structure, properties, and shape of individual layers in a multiple-layer system consisting of high-absorption components.

The following components are chosen for experimental studies:

- high-density polyethylene (HDPE);
- high-density polyethylene with a black pigment additive;
- a barrier layer consisting of a polymer ethylene-vinyl-alcohol resin; and
- adhesive layers comprising resins based on polyethylene that cover the barrier layer on both sides.

Table 7.6 lists the characteristics of all layers of the systems studied.

The total thickness of the multi-layer system varied from 2 to 10 mm with the ratio of thickness of individual layers being kept constant. Ultrasound short-pulse instruments are known in many cases to provide good data on the state of the internal material structure [81, 202, 313, 322].

Table 7.6 Structure of the multilayer polymer system (the first layer is the inner one and the sixth layer is the outer one).

Layer	Material	Layer thickness %	Speed of sound c_L 10^3 m s^{-1}
1	HDPE	40	2.36
2	Adhesion layer	3	2.16
3	Barrier layer	4	3.20
4	Adhesion layer	3	2.16
5	HDPE (ground)	38	2.76
6	HDPE	12	2.68

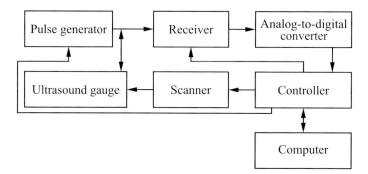

Figure 7.16 Block-diagram of a portable ultrasound gauge of thickness measurements.

Trial investigations have demonstrated that the boundaries of individual layers are insufficiently clear-cut and show poor contrast in the images. This required a compromise in the value of the frequency of the probing ultrasound signal in order to attain adequate resolution at large beam penetration depths [65, 81, 82, 313].

A special measuring instrument was developed to solve this problem. It was comprised of a wide-band piezoelectric transducer, precise scanner, and appropriate software and is shown schematically in Figure 7.16.

The frequency used was 15 MHz, the scanning field length was 10 mm, and the scan steps were 0.05 mm.

To minimize the effect of sample surface tilt, a focused ultrasound transducer with a 15-millimeter focus was used.

The ultrasound gauge unit was equipped with a small immersion chamber manufactured from an acoustically transparent material. The immersion liquid was poured into the chamber and the whole system was in contact with the sample surface via an acoustic gel. The layer thickness and absorption value were assessed from the delay time and the change in the acoustic signal amplitude. Parameters of the reflected signal were compared with the appropriate parameters of the signal reflected from the bottom free surface of a reference sample of desired thickness. A B-scan of the system is shown in Figure 7.17.

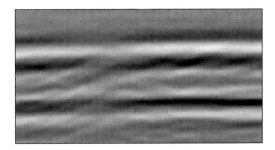

Figure 7.17 B-scan of the boundary layer in a multilayer polymer sample.

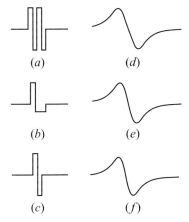

Figure 7.18 Modeled wave shapes of probing pulses (a)–(c) and respective resulting signals (d)–(f). Modeled signals demonstrate that the shape of the received signals is virtually independent of the incident acoustic signal shape [82].

The exponential attenuation coefficient was calculated using the tabulated ultrasound absorption values and approximate layer thickness values were calculated using the delay times between the signal reflected from the front surface and the subsequent reflected signal spike. The high absorption value, which linearly depends on frequency, substantially simplifies the output signal shape. This provides good predictability of the signal shape.

Figure 7.18 represents three wave shape variants of incident signals (similar to the probing signal) as well as signals after they have passed through the filter (equivalent to transmission through a medium with a linear frequency dependence of the absorption value). It can be seen that the resulting signal shapes are nearly identical, independent of the initial acoustic pulse shape. When the time of flight of the ultrasonic pulse in the interior layers is commensurate with the pulse width, the returning signals can significantly overlap. In this case, one can employ the inverse algorithm in the temporal domain. The algorithm is based on minimization of discrepancies between the measured and calculated signal values [322]. The calculated signal is specified as a linear combination of response signals. The reference signal shape was assessed as depicted above. The resulting and recovered signals are exemplified in Figure 7.19.

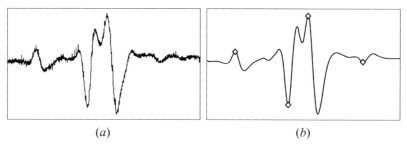

Figure 7.19 An ultrasound signal reflected from internal layers of a sample (*a*) and recovered signal (*b*). The diamonds indicate calculated subsequent delays of the signal between boundaries of the layers 5 and 4, 4 and 3, 3 and 2, and 2 and 1 [82].

The delay time and initially expected amplitude are calculated using the time of the signal's return and the values of the subsequent maximums and minimums of the extended reverse signal. The measurement error did not exceed 5% in the majority of tests. Moreover, reliability of the inversion method improves when changes in the layer thickness in the direction of mechanical scanning are insignificant.

Thus, the aforesaid results demonstrate the feasibility of assessing the thickness of thin (100–300 µm) internal layers in a multilayer (up to 2–100 mm thick) polymer system that consists of intensely absorbing components. This method is based on the solution of the inverse problem of recovering the internal structure from the reflected signal shape. Also confirmed is the possibility of obtaining reliable data on the layer thickness values in experimental studies of samples with curved surfaces, provided that the radius of curvature is much larger than the sample thickness.

7.11
Using the Short-pulse Ultrasound Scanning Technique to Measure the Thickness of Individual Components of Multi-layer Polymer Systems

In the majority of ultrasound diagnostic techniques used currently, the geometry of the material samples studied is not of great importance because, as a rule, measurements are performed with samples and structures whose surfaces are plane-parallel. However, in practice, the necessity for inspection of objects with uneven and curved surfaces arises much more frequently [323].

The previous section discussed the results of investigations in which A- and B-scanning regimes of short-pulse acoustic microscope operation are used to visualize and measure the thickness of internal layers in plane-parallel samples of a multi-layer, intensely absorbing, polymer system.

It seems expedient to investigate the samples with uneven surfaces, particularly in the case when the radius of curvature of the surface is much larger than the ultrasound wavelength. As an example, we consider the problem of measuring the

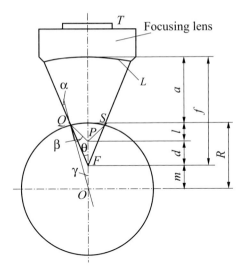

Figure 7.20 Pattern of rays in a focusing lens with a concave structure.

thickness of an internal layer in a multi-layer polymer system with a curved surface. The ultrasound signal frequency values were 15 and 20 MHz.

The short-pulse acoustic microscopy focus lens is sketched schematically in Figure 7.20. The transducer T is mounted at one end of the acoustic waveguide; at the other end a spherical recess is machined. The ultrasound wave spreads normal to the sample surface. L is the spherical surface focusing the ultrasound signal in the immersion liquid. The focused acoustic signal is reflected from the sample surface located at a distance a from the real focus of the spherical lens, P. F is the focal point in the immersion liquid, $(f-a)$ is the focal distance of the focusing lens. A short pulse with central frequency T serves as an input signal and the time gate technique is used to distinguish the signal reflected from the particular depth of the sample structure.

We consider a case when the sample surface is circular with radius R and center O, m is the distance between the circle center and the geometric focal point, and l is the distance from the sample surface to the actual lens focus.

As shown in Figure 7.20:

$$m = R + a - f \tag{7.24}$$

We consider an incident beam at point Q on the sample surface. According to Snell's law,

$$\frac{\sin \alpha}{\sin \beta} = \frac{c_W}{c_S} \tag{7.25}$$

where c_W is the speed of ultrasound in liquid (water) and c_S is the speed of ultrasound in the sample.

In triangle $\triangle OQF$,

$$\frac{R}{\sin(\pi - \theta)} = \frac{|OF|}{\sin \alpha} = \frac{m}{\sin \alpha} \tag{7.26}$$

In triangle $\triangle OQP$,

$$\frac{\sin \beta}{\sin \gamma} = \frac{|OP|}{|QP|} = \frac{m+d}{|QP|} \tag{7.27}$$

$$\sin \gamma = \sin(\theta - \alpha) \tag{7.28}$$

In triangle $\triangle QPF$,

$$\frac{|PF|}{\sin(\beta - \alpha)} = \frac{|QP|}{\sin \theta} \tag{7.29}$$

Multiplying these equations yields:

$$\frac{m+d}{d} = \frac{\sin \theta \sin \beta}{\sin \gamma \sin(\beta - \alpha)} = \frac{\sin \theta \sin \beta}{\sin(\theta - \alpha) \sin(\beta - \alpha)} \tag{7.30}$$

$$d = \left(m\left(k \sin \theta (1-p^2) - kp \cos \theta \sqrt{1-p^2} - \sin \theta \sqrt{(1-p^2)(1-k^2p^2)}\right.\right.$$

$$\left.\left. + p \cos \sqrt{1-k^2p^2}\right)\right) / \left(kp^2 \sin \theta + kp \cos \theta \sqrt{1-p^2}\right.$$

$$\left. + \sin \theta \sqrt{(1-p^2)(1-k^2p^2)} - p \cos \sqrt{1-k^2p^2}\right), \tag{7.31}$$

where

$$k = \frac{c_S}{c_W}$$

$$p = \frac{m}{R} \sin \theta \tag{7.32}$$

$$q = \left(kp\sqrt{1-p^2} - p\sqrt{1-k^2p^2}\right)\left(\sin \theta \sqrt{1-p^2} - p \cos \theta\right) \tag{7.33}$$

Thus, we have

$$l = f - a - d \tag{7.34}$$

and arrive at

$$l = f - a - \frac{mq}{kp \sin \theta - q} \tag{7.35}$$

Given the a value, the distance from the lens to the sample surface, the lens focal distance, the angle θ, and the radius of curvature of the surface, distance l can be

7.11 Using Short-pulse Scanning Technique to Measure Component Thickness

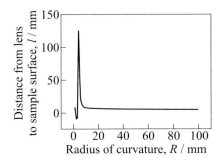

Figure 7.21 Distance from the lens surface to the real focus point vs the radius of the sample curvature ($a = 10$ mm and $f = 10$ mm).

evaluated. The calculated dependence is plotted in Figure 7.21. It is shown that at certain values of the radius of surface curvature R, the distance to the sample surface is a linear function of R. The radius value in Figure 7.21 exceeds 9 mm and $a = 10$ mm.

We rearrange Equation (7.35) using Equation (7.26):

$$p = \sin\theta - \frac{f-a}{R}\sin\theta \tag{7.36}$$

The angle α, which is the lens aperture, is $\leq 15°$ (for lenses at a frequency of 20 MHz); therefore, the most plausible value for the radius of curvature, in the case considered, ranges from 40 to 59 mm. The focal distance for the lenses used was 19 mm for a lens with a frequency of 20 MHz and 12.36 mm for a lens with a frequency of 15 MHz.

In the majority of cases, when $R > f$, the distance from the lens to the sample surface a can vary within the interval $0 < a < f$ until its optimum value is attained. Hence, $p \ll 1$. If $k = c_S/c_W \approx 1.5$, $k_2 p_2 \ll 1$. On the other hand, if angle α is very small, $\cos\alpha \to 1$ and $p\cos\alpha \to p$.

At $p \ll 1$, $1 - p^2 \to 1$, $1 - k^2 p^2 \to 1$, and $p\cos\alpha$ in Equation (7.31) can be neglected; thus, (7.31) assumes the following form:

$$d = \frac{m(k-1)}{k+1} = \frac{R(k-1) - (f-a)(k-1)}{k+1} \tag{7.37}$$

Substituting Equation (7.37) again into Equation (7.35) yields:

$$l = (f-a)\left(1 + \frac{1}{k+1}\right) - \frac{k-1}{k+1}R \tag{7.38}$$

Suppose

$$U = (f-a)\left(1 + \frac{1}{k+1}\right) \tag{7.39}$$

$$V = \frac{k-1}{k+1} \tag{7.40}$$

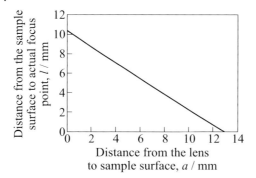

Figure 7.22 Optimal distance between the lens and the sample surface. $R = 60$ mm and $F = 15$ MHz.

Then

$$l = U - VR \tag{7.41}$$

Equation (7.41) becomes a typical linear function $l = f(R)$. This signifies that the sample surface curvature depends linearly on the distance between the surface and the lens. Hence, in every case we can check, with the aid of this equation, the accuracy of inspection of a multilayer curved system and in turn optimize the measurement results choosing a proper focal distance.

The data were substituted in (7.35) and processed with the aid of the MatLab code. The results are illustrated in Figure 7.22. The plotted graph confirms the validity of (7.41) and of our suggestion that, knowing the sample curvature and its approximate thickness, correct results can always be obtained by changing the distance between the lens and sample.

If, for example, the overall sample thickness is about 4.5 mm and the radius of its surface curvature is about 60 mm, the graph in Figure 7.22 suggests that, at 15-megahertz frequency, the optimum distance l from the sample surface is close to 7 mm. Figure 7.23(a) demonstrates the signal recorded in this case on the oscilloscope display. As seen, if the lens is positioned 7 mm away from the sample surface, the recorded signal is quite clear-cut. The actual focus point is positioned to the left of the focal point in the sample. Thus, we record strong signals reflected from each boundary between the layers. As the distance between the lens and sample surface changes, so does the signal pattern.

Figure 7.23(b) shows that for a distance between the lens and sample surface equal to 12 mm, the signal from the internal layer is clear-cut. On the other hand, the signal from the bottom layer is very weak because the focal point is located near the interface between the water and the sample; hence, ultrasound energy is absorbed in the internal layer to a lesser extent and the signal reflected from the water–polymer interface is very weak. As the lens approaches the object, the result is also unsatisfactory (Figure 7.23c). The distance between the lens and sample surface is 4 mm.

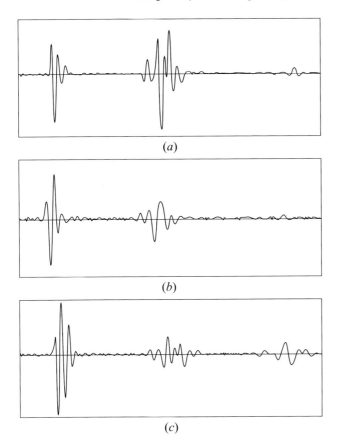

Figure 7.23 Oscilloscopic trace of the ultrasound signal. Radius of curvature $R = 60$ mm, frequency is 15 MHz. Distance from the lens to the sample surface is 7 mm (a), 12 (b), and 4 mm (c).

It can be seen that the resulting signal is unclear; the spikes pertaining to the echo reflected from the adhesion and internal layers are difficult to distinguish. If the lens is positioned very close to the sample surface the beam is focused beyond the sample; therefore, the signal reflected from the interface between the bottom polymer end and the water is clear-cut, whereas the signal reflected from the adhesion layer – internal layer interface is too weak. Measurements using this investigation technique yield good results on the total material thickness but are incapable of revealing microstructure peculiarities in the sample bulk.

In experimental studies, samples were placed in a specially designed device that provided controlled flexure. The whole device, including a bent sample, was placed in water. The transducer was mounted perpendicular to the surface and operated as an emitter and receiver. An ultrasound signal passed through the immersion liquid and was partially reflected at the boundary between water and the upper

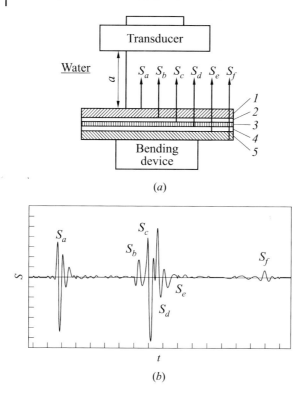

Figure 7.24 (a) Structure of the ultrasound echo-pulse technique: 1 – upper polymer layer, 2 – upper adhesion layer, 3 – internal layer, 4 – second adhesion layer, and 5 – bottom polymer layer. (b) Distribution of echo signals over time.

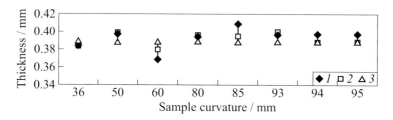

Figure 7.25 Thickness of the internal (barrier) layer: 1 – barrier (15) 0.384, 2 – barrier (20) 0.36, and 3 – barrier (plane case) 0.382.

sample surface (S_a peak); the signal portion transmitted in the sample bulk was reflected from the second interface between polymer and adhesion layer (S_b). The third peak of the reflected signal pertains to the echo reflected from the adhesion layer–internal layer interface (S_c). The fourth echo (S_d) corresponds to a signal reflected from the boundary between the internal and second adhesion layers. Echo S_e arrived from the fifth interface between the second adhesion and bottom poly-

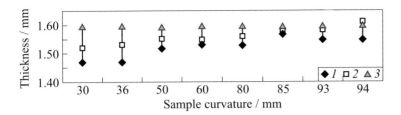

Figure 7.26 Thickness of the topmost polymer layer:
1 – polymer 1 (15), *2* – polymer 1 (20), and *3* – polymer 1
(plane case).

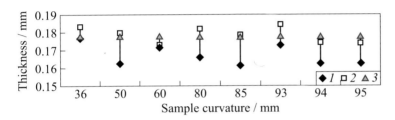

Figure 7.27 Thickness of the upper adhesion layer:
1 – adhesion 1 (15) 0.173, *2* – adhesion 1 (20) 0.179, and
3 – adhesion 1 (plane case) 0.174.

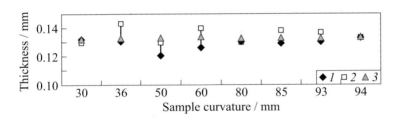

Figure 7.28 Thickness of the second adhesion layer:
1 – adhesion 2 (15), *2* – adhesion 2 (20), and *3* – adhesion 2
(plane case).

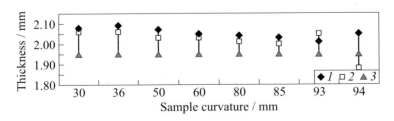

Figure 7.29 Thickness of the bottom polymer layer:
1 – polymer 2 (15), *2* – polymer 2 (20), and *3* – polymer 2
(plane case).

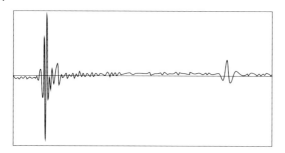

Figure 7.30 Sample with no barrier layer.

mer layers. The last, sixth, echo (S_f) is ascribed to the reflection from the bottom sample surface–immersion liquid (water) interface (Figure 7.24).

When two subsequent reflected signals are split in time, the measured time delays between them makes it possible to evaluate the layer thickness. Thus, $\Delta\tau_{ab}$, $\Delta\tau_{bc}$, $\Delta\tau_{cd}$, $\Delta\tau_{dc}$, and $\Delta\tau_{ef}$ are the delays of the short-pulse signals: S_a, S_b, S_c, S_d, S_e, and S_f, respectively.

Figures 7.25 to 7.29 demonstrate the calculated thickness values of five layers. "Plane case" means that measurements were carried out on the same sample before its bending and the results obtained, with the aid of lenses operating at frequencies of 15 and 20 MHz, were averaged. It can be seen that theoretical calculations are fully consistent with measurements. An optimization algorithm previously used to assess the distance between the lens and sample surface was employed in studying multi-layer structures.

Application of the aforesaid approach is demonstrated to furnish fairly accurate data on the number of internal layers in a material and on the thickness of individual layers in multi-layer structures of samples with a curved surface.

Figure 7.30 provides an additional example, illustrating the results of investigations of a sample containing no internal layer.

8
Investigation of the Microstructure and Physical–Mechanical Properties of Biological Tissues

8.1
Application of Acoustic Microscopy Methods in Studies of Biological Objects

The founders of scanning acoustic microscopy, Lemons and Quate, were the first to obtain acoustic signatures of biological tissues [10, 302, 324]. Even at that time, the benefits offered by acoustic microscopy such as the possibility of exploring native tissues and living cells, a high contrast of acoustic images, and the opportunity to watch biological processes on a real timescale were obvious. But what is of major importance, distinct from optical and electron microscopy images, is that acoustic signatures reveal the mechanical microstructure of the object studied.

Until the latter past of last century, internal tissues were studied mostly by scanning laser or transmission scanning acoustic microscopy, which necessitated the fabrication of fairly thin microscopic sections. The technique for preparing samples for these studies had much in common with methods of classical histological microfabrication. Thin sections were fabricated either after fixation, water removal, and enclosure in paraffin or polymer resins; or after freezing the sample on a microtome. This complex procedure was dictated by the necessity of providing sample rigidity and obtaining regular plane-parallel preparations.

The authors of various methodological works [325–327] have demonstrated that techniques of fixation and the subsequent treatment of samples have a different effect on the acoustic properties of biological tissues, particularly on sound absorption. Specifically, a decrease in the velocity of sound was reported after fixation with formalin: by 0.9–1.5% [326, 328] in liver; and by 0.6% [329] in brain tissue. The paraffin enclosure also significantly reduced the speed of sound and drastically enhanced absorption. Additional sample fixation (with ethanol) of samples fabricated in a freezing microtome improved the quality of acoustic images at a 1.2-megahertz frequency, moreso than does fixation with formalin. Only few works dealt with nonfixed and nonfrozen samples coated with a Mylar or polyethylene film [330]. A great many acoustic signatures of various human and animal tissues were obtained during that period: fixed tissues of normal lung [302], kidney (normal and after long-time dialysis) [331], liver, testicle, and muscle tissues [332]. Development of quantitative investigation techniques made it possible to obtain more accurate data on the acoustic properties of some tissue structures and also to measure their most important acoustic parameters, such as absorption, speed of sound, and acoustic impedance [9, 333–337].

Correlation between the composition of a tissue and its acoustic properties is the most important problem to be solved in tissue studies. Mechanical properties on the tissue level are controlled by water content, conjunctive tissue properties, the pressure in the blood vessels and of the interstitial liquid, peculiarities in the intercellular material, and also by aggregation and interaction of cells. O'Brien et al. [338] have shown that the ultrasound velocity in a liver tissue depends directly on the water content and inversely on the fat content, whereas the protein content (percentage of wet weight) affects its value only slightly. Direct correlation between ultrasound absorption and glycogen content was found in [339].

According to [340, 341], fibrillar albumen plays an important part in ultrasound interaction with tissues. As follows from histological and optical microscopy investigations, malignant growth is associated with collagen growth [342]. Therefore, it is supposed that such tissue abnormality could readily be revealed with the aid of acoustic microscopy. Again, Lemons and Quate [302], who obtained high-contrast signatures of 5-micron fixed sections of a mamma gland tumor on an acoustic microscope operating at 600 MHz, were the first to explore this problem. In fact, they have demonstrated with the aid of acoustic microscopy that one can quickly discover pathologic malformations without using laborious staining techniques [54, 302]. The dynamics of changes in the acoustic parameters and microstructure of a dog's skin in the zone of a maturing wound was studied in [343, 344] using a scanning laser acoustic microscope. Sections fabricated on a freezing microtome were examined. Both the acoustic velocity and absorption were found to increase (the velocity increased from 1540 to 1700–2000 $m\,s^{-1}$) during the course of scar formation; this corresponded to a local increase in the collagen content in this zone.

An acoustic microscope operating at a frequency ranging from 100 to 400 MHz was demonstrated to be an effective tool for differential medical diagnosis of liver carcinoma [345] and gastric cancer tissues [346]. An infarction-affected cardiac muscle was another model used to explore opportunities for acoustic microscopy. Inasmuch as necrotized zones in a cardiac muscle are replaced by a conjunctive tissue, the spot affected by infarction is distinguished in the acoustic microscope due to an increased concentration of fibrillar albumen [347, 348]. A similar mechanism underlies contrast formation in acoustic signatures of atherosclerotic changes in the human aorta [349]. The dynamics of cardiac allograft rejection were studied in [350]. It was found that the acoustic properties of the cardiac muscle tissue changed as a result of edema, cellular infiltration, hemorrhage, and destruction of cardiomyocytes. The ultrasound velocity increased from 1602 $m\,s^{-1}$ in the normal tissue to 1650 and 1640 $m\,s^{-1}$ in zones of lymphocyte accumulation and of near-vascular fibrosis. Ultrasound absorption in the zones of pathological alterations also increased, attaining 1.6 and 1.5 $dB\,mm^{-1}\,MHz^{-1}$ against

1.1 dB mm^{-1} MHz^{-1} in normal tissue. Myocardium anisotropy observed in dog's myocardium tissue in [351] also is deserving of special attention.

The authors of [352] have shown, by way of example, that acoustic images obtained at a 1-gigahertz frequency in the human retina and iris, apart from fibrillar albumen, the image contrast is affected by melanin. Further, anisotropy due to the orientation of collagen fibers in the eye sclera and human skin was disclosed.

Investigations performed with thin tissue sections are of invaluable importance for theoretically grounding the techniques of clinical ultrasound diagnostics and for understanding the imaging mechanisms in ultrasound visualization of internal organs.

Investigations into the microstructure and mechanical properties of individual cells and cell cultures are, at present, one of the most interesting and fruitfully developing areas of application of acoustic microscopy. Studies in this area were launched by Lemons and Quate [54, 302]. They have demonstrated the opportunities offered by acoustic visualization of the microstructure of individual cells, taking as an example acoustic images of erythrocytes and fibroblasts fixed with methanol.

The use of cryogenic liquids as an immersion medium allowed a significant improvement in microscope resolution to be attained in studies of fixed cells. Thus, images of metaphase human chromosomes were obtained in liquid argon at a 1.8-gigahertz frequency (with 0.38-micrometer resolution) [353]. A signature of an internal structure of fibroblasts from chicken heart was obtained in liquid nitrogen at a 2.2-gigahertz frequency (with a resolution of about 0.1 µm). The signature clearly shows nuclei and mitochondria [354]. In [56], bacterium was visualized in liquid helium at 8 GHz ($\lambda = 300$ Å, resolution of about 250 Å). As in electron microscopy, biological objects studied by cryogenic acoustic microscopy are always fixed. Thus, the acoustic images of biological objects were shown to exhibit a high contrast without staining, which offers a unique opportunity to study cells *in vivo*.

It should be pointed out that because the acoustic properties of water or a nutrient medium differ only slightly from those of distilled water the acoustic microscope, like the optical microscope, is quite well suited for operation with living cells. It was established that neither high-frequency radiation nor mechanical scanning causes damage or significant changes in the cell behavior; cells retain their ability to move in a normal way and remain well spread over the substrate [253, 303, 355]. Cell damage by ultrasound radiation at low frequencies (20 kHz, 5 MHz) was observed in [356]. However, significantly greater ultrasound intensities and longer time exposures were used in that work.

A number of images of cells during the course of their motion over the substrate [253] and of their fission [357] were obtained with the aid of SAM at a frequency of 1.3 GHz. A sequence of these images was used to animate the dynamics of changes in the cytoskeleton and adhesion during cell motion over the substrate, and of changes in the cell undergoing mitotic fission.

Cells studied under acoustic microscopes are usually placed on a flat substrate. In this case, images are more difficult to interpret than they are in studies of solid surfaces because they result from the interference of the waves reflected from the substrate surface and also from the top and bottom cell surfaces. In addition, the internal cell structures also contribute to imaging [355, 358]. The authors of [355] studied cells from a chicken embryo heart on various supports: quartz, glass, and polystyrene, at a 1.7-gigahertz frequency (resolution of about 0.72 µm). They have shown that the image pattern depends heavily on the substrate acoustic properties. Focusing of the acoustic lens on the cell surface yields information about the cell impedance, whereas when the lens is focused on the substrate, surface imaging provides data on ultrasound attenuation in the cell. The aforementioned parameters characterize the elastic state of the cell cytoskeleton that consists of filament proteins. Specifically, the impedance value is significantly greater at the cell edge, which is accounted for by a more rigid internal structure in this cell zone. Acoustic microscopy was applied in [253, 359, 360] to investigate the effect of various physical and chemical factors on a cell cytoskeleton. The ultrasound velocity was measured in various cell structures, and it was shown that the ultrasound velocity in cytoplasm depended on the distribution of actin fibrils [361, 362]. The strained cytoplasm state, through contraction of the actinomyosin complex, increased the ultrasound velocity from 1600 to 1800 m s^{-1} [363, 364]. Zones with different values of absorption, acoustic velocity, and cytoplasm thickness in a cell were mapped in 1992 [365]. The topographic distribution of ultrasound velocity values is aligned with the direction of forces supporting the cell shape [366, 367]. Variations in the acoustic velocity, viscosity, and absorption in cytoplasm can be explained by changes in the functional cell state [368].

One of the most interesting problems arising in cell mechanics and associated with cellular mobility is attachment of the cells to a substrate. This problem can also be solved by acoustic microscopy methods [358]. The contact sites exhibit a higher acoustic impedance that can change as the cell moves. These sites effectively control the contrast of acoustic images.

It has been demonstrated that coarse cells in a culture can also be explored at lower frequencies. Model experiments have demonstrated the efficiency of acoustic microscopy in controlling the effect of heating on a tissue fragment or of an antitumor preparation on tumor fragments in the form of cellular spheroids [369–371].

Studies of endothellium cells from a cultured bovine aorta [372] (a frequency of 100 and 210 MHz) revealed that under mechanical stress the cells changed their shape, the viscosity of their cytoplasm increased, and the speed of sound in the cytoplasm increased from 1570 to 1590 m s^{-1}, whereas its value near the cell nucleus did not change (1610 ± 10 m s^{-1}). The feasibility of distinguishing viable, dead (due to overheating), and apoptotic cells with the aid of an acoustic microscope operating at 50 MHz was demonstrated in [373].

No doubt, supravital noninvasive investigation of tissues and organs is the most interesting and practically important application of acoustic microscopy. The depth of ultrasonic signal permeation is limited to 3 or 5 mm; for this reason, skin and eye tissues were the first objects that attracted the attention of researchers.

8.2
Selection of Immersion Media for Acoustic Microscopy Studies of Biological Objects

The selection of an appropriate immersion liquid for any particular group of objects and problem to be solved is an independent task in acoustic microscopy. The choice of the liquid depends on several parameters. These are: the speed of sound, the attenuation coefficient, and low chemical or physical impact on the object.

The set of liquids to be used as an immersion is, in the first place, limited by the value of their ultrasound absorption coefficient. In the majority of liquids, absorption at frequencies used in acoustic microscopy is so large [286] that the receiving transducer fails to record the signal after its passage through the lens system. The immersion liquid used in the initial acoustic microscope version by Quate and Lemons [10, 324] was water. Water still remains the most popular immersion liquid [374].

Absorption in water, at frequencies ranging up to very high values, obeys the classical quadratic dependence [25]:

$$\frac{\alpha}{f^2} = \frac{2\pi^2}{\rho c^3}\left(\eta_V + \frac{4}{3}\eta_S\right)$$

where ρ is the liquid density, c is the speed of sound, η_V is the volumetric viscosity, and η_S is the shear viscosity. The ultrasound absorption coefficient of water ($\alpha/f^2 = 22 \times 10^{-17}$ s^2 cm^{-1} at $T = 25\,°C$) is less than that of the majority of other liquids. Nonetheless, at high frequencies (close to 1.5 GHz), ultrasound waves are intensely attenuated in it. However, it is known that the attenuation coefficient as a function of temperature passes through a minimum. Therefore, preheating water to 60 °C one can reduce the attenuation ($\alpha/f^2 = 12 \times 10^{-17}$ s^2 cm^{-1}) and thereby increase the operation frequency of the microscope to 3 or 4 GHz [182, 375].

To choose an immersion fluid in which ultrasound attenuation is less than that in water, the authors of [376] investigated the acoustic properties of aqueous solutions of electrolytes at frequencies ranging from 250 MHz to 4 GHz. They have found that as a rule attenuation diminishes as the electrolyte concentration increases. However, attenuation in NaCl solutions as a function of the salt concentration exhibits a minimum. Further, salts with cation K$^+$, among them KCN (in which $\alpha/f^2 = 12 \times 10^{-17}$ s^2 cm^{-1}), have the lowest attenuation coefficient values.

The possibility of using hydrogen peroxide at temperatures from 15 to 50 °C as an immersion liquid was explored in [376]. Distinct from water, the absorption in hydrogen peroxide depends only slightly on temperature. The quadratic frequency dependence was observed in a wide frequency range from 500 MHz to 2 GHz. The α/f^2 value in peroxide equals 14×10^{-17} s^2 cm^{-1} at $T = 25\,°C$, which is twice as low as the value in water.

Some organic liquids are known to have fairly long relaxation times [25]. The frequency dependence of parameter $\alpha\lambda$ would be expected to go through a minimum in such liquids at the frequencies used in acoustic microscopy. Absorption in only a small number of organic liquids (such as peroxide, alcohols, and formic acid) is either lower than, or close to, that in water. Liquid carbon sulfide CS_2, in which the αf^2 value drops from 3000×10^{-17} s^2 cm^{-1} at the vibrational frequency value of 78 MHz to 8.5×10^{-17} s^2 cm^{-1} at a frequency of 4 GHz [376], is of particular interest in that respect. At present, this is the only liquid known to have such a low ultrasound attenuation at room temperature in the gigahertz frequency range.

Liquid metals (mercury and gallium at room temperature) are repeatedly mentioned in the literature [377–379] as promising acoustic immersions. Ultrasound attenuation in mercury is half that in water: $\alpha/f^2 = 5.8 \times 10^{-17}$ s^2 cm^{-1}. However, it is very difficult to provide close acoustic contact when mercury is used for immersion because of poor wetting of both the lens and objective by mercury. Losses due to absorption in gallium (melting point $T = 29\,°C$) are extremely low. This material was successfully used as an immersion for the acoustic visualization of the front and rear surfaces of a quartz plate in a reflection microscope [377]. The high adhesion of gallium are a significant drawback as it stains both samples and acoustic lenses. However, one must take into account one very critical issue which makes mercury and gallium almost impossible to be used as a coupling – both of them are hazardous substances.

Table 8.1 lists liquids with ultrasound absorption values which allowed them to be considered as possible acoustic immersions. It also lists the acoustic parameters of those liquids (ultrasound absorption, density, speed of sound, and acoustic impedance) which specify them as candidates for immersions.

The sound speed value c of an immersion liquid specifies the wavelength at the operation frequency of a microscope and, hence, its resolution. Liquids with the lowest possible sound speed values (e.g., alcohols at room temperatures and helium in cryogenic microscopy) are usually used to improve microscope resolution. However, the value of the sound speed velocity in an immersion is also of importance for controlling the image contrast. The ratio between the ultrasound velocities in the object and immersion characterizes peculiarities of wave refraction and reflection at the interface between the object and immersion, producing critical angles (longitudinal, transverse, and Rayleigh) and controlling their value. The mechanism of the output signal formation in an acoustic microscope depends on whether or not a critical angle falls in the acoustic lens aperture [59]. Acoustic images of the surface and near-surface layer of a solid object arise in a reflection microscope when the Rayleigh angle falls in the angular aperture of the lens and when Rayleigh waves take part in signal formation [60, 104, 108]. Therefore, obtaining high-contrast images in reflection microscopy necessitates the use of immersions in which the speed of sound differs appreciably from the transverse speed of sound in the sample.

When visualizing objects in a transmission microscope, one should use an immersion medium for which the speed of sound is close to that in the object, in order to reduce aberrations arising due to refraction of the focused beam at the immersion liquid–sample interface [381]. In this respect, water or aqueous solutions are optimal immersions for studying biological tissues and cells by acoustic microscopy.

Table 8.1 Acoustic properties of liquids.

Liquid	T	Density g cm^{-3}	Speed of sound km s^{-1}	Impedance 10^5 g (cm^2 s)$^{-1}$	Attenuation α/f^2 10^{-17} s^2 cm^{-1}
		Liquids at room temperature [378]			
Water	30 °C	1.000	1.509	1.509	19.1
Methanol	30 °C	0.796	1.088	0.866	30.2
Ethanol	30 °C	0.789	1.127	0.890	48.5
Acetone	30 °C	0.791	1.158	0.916	54.0
Carbon-tetrachloride	25 °C	1.594	0.93	1.482	538.0
Mercury	23.8 °C	13.59	1.449	19.69	5.8
Gallium	30 °C	6.09	2.87	1748	1.58
Hydrogen peroxide	25 °C	1.464	1.545	2.26	10.0
Carbon sulfide	25 °C	1.261	1.310	1.65	10.0 (3 GHz)
Rubidium iodide (8-M solution)	25 °C	–	1.45	–	14.8
Isopentane	20 °C	0.62	1.016	0.63	–
Saturated NaCl solution	20 °C	1.2	1.65	1.98	–
Formic acid	20 °C	1.22	1.4	1.708	22.7
		Cryogenic liquids [380]			
Oxygen	90 K	1.140	0.9	1	9.9
Nitrogen	77 K	0.8	0.85	0.2	13.82
Hydrogen	20 K	0.07	1.19	0.08	5.65
Xenon	166 K	2.9	0.63	1.8	22
Argon	87 K	1.4	0.84	1.2	15.2
Neon	27 K	0.147	0.6	0.72	23.16
Helium	4.2 K	0.147	0.183	0.027	226.5

The difference between the acoustic impedance ρc values in the immersion and sample specifies the fractions of the acoustic energy reflected from the sample and transmitted into it. By selecting an immersion whose acoustic impedance is close to the object impedance, one can enhance the efficiency of penetration of a converging acoustic beam in the sample bulk. This introduces the possibility of studying interior sample structures by reflection microscopy [253, 377] and of thicker samples by transmission microscopy. The majority of crystalline materials have large acoustic impedance values. Hence, of all the conceivable immersions, liquid metals, whose impedance values are very large ($\rho c = 19.69 \times 10^5$ g(s cm^2)$^{-1}$ for mercury and $\rho c = 17.48 \times 10^5$ g(s cm^2)$^{-1}$ for gallium), are best suited for investigations of the interior structures of such samples.

The inactivity of an immersion with respect to the sample material is an important criterion in the selection of an immersion. This problem is particularly important in studies of biological objects because the overwhelming majority of liquids, the acoustic properties of which meet well the conditions imposed on the immersion, will affect the structure and properties of biological materials. For example, formic acid denatures proteins, while alcohols dehydrate biological objects [382]. Obviously, these liquids cannot be used in studies of the acoustic properties of native tissues. Liquid metals are inactive with respect to biological media; however, as a rule, the difference between the acoustic impedance values of the immersion and object is too large.

Cryogenic liquids have very low values of the speed of sound and absorption. Therefore, they were used in acoustic microscopy to attain high resolution, comparable with the resolution of electron microscopes. Liquid argon ($T = 85$ K) was the first cryogenic liquid used as an immersion medium in acoustic microscopy [380]. It was chosen because of its inactivity with the objects, fairly high impedance, and low absorption of ultrasound. At present, liquid nitrogen ($T = 77$ K) is used more often because of its availability. Its acoustic properties, to some extent, are similar to those of water. The absorption of acoustic radiation in it approximately equals absorption in water at $T = 37\,°C$. Helium in the superfluid state at $T < 0.2$ K is considered as an ideal liquid for application in high-resolution acoustic microscopy. The maximum resolution of < 250 Å was obtained in liquid helium at an 8-gigahertz frequency [56]. The authors of [383] investigated the temperature dependence of ultrasound attenuation in liquid propane (from its melting point at 83 K to its boiling point at 231 K). This liquid was not considered as a candidate for an immersion medium but studies of its acoustic properties were of special interest.

In conclusion, we note that acoustic microscopy is now used as an independent technique in studies of the acoustic properties of liquids at high frequencies.

8.3
Imaging and Quantitative Data Acquisition of Biological Cells and Soft Tissues with Scanning Acoustic Microscopy

8.3.1
Introduction

In the biomedical field, many new treatment methods and new medications have been developed. Obviously, their effects and side effects need to be studied qualitatively and quantitatively. Conventional optical microscopes are often used for obtaining such data from tissues or cells taken from patients as specimens. However, the tissues or the cells need to be chemically stained and/or fixed for optical microscopic observation. The staining and/or fixation usually kill the cells. Therefore, when those techniques are applied, it is difficult to understand the real effects of the treatment or the medicine, because the transformation in the tissue may only be visible in living cells. Optical fluorescence microscopy is conventionally used to visualize the living cells. However, a fluorescence stains only a portion of the cellular structure. In addition, the fluorescence fades with time. Therefore, it is necessary to use various types of fluorescent agents, and complete complicated procedures for calibration. Furthermore, a cell may not absorb a fluorescence agent enough to show critical information characterizing the behavior of the cell.

An acoustic image obtained by a mechanical scanning reflection acoustic microscope [109, 378] is formed by contrast mechanisms different from the optical one. Therefore, the staining and/or fixation to the specimen are not required for the SAM. Hence, the living cells can be observed by the SAM [303, 359, 363, 384–386]. Furthermore, the SAM can nondestructively observe not only the surface but also the internal structure of the biological specimen with submicron resolution [387]. The SAM also has the capability to measure mechanical properties (e.g., attenuation, thickness, and velocity) of the cells and/or tissues [355, 360, 366, 388–394]. This article reports the unique applications of the SAM.

8.3.2
Brief Description of the System

Figure 8.1 is the schematic diagram of the modification of the SAM (Olympus Optical Co., Ltd., Model: UH3). Referring to Figure 8.1, the imaging mechanism of the SAM is described below.

The SAM is comprised of a transmitting/receiving section, an $X-Y$ scanning section, a Z scanning section, a computer section, and a display section. The transmitting/receiving section includes a transmitter, a receiver (with an amplifier), a circulator, and an acoustic lens having a piezoelectric transducer (material: zinc oxide), and a buffer rod (material: sapphire). The $X-Y$ scanning section consists of an $X-Y$ stage, and a chamber for containing a specimen and a coupling medium. The electric signal (i.e., a tone-burst wave) generated from the transmitter travels to a piezoelectric transducer, located on the top of the buffer rod, and through the circulator, which has a function permitting the electric signal to be transmitted in only

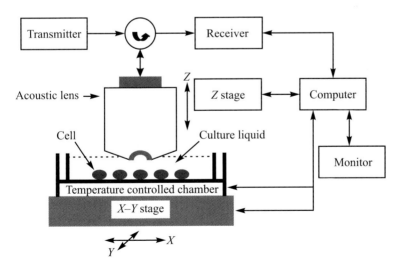

Figure 8.1 Schematic diagram of the SAM designed for biomedical specimens [395].

one direction (i.e., to the piezoelectric transducer or from the piezoelectric transducer to the receiver). The piezoelectric transducer has a function which subjects the transmitted/received signal to electroacoustic conversion. Therefore, the electric signal from the transmitter is converted to an acoustic signal (i.e., ultrasonic plane wave) at the transducer. The ultrasonic plane wave travels through the buffer rod to a spherical recess (hereafter called simply the "lens") located at the bottom of the buffer rod. The lens, coated by an acoustic impedance matching layer, or so-called acoustic anti-reflection coating (AARC), converts the ultrasonic plane wave to an ultrasonic spherical wave (i.e., ultrasonic beam). The ultrasonic beam is focused within the specimen and reflected from the specimen, the specimen being the coupling medium included in the chamber located on the X–Y stage. The reflected beam carries acoustic information of the specimen and is again converted to an ultrasonic plane wave by the lens. The ultrasonic plane wave returns to the transducer through the buffer rod, wherein it is again converted to an electric signal at the transducer. The voltage of the electric signal ranges from 300 mV to 1.0 V. When the operating frequencies range from 100 MHz to 1.0 GHz, the corresponding insertion loss is approximately 30 dB to 80 dB. Therefore, the electric signal must be amplified by 30 to 80 dB at the amplifier.

Furthermore, the electric signal is comprised of transmission leaks, internal reflections from the interface between the lens and the AARC, and reflections from the specimen (Figure 8.2). Therefore, the reflections must be selected by a rectangular signal from a double balanced mixer (DMB), called the first gate. Then, the peak of the amplitude of the electric signal is detected by a peak detector, which includes a diode and a capacitor. The gate noise is removed by using the second gate which exists within the first gate (i.e., the blanking technique).

The peak-detected signal is stored into memory through an analog-to-digital signal (hereinafter simply called the "A/D") converter. The stored signal is again converted into an analog signal by a digital-to-analog signal (referred to as the "D/A") converter to be displayed as intensity on the monitor by means of the display section. This flow of processes gathers the information that is collected at a single spot on a specimen. In order to form a 2D acoustic image, an acoustic lens and/or the

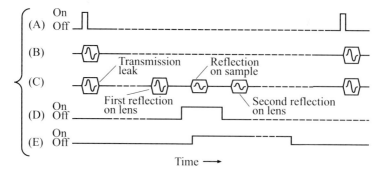

Figure 8.2 Timing chart of the signals for the RF and acoustic lens.

X–Y stage is mechanically scanned across a certain area of the specimen by means of the X–Y scanning section. The acoustic lens is able to translate axially along the Z direction to vary the distance between the specimen and the lens for subsurface visualization by means of a Z scanning section. That is, when visualizing the surface of the specimen, the acoustic lens is focused on the specimen surface (we note $Z = 0$ μm), and when visualizing a subsurface of the specimen, the acoustic lens is mechanically defocused toward the specimen (we note $Z = -x$ μm, where x is the defocused distance). The above described processes are controlled by means of the computer section.

8.3.3
Contrast Factor for Acoustic Imaging of Biological Cells and Tissues

Biological cells and tissues have acoustic impedances close to those of the culture liquid, and virtually no contrast caused by the difference in reflection coefficient can be displayed. However, the contrast in the acoustic images can be generated from the difference in attenuation. When using a background composed of highly reflective materials, the difference in attenuation of the biological specimens can be maximized in the image. This method is useful, when operating SAM with frequency at 1.0 GHz or more. Sapphire, with an acoustic impedance of 44.3×10^6 kg m^{-2}, was selected to be the substrate acting as the background [303]. When operating SAM with frequencies ranging from 200 to 600 MHz for visualizing the biological cells and tissues, a glass slide made of silica glass, commonly used for a conventional optical microscope, was selected as the substrate. The scanning reflection acoustic microscope can also shed light on the adhesive condition between the cells and the substrate (Figure 8.3). The corresponding contrast mechanism depends upon the $V(z)$ curve [59, 61, 108, 112–114, 397–401].

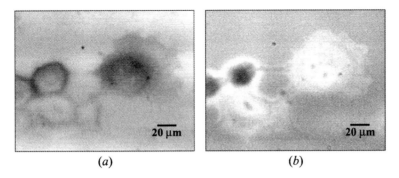

Figure 8.3 The 1.0-gigahertz image in two different lens positions: (a) $Z = 0$ μm; and (b) $Z = -1.6$ μm, gives the adhesive condition between the human skin cells (HaCat) and the substrate (Sapphire). The temperature of the culturing medium was 42.5 °C [396].

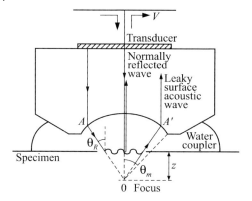

Figure 8.4 Cross-sectional geometry of spherical acoustic lens to explain the mechanism of the $V(z)$ curves.

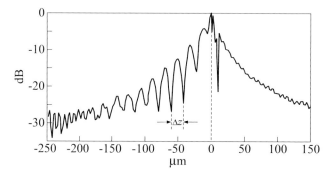

Figure 8.5 Curve $V(z)$ for fused quartz. Specimen: fused quartz; coupling medium: distilled water; temperature of the coupling medium: 22.3 °C (change less than ±0.1 °C). Parameters of the acoustic lens are as follows: frequency = 400 MHz; aperture angle = 120°; and working distance = 310 μm [395].

$V(z)$ Curve. The transducer output voltage can be periodic with axial motion as the acoustic lens advances from the focal plane toward the specimen. The period of this variation is characteristic of the specimen's elastic material properties and results from interaction between the two ray components, that radiate into the liquid from the solid–liquid interface shown in Figure 8.4. Figure 8.5 is an example of a $V(z)$ curve.

The $V(z)$ curve is expressed as follows:

$$V(z) = C^{-1} \int_0^\infty u^2(r) P^2(r) R\left(\frac{r}{f}\right) \exp\left\{i2kz\sqrt{1-\left(\frac{r}{f}\right)^2}\right\} r\, dr \qquad (8.1)$$

$$C = \int_0^\infty u^2(r) P^2(r) r\, dr \qquad (8.2)$$

where u is the acoustic field, P is the pupil function defined by the lens geometry, R is the reflectance function, k is the wave number in the coupling medium, and f is the focal length.

Equations (8.1) and (8.2) are expressed by using $r = f \sin\theta$ as follows:

$$V(z) = C^{-1} \int_0^\infty u^2(\theta) P^2(\theta) R(\theta) \exp(i2kz\cos\theta) \sin\theta \cos\theta \, d\theta \qquad (8.3)$$

$$C = \int_0^{\theta_0} u^2(\theta) P^2(\theta) \sin\theta \cos\theta \, d\theta \qquad (8.4)$$

where θ is half the aperture angle of the lens.

When using $k_z = k\cos\theta$, (8.3) and (8.4) are expressed as follows:

$$V(z) = C^{-1} \int_k^{k\cos\theta_0} Q^2(k_z) R(k_z) \exp(i2k_z z) \, dk_z \qquad (8.5)$$

where

$$C = \int_k^{k\cos\theta_0} Q^2(k_z) \, dk_z \qquad Q^2(k_2) = u^2(k_z) P^2(k_z) k_z$$

From Equation (8.5), the following equation is obtained:

$$F^{-1}\{V(z)\} = C^{-1} Q^2(k_z) R(k_z)$$

where $F^{-1}\{\}$ is the inverse Fourier transform. In this case, the contrast caused by the reflectance function depends upon the cell and its peripheral conditions. When an acoustic beam is emitted from the acoustic lens onto the healthy cell grown on the substrate in the culturing medium, the reflectance function is determined by the layered media comprising the culturing medium, the cell, and the substrate (Figure 8.6a). On the other hand, when analyzing the image of the injured cell, the reflectance function is determined by the layered media comprising the culturing medium, the cell, the fluid, and the substrate (Figure 8.6b). The $V(z)$ curve with frequency at 1.0 GHz was sensitive enough to capture the minute change in the layered media.

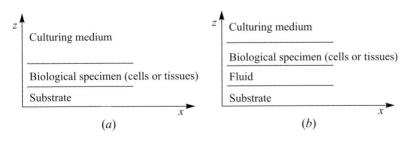

Figure 8.6 The structures of the specimens and their periphery conditions.

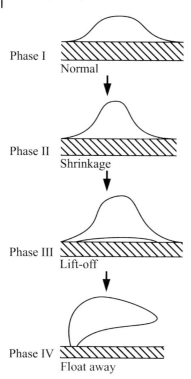

Figure 8.7 The processes of physical deformation (i.e., shrinkage of the cells and decrease in the contact area of the cells), lift-off, and float away due to the elevation of temperature in the culturing medium [395].

8.3.4
Thermal Insult

Processes of Physical Deformation of a Cell. During exposure of living, healthy cells to gradual increases in heat, the cells typically undergo several stages before their death: (*1*) shrinkage of the cells; (*2*) decrease in the contact area of the cells and the substrate; (*3*) lift-off; and (*4*) death as manifested by float away (Figure 8.7). The lift-off cells were found to be badly injured, and statistically never recovered. Therefore, the onset of lift-off was judged to be a convenient threshold between dead and living cells.

The human skin cells (i.e., HaCat) and the human skin cancer cells (i.e., Melanoma) were grown on the substrate (i.e., sapphire) mounted on the bottom of the well with a culturing medium (i.e., Dulebecco's modification of Eagle's medium: 1 × (Mod)). The well was mounted into the chamber having the function of a heating plate with temperature controller, and in turn located on the X–Y stage of the SAM. The temperature of the culturing medium was monitored by thermocouples equipped within the chamber.

The changes in the behavior of the living cells due to the elevation of temperature in the culturing medium are shown in Figures 8.8 to 8.10. The SAM images show the clear effects of increasing the temperature from 37.5 to 44.5 °C in terms of shrinkage, lift-off, and death.

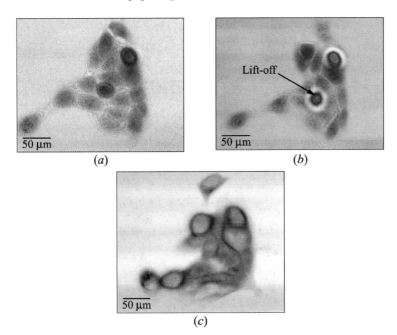

Figure 8.8 The SAM images show the clear effects of increasing the temperature from 37.5 to 44.5 °C. The images of shrinkage, deformation, and death of human skin cells are obtained. (a) The image of healthy cells at 37.5 °C; (b) the image of the shrunken and lifted off cells at 42.5 °C; and (c) the image of the cells, wherein some dead cells moved from the group at 44.5 °C. The images were formed with frequency of 1.0 GHz. The acoustic lens was defocused at $z = -1.6$ μm [396].

Shrinkage. Another set of human skin cells (Figure 8.11) were grown on the silica glass located on the bottom of the well with 1 × (Mod) of Dulebecco's modification of Eagle's medium to find the shrinkage rate in accordance with the elevation of temperature of the medium. The temperature was increased from 38.5 to 52.5 °C. The image was formed with a frequency of 0.6 GHz. The greater the temperature increase, the more cells shrink (Figure 8.12 and Table 8.2). All normal skin cells were subsequently detached from the substrate and floated away in the culturing medium when the temperature was 52.5 °C. Most of the cancer cells were detached before the temperature reached 52.5 °C.

8.3.5
Shock Wave Insult

A laser induced ultrasonic shock wave system (Figure 8.13) was used for shock wave insult. Here a Q-switched Nd:YAG laser is pulsed and emitted onto a flier plate made of tungsten. Plasma is created due to the interaction between the laser and the flier plate. Then, a shock wave is generated and emits onto the human skin cells (HaCat) located under the flier plate, where the strength of the impact corresponds to the conservation of momentum of the plate. Note that the acoustic

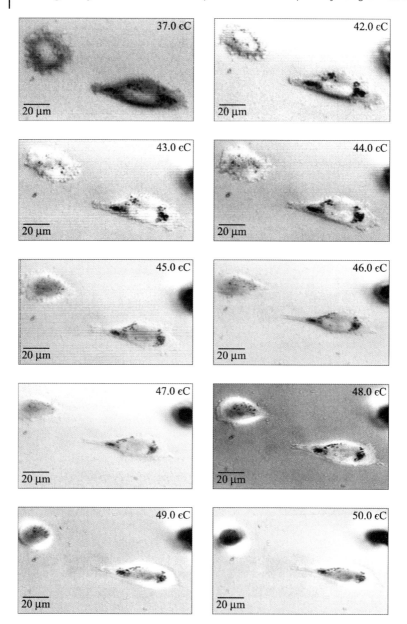

Figure 8.9 Acoustic images of Melanoma cells with temperature increase [395].

impedance of tungsten is 101.0×10^6 kg m^{-2}. The force generated by the system was 750 N. The temperature was monitored by thermocouples to make sure that the heat generated was not significant.

Figures 8.14(a) and (b) show the healthy cells and the cells impacted by the shock waves, respectively. As can be seen in Figure 8.14(b), all impacted cells in the region

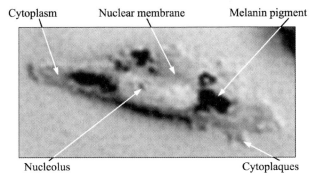

Figure 8.10 Schematic diagram of a melanoma cell. The melanoma is a malignancy of melanocytes which leads to the presence of melanosomes in the cytoplasm which gives the characteristic melanin pigment. The melanoma manifests itself in the variation of the melanin pigment. The cancer cells exhibit pleomorphism with many different shapes varying from oval to nearly columnar, and secondary degeneration [402].

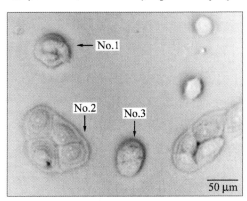

Figure 8.11 The image was formed with frequency of 0.6 GHz. The acoustic lens was defocused at $z = -5.0$ μm. The temperature of the culturing medium was 38.5 °C [403].

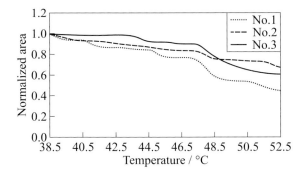

Figure 8.12 Shrinkage rate of the cells [403].

Table 8.2 Area calculated from each acoustic image of the cells at each temperature [403].

Temperature °C	No. 1 µm²	No. 2 µm²	No. 3 µm²
38.5	840	3050	790
39.5	794	3044	760
40.5	790	3017	739
41.5	738	3015	735
42.5	732	3006	717
43.5	719	2991	704
44.5	711	2825	692
45.5	654	2807	671
46.5	650	2757	662
47.5	620	2725	655
48.5	499	2388	599
49.5	456	2144	592
50.5	452	2002	580
51.5	407	1886	575
52.5	375	1841	531

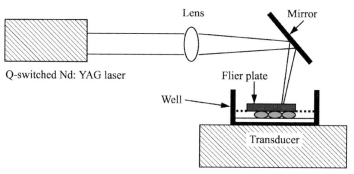

Figure 8.13 A Q-switched Nd:YAG laser is pulsed and emitted onto a flier plate made of tungsten. Plasma is created due to the interaction between the laser and the flier plate. Then a shock wave is generated and impacts human skin cells (HaCat) located under the flier plate, where the strength of the impact corresponds to the conservation of momentum of the plate. A tungsten base plate was located under the well. Note that the acoustic impedance of tungsten is $101.0 \times 10^6 \, \mathrm{kg \, m^{-2}}$. The force generated by the system was 750 N [396].

were lifted off. Fifty pulses in a ten-second interval at full power were necessary to destroy the cells. We stained the lift-off cells with TRYPAN-BLUE (0.4%) solution. Then we confirmed that the lift-off cells were dead by means of a laser scanning microscope (Olympus Optical Co., Ltd.). The SAM images show that cells subjected to shock wave insult (i.e., mechanical insult) had little physical deformation or shrinkage and were damaged to the point of complete lift-off.

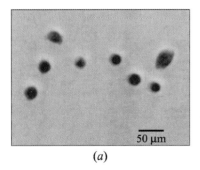
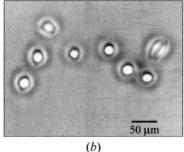

(a) (b)

Figure 8.14 The SAM images show that the cells subjected to mechanical insults remained flat on the substrate. The acoustic lens was defocused at $z = -1.6$ μm to obtain both images: (a) healthy cells at 37.5 °C; and (b) impacted cells at 37.5 °C [396].

8.3.6
Velocity Measurement for Biological Tissue

Ever since the advent of SAM, the key objectives have been quantitative data acquisition and the enhancement of the resolution in the acoustic images. For quantitative data acquisition with the SAM, $V(z)$ curve techniques have been developed and applied to various materials with fruitful results [59, 108, 112–114, 366, 390–394]. With the $V(z)$ curve technique, we can obtain the velocity of the surface acoustic wave (e.g., Rayleigh wave) of the small area of the specimen.

The velocity of the SAW, such as a Rayleigh wave, is expressed as:

$$V_{saw} = \frac{V_w}{\sqrt{1 - (1 - V_w/(2\Delta z f))^2}} \quad (8.6)$$

where V_w is the velocity of the longitudinal wave of the coupling medium, f is the frequency of the ultrasonic wave emitted from the acoustic lens, and Δz is the spacing of the minima in the oscillations after removing the background of the $V(z)$ curve.

However, it is not easy to characterize soft materials (e.g., biological tissues) by the $V(z)$ curve technique. First, the critical angle of the Rayleigh wave generation of the soft materials is generally high. Therefore, the Rayleigh wave is often not generated within a specimen, even though an acoustic lens with a high numerical aperture (e.g., 120°) is used. This means that no $V(z)$ curve is formed. Second, the relative attenuation of soft materials is generally high. Therefore, the $V(z)$ curve may not have enough oscillations for the FFT analysis to measure the Rayleigh wave velocities accurately, even though the Rayleigh wave is generated. However, the attenuation is typically frequency dependent. When an acoustic lens, with a high numerical aperture, a long working distance, and a low frequency (e.g., 10 MHz or less) is used for the soft materials, a $V(z)$ curve might be formed. Unfortunately, in this case, the advantage of using the $V(z)$ curve technique for characterizing a small area of materials is lost. A simple solution for applying the $V(z)$ curve techniques to soft materials is to approximate them as a thin film coated on an isotropic substrate (e.g., sapphire, fused quartz, silica glass, or the like). In this case, the reflectance function can be obtained by using the theory of ultrasonic propagation in layered media.

Reflectance Function. When an acoustic beam is emitted from the acoustic lens onto the biological tissue mounted on the substrate in the culturing medium, the reflectance function is determined by the layered media comprising the culturing medium, the biological tissue, and the substrate. Waves traveling within the biological tissues behave very much like those in water. Hence, the boundary condition for this case (i.e., coupling medium–specimen–substrate) is considered in the same way as that of a liquid–liquid–solid [61].

When a longitudinal wave is incident from water onto the specimen waves in water, the biological tissue, and the substrate can be expressed by using potentials as follows:

Water (Medium I)

$$\phi^{I^-} = \Phi^{I^-} \exp(i\alpha^I z) \exp\{-i(sx - \omega t)\}$$

$$\phi^{I^+} = \Phi^{I^-} \exp(-i\alpha^I z) \exp\{-i(sx - \omega t)\}$$

Biological Specimen (Medium II)

$$\phi^{II^-} = \Phi^{II^-} \exp\{i\alpha^{II}(z-d)\} \exp\{-i(sx - \omega t)\}$$

$$\phi^{II^+} = \Phi^{II^-} \exp\{-i\alpha^{II}(z-d)\} \exp\{-i(sx - \omega t)\}$$

Substrate (Medium III)

$$\phi^{III^+} = \Phi^{III^+} \exp\{-i\alpha^{III}(z-d)\} \exp\{-i(sx - \omega t)\}$$

$$\psi^{III^+} = \Psi^{III^+} \exp\{-i\beta^{III}(z-d)\} \exp\{-i(sx - \omega t)\}$$

where ϕ is the potential of the longitudinal wave, ψ is the potential of the shear wave, Φ is the amplitude of the potential of the longitudinal wave, Ψ is the amplitude of the potential of the shear wave, the superscript "+" shows waves propagating in the positive direction of the z-axis, the superscript "−" shows waves propagating in the negative direction of the z-axis, and the Roman numerals *I*, *II*, and *III* indicate the media for wave propagation.

By Snell's law, we obtain the following equations:

$$k_L^I \sin\theta_L^I = k_L^{II} \sin\theta_L^{II} = k_L^{III} \sin\theta_L^{III} = k_s^{III} \sin\theta_s^{III} = s$$

$$\alpha = k_L \cos\theta_L \qquad \beta = k_s \cos\theta_s$$

where θ is an incident angle or a reflected angle, a subscript L indicates a longitudinal wave, and a subscript s indicates a shear wave.

The boundary conditions for the present case are the particle velocity in the direction of the z-axis and the continuity of stress. These are expressed as follows:

$$\sigma_z^I = \sigma_z^{II} \qquad \qquad v_z^I = v_z^{II} \quad \text{at } z = 0 \tag{8.7}$$

$$\sigma_z^{II} = \sigma_z^{III} \qquad \tau_{xz}^{II} = \tau_{xz}^{III} \qquad v_z^{II} = v_z^{III} \quad \text{at } z = 0 \tag{8.8}$$

From Equations (8.7) and (8.8), we obtain the sets of the first order of simultaneous equations relating to the potentials:

$$\begin{bmatrix} -\sigma_z \\ v_z/\omega \end{bmatrix}\bigg|_{z=d} = \begin{bmatrix} \lambda^I k_L^{I^2} & 0 \\ 0 & \alpha^I \end{bmatrix} \begin{bmatrix} \Phi^{I+} + \Phi^{I-} \\ \Phi^{I+} - \Phi^{I-} \end{bmatrix}$$

$$= \begin{bmatrix} \lambda^{II} k_L^{II^2} & 0 \\ 0 & \alpha^{II} \end{bmatrix} \begin{bmatrix} \Phi^{II+} + \Phi^{II-} \\ \Phi^{II+} - \Phi^{II-} \end{bmatrix} \tag{8.9}$$

$$\begin{bmatrix} -\sigma_z \\ v_z/\omega \end{bmatrix}\bigg|_{z=d} = \begin{bmatrix} \lambda^{II} k_L^{II^2} \cos\alpha^{II} d & i\lambda^{II} k_L^{II^2} \sin\alpha^{II} d \\ i\alpha^{II} \sin\alpha^{II} & d\alpha^{II} \cos\alpha^{II} d \end{bmatrix} \begin{bmatrix} \Phi^{II+} + \Phi^{II-} \\ \Phi^{II+} - \Phi^{II-} \end{bmatrix}$$

$$= \begin{bmatrix} \mu^{III}(\beta^{III^2} - s^2) & 2\mu^{III} s\beta^{III} \\ \alpha^{III} & s \end{bmatrix} \begin{bmatrix} \Phi^{III+} \\ \Phi^{III+} \end{bmatrix} \tag{8.10}$$

$$2s\alpha^{III}\Phi^{III+} + (s^2 - \beta^{III^2})\Psi^{III+} = 0 \tag{8.11}$$

From Equations (8.9)–(8.11), the reflectance function of the amplitude is expressed as follows:

$$\frac{\Phi^{I-}}{\Phi^{I+}} = \frac{(B_1 - \gamma B_2)\cos\alpha^{II}d + i(\gamma B_1 - B_2)\sin\alpha^{II}d}{(B_1 + \gamma B_2)\cos\alpha^{II}d - i(\gamma B_1 + B_2)\sin\alpha^{II}d} \tag{8.12}$$

$$\gamma = \frac{\rho^I \alpha^{II}}{\rho^{II} \alpha^I} \tag{8.13}$$

$$B_1 = \left(k_s^{III^2} - 2s^2\right)^2 + 4s^2 \alpha^{III} \beta^{III}$$

$$B_2 = \frac{\rho^{II} \alpha^{III}}{\rho^{III} \alpha^{II}} k_s^{III^4} \tag{8.14}$$

The $V(z)$ curve can be simulated after determining the reflectance function by Equations (8.12)–(8.14).

Using the reflectance function, the $V(z)$ curve for the thin soft material mounted onto the substrate can be simulated. In the simulation, only the velocity of the longitudinal wave of the soft material is set by estimation. The actual velocity of the longitudinal wave of the soft material can be obtained by matching the $V(z)$ curve obtained from the experiment by means of an iterative procedure.

Simulation of $V(z)$ Curve. The algorithm of the $V(z)$ curve simulation is described in the following steps. First, initialize the parameters of the acoustic lens, specimen (i.e., biological tissue), and substrate. Second, calculate the parameters of the acoustic field at the back focal plane, the pupil function of the lens, and the reflectance function. Third, calculate and draw the $V(z)$ curve. The simulation parameters of the acoustic lens are shown in Table 8.3. The velocity of the longitudinal wave of water was set as 1438 m s^{-1}. A human kidney (thickness 3 µm) was chosen as the soft material, and its longitudinal velocity was set as 1560 m s^{-1}. Fused quartz was chosen as the substrate, and its velocities of longitudinal and shear waves were 5970 and 3760 m s^{-1}, respectively.

Figure 8.15 is the result of the $V(z)$ curve simulation. It shows that the $V(z)$ curve for the kidney mounted on the fused quartz has different periods than the $V(z)$ curve for the fused quartz only. This means that the $V(z)$ curve technique can measure surface acoustic velocity of a thin biological specimen mounted on a fused quartz substrate.

Experiment of Velocity Measurement with $V(z)$ Curve. A human kidney was chosen as the biological tissue. The kidney was preserved in formalin, and then wrapped by paraffin. The wrapped specimen was sliced by a microtome. The thickness of the

Table 8.3 Parameters of acoustic lens.

Parameters	Conditions
Transducer	ZnO
Radius of transducer	383.00 µm
Frequency	400 MHz
Buffer rod	Al_2O_3
Aperture angle of lens	120°
Focal distance	577.52 µm

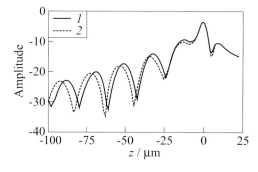

Figure 8.15 Simulation of $V(z)$ curve: 1 – fused quartz and 2 – kidney (3 µm) fused quartz [404].

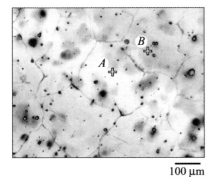

Figure 8.16 The acoustic image is the surface of the kidney (3 μm) fused quartz, and shows the positions where the longitudinal velocities were measured by the $V(z)$ curves technique. The image was formed with the frequency at 400 MHz [404].

100 μm

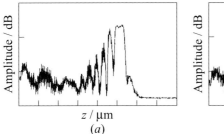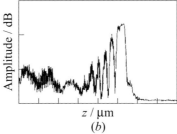

Figure 8.17 $V(z)$ curves of a kidney mounted on the fused quartz substrate: (a) point A and (b) point B.

sliced specimen was 3 μm. The sliced specimen was located on the fused quartz glass. The positions for applying the $V(z)$ curve technique were determined by the spherical acoustic lens (Figure 8.16).

Figures 8.17(a) and (b) are the $V(z)$ curves obtained at points indicated in Figure 8.16. From the $V(z)$ curves in Figures 8.17(a) and (b), the surface acoustic waves were measured as 3351 and 3277 m s^{-1}, respectively. Changing the values of the longitudinal wave velocity, the simulations were continued until the velocities of the surface acoustic waves coincided substantially with the experimental values. The longitudinal wave velocities were obtained as 1820 and 1765 m s^{-1}, respectively. These values are large compared to the longitudinal wave velocity obtained from the literature (1560 m s^{-1}).

Error Analysis. Assuming $1 \gg V_w/(2f\Delta z)$, Equation (8.6) is approximately expressed as follows:

$$V_R \cong \sqrt{V_w f \Delta z} \qquad (8.15)$$

In order to enhance the precision of the SAW velocity measurement by the $V(z)$ curve technique, both sides of (8.15) were differentiated after taking logarithms of both sides. After doing this, the following equation was obtained:

$$\frac{dV_R}{V_R} = \frac{1}{2}\frac{dV_w}{V_w} + \frac{1}{2}\frac{df}{f} + \frac{1}{2}\frac{d\Delta z}{\Delta z} \tag{8.16}$$

Equation (8.16) shows that the errors in the measurement of the SAW velocity are the sum of the errors in the values of the velocity of the coupling medium, the frequency of the acoustic wave, and the distance of the period. Therefore, in order to minimize the measurement error, it is necessary to maintain a constant temperature for the coupling medium to stabilize the frequency of the acoustic wave, and accurately measure the movement of the acoustic lens along the z-axis.

The frequency of the electric signal (i.e., tone-burst wave) generated by the transmitter was 400 MHz. The network analyzer (Agilent Technologies; model: 8753D) was used to confirm the stability of the frequency. The specimen was located at the bottom of the glass well containing the coupling medium (i.e., distilled water). The temperature of the coupling medium was maintained at 20 °C, and monitored by the thermocouple. The corresponding velocity of water (V_w) was 1438 m s^{-1}. The distance of movement of the acoustic lens along the z-axis was monitored. The error caused by the nonlinearity of the movement is statistically reduced by repeating the experiment. The V_{saw} value was obtained from the $V(z)$ curve through an FFT treatment.

The surface acoustic velocities of fused quartz obtained by the $V(z)$ curve in the experiments along with the $V(z)$ curves obtained by the simulation and literature were 3443, 3434, and 3430 m s^{-1}, respectively. Both experimental and simulated values are close enough for accuracy. Therefore, the difference in the longitudinal velocities obtained in the experiments and the simulations were considered to occur due to the conditions of the specimen.

8.3.7
Concluding Remarks

The following primary conclusions can be drawn:

(1) In-vivo images observed with a frequency of 0.4 and 1.0 GHz by SAM show the clear effects (i.e., from shrinkage to lift-off) of increasing temperature. The SAM images show that cells subjected to mechanical insult had little physical deformation or shrinkage and were damaged to the point of complete lift-off.
(2) Scanning acoustic microscopy was found to be a useful tool for visualizing injury to skin cells. Clear differences could be observed between thermal and shock wave insults.
(3) $V(z)$ curve for the kidney/fused quartz specimen was simulated by developing a software program. The procedure for obtaining the longitudinal wave velocity of a thin soft material (e.g., biological tissue) was developed by using the $V(z)$

curve technique. Longitudinal wave velocities of kidney tissue were obtained by the above procedure.

8.4
Methods for Tissue Preparation and Investigation

For as long as acoustic microscopy was a new method in biological studies, the development of techniques for exploring native biological tissues in transmission acoustic microscopy has been a top-priority task. The fixation of tissues used to fabricate histological preparations in optical studies interferes in the mechanical arrangement of tissues and cells. Therefore, it is nonfixed native preparations that are the basic biological objects which can be studied by acoustic microscopy.

A technique using thin frozen sections which allows the mechanical tissue structure to be retained and samples of uniform thickness to be fabricated, was chosen. Sections of thickness ranging from 5 to 20 µm were fabricated on a Leitz freezing microtome. To fabricate sections of a good quality, in which the structure inherent in this tissue type is retained, one should cool simultaneously both the knife and support used to fix the object. The temperatures which are optimal for section fabrication were ascertained. Dense tissues, such as sclera and skin, were cut at $T = -20\,°C$, and soft tissues were cut at $T = -10\,°C$.

Fabricated sections were placed in a special device we proposed for concurrent studies in optical and acoustic microscopes. The device consisted of a set of rings between which two-micron Mylar films, which absorb ultrasound only slightly, were stretched. Ultrasound attenuation in Mylar at 450 MHz amounts to $0.0186\,\mu m^{-1}$. The amplitude of a signal after its journey through a two-micron film equals $0.964 A_0$ (A_0 is the amplitude of an incident wave). The major energy is lost at wave reflection. The speed of sound in Mylar equals $2.54\,km\,s^{-1}$; Mylar density is $1.18\,g\,cm^{-3}$ [378]. The reflection coefficient is 0.51. The wave loses about 50% of its energy after it passes through two Mylar films. However, the use of Mylar films is necessary to save the samples and make them strictly plane-parallel. The tissue section to be studied is placed inside the orifice of an electron microscope blind in a drop of water or physiological solution, between two films. The blind is preliminarily glued to the bottom film. The blind was used to identify the tissue area of interest and to the scale images. The size of the inner blind orifice was 2×1 mm. The section studied was mounted so that some free space was left in the blind orifice to allow the ultrasound signal to be measured both in the sample and water. The Mylar films were thoroughly treated with 96 percent ethanol and dried. The object was examined under an optical microscope before studying it in the acoustic microscope in order to reveal its characteristic peculiarities and to compare acoustic and optical images. Only nonfixed and nonstained samples were investigated.

Water and a physiological solution were used in studies of biological objects by acoustic microscopy because the maximum signal amplitudes within them were almost identical. Water as an immersion medium exhibits a number of advantages:

namely, good wettability, high surface tension ($\sigma = 72.75$ dyne cm^{-1} at $T = 20\,°\mathrm{C}$), and low reactivity. The acoustic impedance of water is close to that of the majority of biological tissues; therefore, acoustic waves traverse the interface with virtually no reflection (solid tissues such as tooth and bone tissues are an exception).

The speed of sound in water differs only slightly from that in the majority of soft tissues, which reduces the aberrations caused by refraction of acoustic waves in the sample. Ultrasound attenuation in water at an operating frequency of the microscope of 450 MHz at room temperature ($20\,°\mathrm{C}$) amounted to 50.625 cm^{-1}.

8.5
Acoustic Properties of Biological Tissues and their Effect on the Image Contrast

Currently, the goal of measurements of attenuation and the velocities of ultrasound waves in biological objects concerns the acquisition of data interrelation between the acoustic parameters of tissues and their biological composition, structure, and state. Analysis of the viscoelastic properties of tissues at the microscopic level is progressively becoming essential. Therefore, one of the important fields with acoustic microscopy in biology is the investigation of the microstructure and local acoustic properties of biological tissues. However, unlike traditional optical microscopy in histologic investigations, acoustic microscopy is at an early stage in this area. Works in which the acoustic properties of biological tissues were studied at high frequencies are almost lacking, except for the studies performed recently by laser scanning microscopy [336, 405]. At the same time, a great number of works have dealt with studies on the peculiarities of sound absorption and also with measurements of the speed of sound in tissues and other biological objects at low frequencies (ranging from 1 to 10 MHz) [406, 407].

The data on the speed of sound and attenuation at a frequency of 1 MHz in various organs can be found, for example, in reviews [186, 408]. The acoustic parameters of some tissues are listed in Table 8.4. They will be used below for comparison with measurements made by the transmission microscopy technique.

Studies of the acoustic properties of tissues at low frequencies suggest that the speed of sound in soft tissues differs only slightly from its value in water and is virtually independent of frequency. Significant differences between the sound velocities in water and objects are observed only for solid biological tissues (bones) and tissues containing large concentrations of fibrillar proteins. The attenuation of ultrasound in tissues is much greater than that in water or many other liquids. Studies of the frequency dependence of the attenuation coefficients in tissues are extremely scarce. At the very beginning of acoustic microscopy development it was believed that the application of acoustic microscopes operating at frequencies in a high ultrasound range would permit high-contrast images of native tissues to be obtained. Measurements of local acoustic parameters in tissues would make it possible to interpret these images. This belief has been recently supported by [409, 410]. Thus, the high acoustic contrast of the majority of biological tissues stems from ultrasound attenuation within them, which is more intense than that in wa-

Table 8.4 Attenuation coefficient and speed of sound values in biological tissues [408].

Tissue	Attenuation coefficient at 1 MHz cm^{-1}	Speed of sound m s^{-1}
Liver	0.07–0.13	1553–1607
Myocardium	0.25–0.38	1570–1585
Kidney	0.09–0.13	1558–1568
Brain	0.09–0.15	1510–1565
Spleen	0.06	1520–1591
Skin	0.66–1	1498–2030
Bone	1–5	3360–4500
Water	0.0003	1500

ter, and from the presence of fibrillar albumen, the acoustic impedance of which is greater than that of the other tissue components.

Melanin [352, 409], the walls of vegetable cells consisting of a dense cellulose network [302, 410], and chitin contained in wings of various insects, were shown to give rise to the most contrast signatures of biological objects.

A great number of acoustic images were collected and analyzed by the team of researchers headed by Quate [302, 378]. Mostly fixed unstained tissues were studied. It is unknown how much fixation changes the high-frequency properties of biological tissues. This problem calls for special investigation. Low-frequency studies (in the 1- to 7-megahertz range) of acoustic parameters of tissues (attenuation coefficients, inverse scattering, and speed of sound) [328] have demonstrated that fixation in 4% formalin solution and 5% potassium dichromate solution changes these parameters only slightly (variations amount to a few percent of their values for the unfixed material); however, histochemical fixation in absolute alcohol causes large changes in the acoustic properties of tissues and, hence, their mechanical structure.

8.6
Investigation of Soft Tissue Sections

8.6.1
Skin

Researchers working in this area have paid great attention to studies of the microstructure of healthy tissues. Skin carries out extremely important biochemical functions in the human organism and possesses a complex mechanical microstructure. Plentiful material on microscopic skin structure has been accumulated [342], allowing one to perform a comparative analysis of acoustic and optical images.

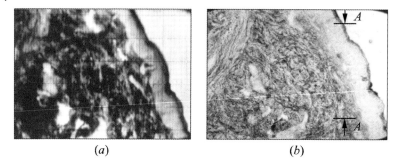

Figure 8.18 Acoustic (a) and optical (b) images of an unfixed, unstained section of human skin (7 μm thick).

The goal of the first period of investigation was to obtain high-quality images of thin (7 or 10 μm thick) native sections of human skin [410]. Figure 8.18 shows acoustic and optical images of the same unstained and unfixed skin section. The general similarity of the two images is noteworthy. Three layers can be clearly distinguished in the acoustic image; namely, a corneous epidermis layer seen as a dark strip on the skin surface and consisting of dense fibrillar keratin albumen; a layer of epithelial cells containing no fibrillar albumen, therefore being more transparent; and finally a skin layer, called the derma, of a very complicated structure which, most importantly, consists of a dense network of entangled collagen and elastin fibers immersed in a gel of intercellular fluid [342]. Acoustic images exhibit variations in the amplitude of the transmitted signal that arise due both to the difference in absorption of individual structures and to a difference in sound wave reflection and scatter at the interfaces. Areas weakly absorbing ultrasound; e.g., the epidermis and some derma spots occupied by the intertissue fluid, appear light in the microphotograph. The epidermis consists of cells 1 to 7 μm in size, with acoustic properties which differ only insignificantly (by 1% or 3%) from those of water. Individual cells are not discernible in the acoustic microphotographs. However, the section thickness could be too thin for keeping the cellular membranes intact. It is the scatter of ultrasound radiation at the cellular structures which is mostly responsible for attenuation of the acoustic waves in the epidermis. The sound reflection at the epidermis–water interface is negligibly small.

The corneous epidermis layer and some derma areas, in which ultrasound attenuation is low, appear dark in the acoustic image. The degree of darkening of individual derma areas in the microphotograph is proportional to the local collagen fiber density because domains containing more collagen absorb and reflect ultrasound more intensely. The fiber size in derma varies from 0.1 to 10 μm. Entangled fibers are oriented in different directions. The corneous layer is 15 μm thick. This layer consists of stacked keratinized cells (scales). From the point of view of acoustic properties, the corneous epidermis layer is a dense system intensely attenuating ultrasound, primarily due to its reflection. The coefficient of sound transmission through the corneous layer is $T = 0.3$. Depending on the area from which the skin sample is taken, the corneous layer thickness can vary from 10 to 20 µm. However, the transmission coefficient value in the corneous layer does not vary from section to section (0.3 ± 0.05). The width of epidermis at different skin areas varies from 100 to 200 µm, and, as in the corneous layer, the transmission coefficient in it remains invariable and equal to 0.8 ± 0.05. The drastic change in the transmitted signal amplitude between the epidermis and derma is caused by the basal membrane. The basal membrane thickness ranges between 0.1 and 1 µm and often is too small to be resolved in acoustic images. However, because of its very high density and close connection with the derma collagen, the boundary between the epidermis and derma is sharply outlined in acoustic images of normal skin. The linkage between the epidermis and derma deteriorates during skin pathology because of changes in the basal membrane structure and this results in a vanishing of the clear-cut boundary in acoustic images.

During the course of scanning, fibers which are going in different directions and which are of various density, and also the boundaries of fibers and the gaps between them, can fall within the ultrasound beam section. Measurements of local coefficients of sound transmission in derma make it possible to study the density and distribution of collagen in derma [304]. The transmission coefficient of derma amounts to 0.3–0.4. The attenuation coefficient of the skin derma at 450 MHz equals 1490 ± 200 cm^{-1}. Comparison of the attenuation coefficient measured in skin with its counterpart linearly extrapolated from the low-frequency region (1450 cm^{-1}) suggests that the frequency dependence of the attenuation coefficient at frequencies above 100 MHz is presumably linear.

Unfortunately, hair fell in the plane of some sections, which permitted some of its characteristics to be assessed. The average hair diameter was 50 µm, the transmission coefficient turned out to be 0.2, which is lower than its value in the corneous layer, also consisting of keratin. According to the data reported in [411], the longitudinal speed of sound in a hair is 3.84 km s^{-1}, the average hair density is 1.4 g cm^{-3}. The reflection coefficient of water is about 0.565 for normal wave incidence. Assessment of the ultrasound attenuation coefficient for a hair at a section thickness of 7 µm (although we acknowledge that the actual thickness of the hair section might be greater) yields 1800 cm^{-1}, which exceeds the average value of sound attenuation in derma.

To quantitatively characterize skin, we used the acoustic parameters of collagen measured at low frequencies. The longitudinal sound velocity in native collagen is equal to 1.73 km s^{-1}, its density is 1.25 g cm^{-3}, and the attenuation coefficient at a frequency of 1 MHz is 1.0 cm^{-1} [412].

8.6.2
Eye Sclera

Human sclera, with a structure more ordered than that of the skin tissue, was also investigated. Fibers 1 or 4 µm in diameter dominate in sclera. They are arranged in the form of plane bundles of a thickness ranging from between 10 and 20 µm and of width varying from 100 to 140 µm [342]. Every bundle contains from 7 to

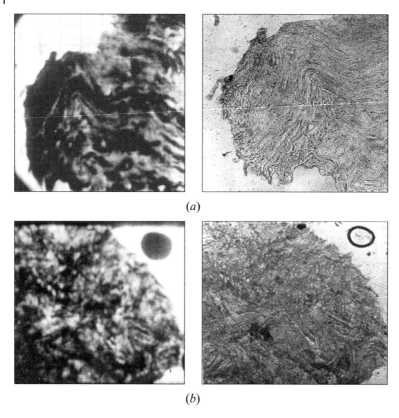

Figure 8.19 Acoustic (on the left) and optical (on the right) images of unfixed and unstained sections from the posterior zone (10 μm thick): (a) section normal to the surface, and (b) section parallel to the surface.

80 fibers in its cross-section. Fibers in the bundle are normally aligned in the same direction. The thickness of fibers, the density of their packing, their mutual orientation, and the size of fiber bundles depend on the age, sex, and position in the sclera. For this reason, the speed of sound varies significantly over the sclera thickness in different zones of a normal eye [413]: it is 1835 m s^{-1} in the anterior zone, 1714 m s^{-1} in the equatorial zone, and 1694 m s^{-1} in the posterior pole.

Sections cut from different zones parallel and perpendicular to the sclera surface were studied under a transmission acoustic microscope. Figure 8.19(a) shows acoustic and optical images of an unfixed and unstained 10-micron section of sclera from the posterior zone cut normally to the surface. Bundles of collagen fibers are clearly seen in the images. The longitudinal axes of nearly all fibers lie in the section plane. In some areas the fibers exhibit a wavy shape. The contrast of an acoustic image depends on the packing density of fibers in the bundle. The transmission coefficient of sclera for an ultrasound signal, incident normal to the fiber axis, varies from 0.2 to 0.4 for different sclera areas. The appropriate attenuation coefficient

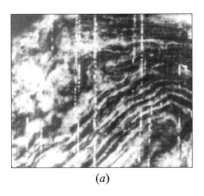 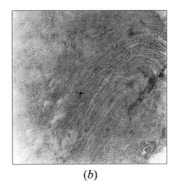

(a)   (b)

Figure 8.20 Acoustic (*a*) and optical (*b*) images of an unfixed and unstained section of sclera from the lateral zone (cut normally to the surface, 10 µm thick).

varies from 930 to 1630 cm^{-1}. Ultrasound is attenuated due mostly to scatter at the boundaries of individual fibers, the reflection coefficient amounts to 0.2.

Figure 8.19(*b*) displays optical and acoustic images of an unfixed and unstained 10-micron section of sclera from the posterior zone cut parallel to the sclera surface. Differently oriented fibers fall in the section plane; therefore, the acoustic image is inhomogeneous as scatter at the boundaries of individual fibers presumably plays an important part in imaging.

Figure 8.20 displays acoustic and optical images of an unfixed and unstained sclera section perpendicular to the surface from the lateral zone.

Resolution of the acoustic microscope permits individual fiber bundles to be distinguished. Closer to the sclera surface, the fibers are aligned in one direction. The regular arrangement of fibers is disturbed in deeper sclera layers.

8.6.3
Liver

This section discusses the results of studies on mouse liver in an acoustic microscope.

The tissue studied belongs to a category of tissues which moderately absorb ultrasound.

Liver sections 20 µm thick, fabricated with the aid of a freezing microtome at $T = -8\,°C$, were examined [410]. The liver structure is quite homogeneous, it consists of cells (hepatocytes) about 20 µm in size [342].

Acoustic and optical images of one and the same unstained and unfixed section of mouse liver are shown in Figure 8.21 (the scanning field is 700 × 700 µm). It should be noted that the images demonstrate general similarity and some common characteristic peculiarities. Dark capillaries and an air bubble are seen in the bottom left corner (the bubble is dark in the optical image and light in the acoustic image). Light large areas in the optical and acoustic microphotographs pertain to vessel gaps. Acoustic images reproduce variations in the intensity of a transmitted

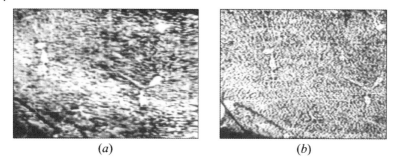

Figure 8.21 Acoustic (*a*) and optical (*b*) images of an unfixed and unstained section of mouse liver (20 µm thick).

ultrasound signal; therefore, light areas belong to sites with low sound attenuation, while dark areas correspond to sites intensely attenuating sound due to its reflection, absorption and scatter from the structural inhomogeneities. The speed of sound in liver is 1.575 km s^{-1} and the acoustic impedance is 1.65×10^5 g(s cm^2)$^{-1}$.

The coefficient of sound wave reflection at the water–liver interface equals 0.048 under the assumption that the sound speed is independent of the frequency, that is, only 0.002% of the wave energy is reflected. The transmission coefficient was virtually identical in various section areas and equal to 0.7 ± 0.03. The attenuation value in liver amounts 220 ± 40 cm^{-1} at a 450-megahertz frequency.

8.6.4
Cardiac Muscle

The cardiac muscle (myocardium) of a rat was the next object of investigation.

Thin (5 or 10 µm thick) unfixed sections were studied. Myocardium consists of cells called muscle fibers. The length of fibers ranges from 1 to 40 mm and their thickness can attain 0.1 mm [342]. Under an optical microscope, a myocardium tissue looks like a body of fibers separated by slit-like gaps filled with the intratissue fluid. Figure 8.22 shows acoustic and optical images of a 10-micron myocardium section. The muscle fibers lie in the section plane. It is worth mentioning that the acoustic images of myocardium sections cut both along and across the fibers can be of a low contrast. Gaps between the fibers can be seen fairly well. Muscle fibers differ only slightly in their acoustic properties from the liquid filling the interslit gaps, which is responsible for the poor contrast of images.

The coefficient of sound wave transmission through a section cut perpendicular to the axis of the muscle fibers is 0.9. The attenuation coefficient in the muscle across the fibers is 250 ± 40 cm^{-1}. The attenuation coefficient measured in a native section cut along the fibers differs only insignificantly from its counterpart measured in the section cut across the fibers. The attenuation coefficient assessed by linear extrapolation equals 171 cm^{-1}.

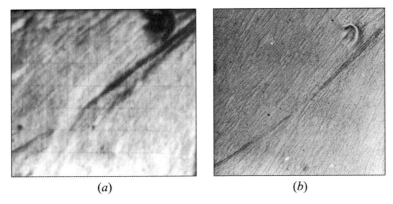

Figure 8.22 Acoustic (*a*) and optical (*b*) images of an unfixed and unstained section of rat myocardium (10 μm thick).

8.7
Investigation of Hard Mineralized Tissues

8.7.1
Bone Tissue and the Bone–Implant System

Bone tissue is a heterogeneous nonuniform material with a complicated spatial structure. On the molecular level, the organic bone component is represented by long collagen molecules packed in fascicles incorporated in collagen fibers, which, in turn, form plates known as lamellas. The mineral bone tissue component consists mostly of hydroxyapatite crystals. The collagen fibers and hydroxyapatite crystals are wrapped with a poorly structured basic substance consisting of mucopolysaccharides and glucoproteides.

Bone lamellas are the major components of the supramolecular bone structure. On the tissue level, the basic components of a compact bone are osteons pierced with Havers' system channels containing blood vessels; spongy bones consist of bone trabeculas.

Following the histological structure in the structural hierarchy, the anatomic structure of each particular bone is considered as an organ. Thus, the biomechanical properties of bone tissue are governed by its structure. The type of collagen fiber packing, which controls the density and orientation of the arrangement of mineral component crystals, makes the major contribution to the structure formation. It should be noted that other bone tissue components, including the cellular mass, vessels and inter-cellular material, also contribute to the mechanical properties of both the tissue and whole organ levels.

Information about various attempts to ascertain the state of bone tissue with the aid of acoustic microscopy is available in the literature. However, the investigations were performed at a low ultrasound signal frequency ranging from 25 to 50 MHz [414]. Such a low frequency did not permit the researchers to attain the resolution needed to observe interaction between an implant and bone tissue.

Figure 8.23 Acoustic signature of the compact material of a maxillary bone.

The authors of [414] have demonstrated, in 1988, the feasibility of observing inhomogeneities of the mechanical properties of a bone at the anatomical level. They used a 20-megahertz frequency microscope to study cross-sections of an unfixed tubular bone. The objective of the investigation was to analyze the whole set of successive steps needed to prepare and carry out correct measurements and to thereby assess, the possibility of using acoustic microscopy in studies of the histological structure of bone tissues in a wide frequency range. A bone tissue sample containing a coated porous implant (stainless steel 316L sputtered with a Co–Cr–Mo powder) was fabricated in the previous study, in which the implant was used to replace an artificial defect in a dog's bone [415]. The experiment demonstrated that a fibrous capsule arises around the implant. The capsule fixes the implant within the bone. The bone block containing the implant was prepared [416], dried, and enclosed in acryl. The sample was sawn in the implant area. The section plane was studied using an acoustic microscope.

Acoustic images were taken on an ELSAM (Leitz, Germany) microscope which allows the surface of an object to be scanned over a wide frequency range: 100, 200, 400 MHz, also 0.8 and 1.3–2.0 GHz. Acoustic images of the histological structure and finer structure units were obtained at frequencies of 200 MHz and 1.3 GHz.

8.7 Investigation of Hard Mineralized Tissues | 221

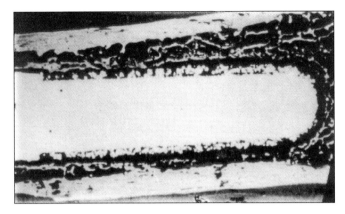

Figure 8.24 Acoustic signature of compact bone tissue with a metal implant. The frequency is 25 MHz and the scanning field size is 12×25 mm.

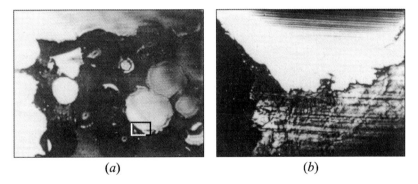

Figure 8.25 Acoustic image of a compact bone tissue in contact with a metal implant: (a) 200-megahertz frequency, image size 1.0×1.2 μm; and (b) the scanned area corresponds to the rectangle indicated in (a); frequency 1.6 GHz.

To ascertain the tissue structure, a set of acoustic images with successively increasing resolution was analyzed. Figure 8.23 demonstrates the acoustic signature of a section of compact bone tissue in contact with a polymer material in which the sample studied has been immersed for convenience of treatment. The scan length was about 10 mm. All the basic components of the bone tissue such as osteons, lacunas of bone cells, and central osteon channels are clearly discernible in the image. The laminated osteon structure formed by bone lamellas provides good contrast owing to the high density and regular collagen orientation.

Figure 8.24 shows the microstructure of the interface between the bone and the implant. The image illustrated in Figure 8.24 is fairly large in size, about 1×2 cm. The implant itself and the fine spherical grain-shaped coating formations on its surface are of a light color. Figure 8.25(b) displays a finer microstructure of the bone-implant interface in the vicinity of one of the metal grains; it is an enlarged image rectangular sector as indicated in Figure 8.25(a). The high resolution image was obtained at 1.6 GHz, the scan size is 100 × 100 μm. No tissue units with a regularly arranged structure were detected nearby the implant boundary. The newly formed bone tissue looks like an isotropic crystalline material.

8.7.2
Dental Tissue

Although attempts to use ultrasound in studies of dental tissues were first undertaken nearly 40 years ago, the results achieved in this field are still rather modest. This is primarily because of the peculiarities of the chemical composition and structure of a tooth as an organ.

Dental tissues are not only the hardest in the human body, they have the highest density impedance, reflection, and refraction coefficients. A tooth has a fairly small size and an irregular uneven surface; moreover, enamel and dentin are arranged in layers of dissimilar thickness and density in various tooth zones. Apart from longitudinal sound waves, surface and transverse acoustic waves make a significant contribution to the formation of the image contrast.

Figure 8.26 shows a longitudinal human tooth section schematically. The tooth portion above the gum, called the corona, is coated with enamel, the root is coated with cement. The tooth root is positioned in the recess of the alveolar jaw bone. The central part of the tooth is filled with pulp, which is a connective tissue pierced with nerve fibers, blood and lymph vessels. The root canal diameter varies from 0.1 to 1.5 mm. Enamel is the hardest tissue in the human organism. Its thickness on the masticatory surface is (1.5–3) mm, while on the lateral surfaces it is much thinner and vanishes toward the cement boundary. Enamel prisms 4–6 μm in diameter are the basic enamel component. The prism length corresponds to the enamel thickness and can even exceed it due to the curved shape of the enamel surface. The interprism enamel material consists of crystals identical to those constituting the prism, however their orientation is different. The organic enamel material has

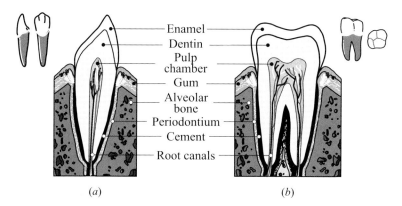

Figure 8.26 Schematic of tooth structure: (a) canine tooth and (b) molar tooth.

a very fine fibrillar structure. Crystals of various apatites (hydroxyapatite (75%); carbonapatite (12%); and chlorapatite (4.39%)) are assumed to be the major prism structure units. Inorganic compounds constitute 95% of the total enamel mass, the rest is water (3.8%), and organic material content amounts to only 1.2%.

Although softer than enamel, dentin is fairly hard compared to the other bone tissues and cement. It contains 70–72% organic substance and 28–30% water and organic components. Potassium phosphate (hydroxyapatite) is the basic component of the inorganic material. The basic dentin component is pierced with many dentin tubes, numbers ranging between 30 000 and 75 000 per 1 mm^2 of dentin. Dentin liquid, which provides the metabolism processes, circulates in the dentin tubes.

Cement is a tissue similar in its composition to normal bone tissue, consisting of 68% inorganic and 32% organic species. Unlike bones, cement contains no blood vessels.

Caries are one of the most widespread diseases of dental tissue. They demineralize the dental tissue and reduces its strength leading, thereby, to destruction of the tooth. Diagnosis of caries often presents difficulties. Visual examination or X-ray diagnostics are usually used. Unfortunately these methods are insufficiently sensitive; therefore, some affected areas may escape detection. Difficulties arise also in detecting caries in their early stages. It should be mentioned that the methods used at present do not allow examination of a tooth coated with materials which are opaque to X-rays, e.g., metal corona.

Only a few experiments in which ultrasound is applied to the examination of healthy and carious tooth tissue have been carried out so far. The first trial investigations of teeth by ultrasound microscopy were conducted in 1963 [417]. Based on the images obtained, the authors of that paper concluded that ultrasound visualization permits dentin and enamel to be distinguished and the pulp cavity to be revealed; however, the diagnosis of caries necessitates the use of ultrasound signals at higher frequencies. The authors found that the image quality depended on the time lapsed after tooth extraction.

Later, in 1966, Kossoff and Sharp published a paper [418], the objective of which was to develop instrumentation for the diagnostics of degenerate pulpitis. The contact technique was applied: one of the sample surfaces was polished to make a flat area to which a plane transducer was clamped. The sound velocity and impedance values were measured. Experiments were conducted with the aid of A-scanning at 18 MHz, both in the reflection and transmission operation modes. Whether or not a gas phase was present in the pulp was determined by whether the ultrasound signal passed through the pulp. A human tooth with cylindrical canals drilled into it was used as a test sample. The authors managed to discriminate between empty and water-filled canals. However, the results of measurements lacked stability and clarity because of the large transducer size and fairly long pulse duration.

Soon after the above publication, Barber and co-workers [419] carried out a more thorough investigation into the acoustic properties of various dental tissues. Table 8.5 compares the data from [419] with data reported by other authors.

Table 8.5 Speed of sound in enamel and dentin, m s^{-1}.

Tissue	Pulses at a 18-megahertz frequency [418]	Pulses at a 9-megahertz frequency [420]	Pulses of a 150-nanosecond duration [419]
Enamel	4500	6000	6100
Dentin	4000	4570	3900

Using their own data, Barber et al. calculated the transmission and reflection coefficients for ultrasound waves spreading in the enamel–dentin–pulp system. The authors of [419] decided not to use water as an immersion liquid and replaced it by a solid material. The motivation for their choice was that it was much easier to find a solid with a material acoustic impedance which was close to that of dental tissues. A significant difference in the impedance values limits the penetration of the acoustic waves through the immersion–enamel interface; in addition, it increases the number of repeated wave reflections in the thin enamel layer. A bismuth alloy rod served as an immersion, and was clamped to a pretreated enamel area. A small enamel zone was leveled and made flat and a droplet of castor oil was placed at the contact between the rod and enamel. Measurements were performed on a great number of human and bovine teeth. The possibility of distinguishing the enamel–dentin and dentin–pulp interfaces was demonstrated. The time resolution of the instrument was 75 ns. Rough estimates show that the depth resolution of the system amounts to 0.45 mm in enamel and 0.3 mm in dentin. The drawbacks of the method are obvious: the necessity to pretreat a test zone on the tooth surface, poor resolution along the direction parallel to the enamel surface, and the impossibility of scanning the surface. In 1986, Peck and Briggs performed [421] a comparative analysis of caries signatures taken with the aid of an optical polarization and acoustic microscopes. Plane-parallel sections were studied. The reflection scanning acoustic microscope operated at 380 MHz. The two microscopes disclosed a demineralized domain of approximately the same size and of similar shape. However, the acoustic images contained details and information that the optical polarization microscope and other techniques failed to provide.

Later on, in 1986, Peck et al. [422] used cylindrical lenses to measure the anisotropy of the Rayleigh velocity and Rayleigh absorption in enamel. The results demonstrated that the speed of sound is highest along the longitudinal tooth axis (3124 ± 10 m s^{-1}) and lowest in the transverse direction (2943 ± 10 m s^{-1}). The absorption changes more significantly: from 0.087 ± 0.05 to 0.381 ± 0.05 in the longitudinal and transverse directions, respectively. The anisotropy of the mechanical enamel properties is most likely to stem from the fact that hydroxyapatite crystals are packed into the so-called prisms aligned with the longitudinal tooth axis.

The present section discusses the results of studies of human teeth extracted for orthodontic reasons. Before examination, the teeth tested were held in a physiological solution with a small thymol additive. Semisections and plane-parallel sections

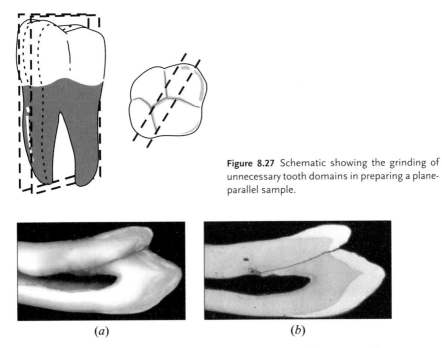

Figure 8.27 Schematic showing the grinding of unnecessary tooth domains in preparing a plane-parallel sample.

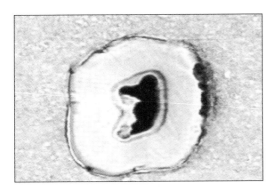

Figure 8.28 Optical (*a*) and acoustic (*b*) 50-megahertz images of a human premolar.

Figure 8.29 Acoustic image of a transverse section of a human tooth at 25 MHz.

were fabricated by grinding off unnecessary parts with a diamond grinding stone followed by polishing. A schematic of the fabricated section is illustrated in Figure 8.27. The procedure permitted plane-parallel plates 1 mm thick to be fabricated which were uneven to within no more than 2%.

Comparison of the optical and acoustic human tooth images (Figure 8.28) demonstrates that dentin and enamel, as well as the boundary of the pulp camera are distinguished to a great extent in the acoustic image. Areas with well expressed interglobular spaces are discernible in dentin.

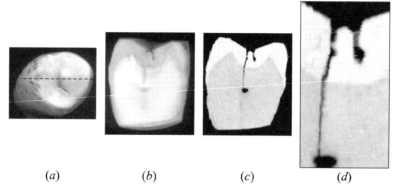

(a) (b) (c) (d)

Figure 8.30 Human premolar with fissure caries: (a) optical image seen from the masticatory surface side; (b) optical image of the longitudinal polished section plane; (c) appropriate acoustic image at 100 MHz; and (d) enlarged part of the image in the area affected by caries.

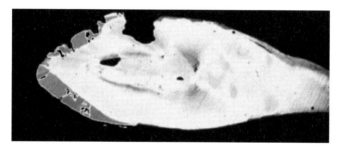

Figure 8.31 Acoustic image of the surface of the longitudinal section of an human incisor affected by deep caries, 50 MHz.

The shape and size of secondary dentin deposits can readily be assessed in the pulp camera cavity (Figure 8.29).

Figure 8.30 shows optical and acoustic images of fissure caries on a longitudinal polished section of a human molar. It can be seen that the acoustic image (Figure 8.30c and d) reveals more structural details than does the optical image; in particular, a darker zone around the caries cavity – the boundary of which quite definitely outlines the area of affected enamel with changing biomechanical properties – is clearly visible.

In teeth with deep caries (Figures 8.31 and 8.32), cavity boundaries can be clearly seen; moreover, the changes in the dentin state in the areas adjacent to the domains affected by caries can be assessed. Various intensity of changes in dentin are reproduced in the acoustic images by different gradations of the white-black scale. As can be seen in Figure 8.31, a so-called transparent dentin, which is denser than the healthy tissue, forms in the area adjacent to the caries cavity (blue-green colored area). At the same time, for the acoustic image of a tooth shown in Figure 8.32, sealing did not stop the spread of the pathological process. In fact, dentine destruc-

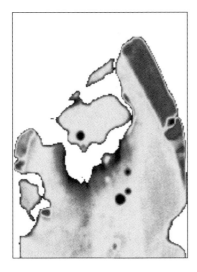

Figure 8.32 Acoustic image of the surface of the longitudinal section of a carious human tooth after unsuccessful healing, 50 MHz.

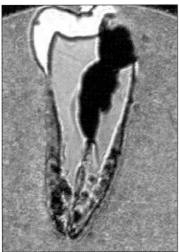

Figure 8.33 Acoustic image of the polished section of a human tooth immersed, before polishing, in diallylphthalate at 90 °C, 25 MHz.

tion was ongoing under the seal, which resulted in tissue softening. The softened tissue shows up in the image as bright red areas of various tints that pertain to zones with a lower intensity of the acoustic signal.

Figure 8.33 demonstrates that warming the tooth to 90 °C, concomitant with the procedure of its immersion in the resin, drastically affects the elastic mechanical properties of the dental tissue; therefore, the image quality has significantly deteriorated.

Investigations at higher frequencies allowed a great many microstructure details to be observed in acoustic images. Figure 8.34 displays nonuniform biomechanical properties of dentin that show up as variations in the intensity of the reflected acoustic signal in the bulk, cape dentin, and in the zone adjacent to the pulp. Alternating oblique striations pertaining to Günter–Schrader striations can be distin-

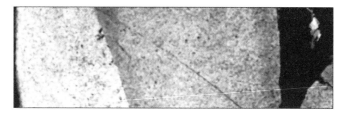

Figure 8.34 Enamel (left), dentin (center), and pulp cavity (right) under acoustic microscope, 4×1.3 mm. ELSAM acoustic microscope, 200 MHz.

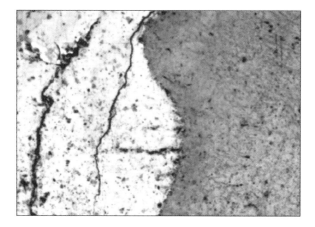

Figure 8.35 The enamel–dentin interface (acoustic image). ELSAM acoustic microscope, $f = 200$ MHz, $\Delta t = 10$ μm, image size 2×1.5 mm.

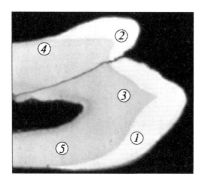

Figure 8.36 Illustration demonstrating the choice of points for measurements of local ultrasound velocity in dental tissue sections. The sample thickness $d = 1.93 \pm 0.03$ mm. Delays of the L and LT pulses were measured in enamel (spots 1 and 2) and in dentin (spots 3, 4, and 5).

guished in the enamel. These striations in longitudinal sections are known to show up, depending on what plane of the curved enamel prisms the section has been cut through. Lighter zones belong to prisms cut transversely while darker ones pertain to longitudinally cut prisms. The scalloped nature of the dentin–enamel joint can be clearly seen in Figure 8.35. This joint type provides better coupling between enamel and dentin under a mechanical load. Narrow dentinal canals going through

Table 8.6 Longitudinal (c_L) and transverse (c_T) sound speeds in enamel and dentin of a healthy and carious tooth measured in the A-scanning mode using the reflected signal.

Sample	c_L m s^{-1}	c_T m s^{-1}
Normal enamel	5900 ± 300	3200 ± 200
Normal dentin	3900 ± 200	1600 ± 100

Table 8.7 Reflection and transmission coefficients for normal incidence of sound waves.

Interface	R %	T %
Water–enamel	72	28
Enamel–dentin	15	85
Dentin–pulp	47	53
Water–silver	86	14
Silver–enamel	14	86

the dentin bulk normal to the dentin–enamel joint can be clearly seen in the top right-hand part of the image.

The ultrasound velocity in dental tissues was measured in plane-parallel sections of thickness ranging between 1.2 and 2 mm. One of these samples is illustrated in Figure 8.36. Measurements were performed in sections of seven healthy human premolars. The calculation results are summarized in Table 8.6.

The results of measurements listed in Table 8.6 are, on the whole, consistent with the literature. The ultrasound velocity in enamel of a healthy tooth is 1.5 times its value in dentin. Based on the ultrasound velocity values measured in enamel and dentin, and the tabulated density values, the specific acoustic impedance (Z) was assessed for these tissues: $1.8 \pm 0.2 \times 10^6$ g(cm^2 s)$^{-1}$ for enamel and $0.8 \pm 0.2 \times 10^6$ g(cm^2 s)$^{-1}$ for dentin. For comparison, the acoustic impedance values for water and silver are 0.15×10^6 g(cm^2 s)$^{-1}$ and 3.9×10^6 g(cm^2 s)$^{-1}$, respectively.

Based on the values obtained, we estimate the transmission and reflection coefficients for the laminated tooth structure. To do this, we consider the case of normal incidence of a plane wave on a flat interface.

The reflection coefficient R, that is the ratio between the intensities of the reflected and incident waves, is calculated by the following formula:

$$R = \left(\frac{Z_2 - Z_1}{Z_2 + Z_1}\right)^2 = \frac{r-1}{r+1}$$

where Z_1 and Z_2 are acoustic impedance values on different sides of the interface and $r = Z_2/Z_1$ is their ratio. The transmission coefficient $T = 1 - R$. The estimated values are listed in Table 8.7.

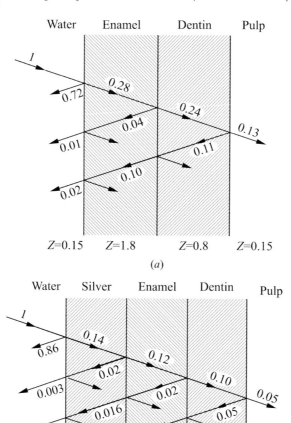

Figure 8.37 Relative intensities of a wave traveling in the water–enamel–dentin–pulp (a) and water–silver–enamel–dentin–pulp (b) laminated systems. The impedance Z values are in the 10^6 g(cm^2 s)$^{-1}$ units.

Assuming that ultrasound is not absorbed, we calculate the overall coefficients for a simple tooth model and a tooth under a silver seal (Figure 8.37). The intensity of an incident acoustic wave equals unity; the remaining intensities are expressed in its units.

The calculations were checked against the available experimental results of studies of a tooth sample with a plastic seal (Figure 8.38). Intense illumination allows

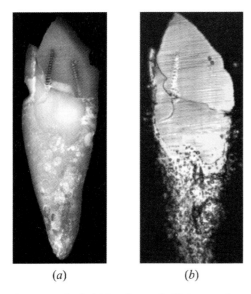

Figure 8.38 Optical (*a*) and acoustic (*b*) images of a tooth with a plastic seal and two metal drills in the dentin bulk; 100 MHz.

the shades of the two drills in the tooth bulk to be seen through the plastic seal; however, it provides no information about the depth of their submersion.

The acoustic image taken through the plastic seal in the zone around one of the drills shows only the drill, the plane of which is traversed by the C-scan. The depth of drill immersion can be assessed by calculating the time it takes for the reflected signal to come back.

Thus, the aforesaid results demonstrate that scanning with a focused ultrasound beam makes it possible to obtain acoustic images of the sample surface which reproduce the actual microstructure and distribution of individual structure units and tissues.

The measured longitudinal and transverse speeds of ultrasound in tooth tissues support the feasibility of a precise quantitative characterization of mineralized tissues and of a comparative assessment of their biomechanical properties.

Theoretical calculations and experimental data confirm the conclusion that acoustic microscopy is a unique tool for visualizing the structure of solid, opaque, multi-layer biological tissues – specifically dental tissue – under metal and plastic seals.

The data discussed in this section serve as a fundamental basis for developing new methods of clinical diagnosis in stomatology with the aid of acoustic microscopy.

8.8
Acoustic Properties of Collagen

Comparison of the acoustic images of different tissues confirms that the orientation of collagen fibers plays one of the most important roles in the formation of the mechanical properties of tissues on the microstructure level. The orientation of collagen fibers also significantly affects acoustic imaging. Therefore, investigations into the physical mechanisms, regarding how the arrangement of collagen fibrils affects the acoustic properties of a tissue containing fibrils, is a basic component and is believed to be very important in interpreting the acoustic images obtained.

8.8.1
The Effect of Collagen Anisotropy on Propagation of an Ultrasound Wave

Collagen is the basic structure component of connective tissues. The overwhelming majority of collagen in tissues is incorporated in collagen fibers. The average diameter of individual collagen fibers ranges from 0.5 to 10 µm. Each fiber consists of thin fibrils that, in turn, were formed by long molecules aligned in one direction. Researchers usually consider the models of tetragonal, hexagonal, cylindrical, and spiral-helix packing of molecules in fibrils [423, 424]. However, all the models assume that each individual fiber and the structures consisting of identically oriented fibers have hexagonal symmetry. Even pioneer studies, in which the speed of sound was measured along with other viscoelastic characteristics of collagen in tissue samples containing well oriented collagen fibers, disclosed an appreciable difference in the acoustic velocity values for waves propagating along and across fibers [412, 425].

Later on, the authors of [411] measured all the modulus-of-elasticity values (C_{11}, C_{12}, C_{33}, C_{13}, C_{44}) in clean, dry samples of a rat tail tendon by employing the Mandelshtam–Brillouin light-scattering method. The elastic collagen properties depend essentially on the water content in the tissue. If collagen is saturated with water, the longitudinal ultrasonic velocity drops drastically (by 30%–40%, from 3.64 to 2.64 km s^{-1} along the fiber and from 2.94 to 1.89 km s^{-1} across the fiber), the transverse waves vanish.

The data of [411] were used to theoretically predict the phase ultrasound velocities in oriented collagen fibers for a large variety of directions with respect to the longitudinal axis of an individual fiber (Table 8.8) [426]. The results are illustrated

Table 8.8 Density ρ and modulus of elasticity C of collagen.

Density ρ, g cm^{-3}	Modulus of elasticity 10^9 N m^{-2}				
	C_{11}	C_{12}	C_{33}	C_{13}	C_{44}
1.35	11.7	5.1	17.9	7.1	3.3

8.8 Acoustic Properties of Collagen

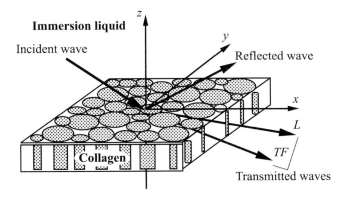

Figure 8.39 Schematic of wave reflection at the collagen–liquid interface [66, 426].

in Figure 8.39. Here, the z-axis is parallel to the long axis of the collagen fiber, the ultrasonic wave spreads at an angle θ with respect to the z-axis and the x-axis lies in the plane of ultrasonic wave propagation.

The three acoustic modes differ in an oriented, packed collagen. Two wave modes propagate in conformity with the polarization vector, aligned with the wave propagation direction; they are quasi-longitudinal waves (L) and fast transverse waves (TF). The angular dependence of the velocity for the quasi-longitudinal mode reads:

$$c_L = \frac{1}{\sqrt{2\rho}} \{(C_{11} \sin^2\theta + C_{33} \cos^2\theta + C_{44}) + [\{(C_{11} - C_{44}) \sin^2\theta \\ + (C_{44} - C_{33}) \cos^2\theta\}^2 + 4(C_{13} + C_{44})^2 \cos^2\theta \sin^2\theta]^{1/2}\}^{1/2} \quad (8.17)$$

The polarization vector $\vec{e}_L$ is specified by three components:

$$\vec{e}_L = \frac{1}{B} \left\{ \sin\theta \quad 0 \quad \frac{C_{11} \sin^2\theta + C_{44} \cos^2\theta - \rho c_L^2}{(C_{13} + C_{44}) \cos\theta} \right\} \quad (8.18)$$

where

$$B = \sqrt{\sin^2\theta + \left(\frac{C_{11} \sin^2\theta + C_{44} \cos^2\theta - \rho c_L^2}{(C_{13} + C_{44}) \cos\theta}\right)^2}$$

is a normalizing factor depending on the direction of the wave propagation vector. A similar formula describes the angular dependence of the phase velocity for the quasi-transverse mode. The angular dependence of the phase velocity is derived by changing the sign of the square-root in (8.17):

$$c_L = \frac{1}{\sqrt{2\rho}} \{(C_{11} \sin^2\theta C_{33} \cos^2\theta + C_{44}) - [\{(C_{11} - C_{44}) \sin^2\theta \\ + (C_{44} - C_{33}) \cos^2\theta\}^2 + 4(C_{13} + C_{44})^2 \cos^2\theta \sin^2\theta]^{1/2}\}^{1/2}$$

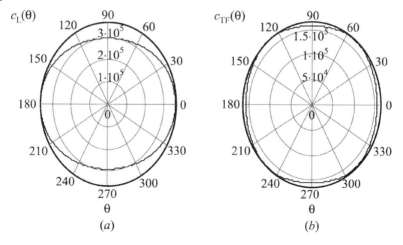

Figure 8.40 Angular dependence of the phase velocities in a dried oriented collagen: (a) quasi-transverse waves (L-mode); and (b) fast quasi-transverse waves (TF-mode) [66, 426].

The polarization vector for the TF mode is derived from Equation (8.18) by changing subscript TF to L. Given the polarization direction of e_L, the e_{TF} vector is also readily determined. The two vectors are orthogonal and lie within the wave propagation plane.

The third mode (TS) is a purely transverse wave, the polarization vector of which is perpendicular to the plane in which the z-axis and propagation direction lie:

$$\vec{e}_{TS} = \{1, 0, 1\}$$

The angular dependence of the phase velocity for the slow transverse mode is

$$c_{TS} = \sqrt{\frac{1}{2\rho}\{(C_{11} - C_{12})\sin^2\theta + C_{44}\cos^2\theta\}}$$

The calculated diagrams of the angular dependence of the L and TF wave velocities are illustrated in Figure 8.40. The maximum and minimum values of the velocities and the c_{min}/c_{max} ratio characterizing the anisotropy of each acoustic mode are listed in Table 8.9.

The longitudinal mode has the highest phase velocity along the fibrils ($c_L^{\parallel} = 3.64 \times 10^5$ cm s^{-1}) and the lowest velocity across the fibrils ($c_L^{\perp} = 2.94 \times 10^5$ cm s^{-1}), its anisotropy coefficient is 0.8. The angle β_L specifying the deviation of vector e_L from the propagation direction of the basic wave is at a maximum: $\varphi_{L\,max} = 8.3°$, near $\theta \approx 40°$. The phase velocity of quasi-transverse waves peaks at about $\theta \approx 49°$ ($c_{TF}^{\theta \approx 49°} = 1.64 \times 10^5$ cm s^{-1}); the lowest velocity corresponds to the same angular positions in the diagram for waves spreading both along and across fibrils ($c_{TF}^{\parallel} = c_{TF}^{\perp} = 1.56 \times 10^5$ cm s^{-1}). The phase velocity of a totally transverse wave (TS-mode) is independent of the propagation direction, in accordance with

Table 8.9 Properties of the longitudinal and fast quasi-transverse waves in a dried oriented collagen [66, 426].

	Longitudinal waves
Maximum velocity $c_{L\,max}$	$c_L^{\parallel} = 3.64 \times 10^5$ cm s^{-1} (spreads along the fibrils)
Minimum velocity $c_{L\,min}$	$c_L^{\perp} = 2.94 \times 10^5$ cm s^{-1} (spreads across the fibrils)
Anisotropy coefficient $K = c_{L\,min}/c_{L\,max}$	$K = 0.8$

	Transverse waves
Maximum velocity $c_{TF\,max}$	$c_{TF}^{\parallel} = 1.64 \times 10^5$ cm s^{-1} (spreads at an angle $\theta \approx 49°$)
Minimum velocity $c_{TF\,min}$	$c_{TF}^{\perp} = 1.56 \times 10^5$ cm s^{-1} (spreads along and across the fibrils)
Anisotropy coefficient $K = c_{L\,min}/c_{L\,max}$	$K = 0.95$

the equation $(C_{11} - C_{12})/2 = C_{44}$ that was deduced from experimental data [411] ($c_{TS} = 1.56 \times 10^5$ cm s^{-1}).

The anisotropy of a material is most conveniently characterized by the slowness surfaces (surfaces of reciprocal sonic velocities $1/c_L$, $1/c_{TF}$, and $1/c_{TS}$, or equivalent surfaces of wave vectors). Propagation (concentration) of the energy depends on the topology of the surfaces, namely, their smoothness, the presence of steady points, and plane areas [427].

For each wave vector, the normal to the surface at every appropriate point specifies the direction of the beam (group) velocity, i.e., the direction along which a group of waves belonging to a particular mode with a given wave number spreads. The beam velocity direction in an anisotropic medium does not coincide with the direction of the appropriate wave vector. Moreover, for a certain rank of wave vectors, the beam velocity direction can be dictated by the appropriate wave vector; the energy of many waves finding themselves in the angular spectrum, spreads along a certain direction. As a result, the energy of a pulse incident on a surface is redistributed in the bulk; thus, one can only indicate directions of the predominant spread (concentration) of the acoustic energy. Similar energy concentration is observed for wave vectors pertaining to the flat areas of the slowness surface.

Based on calculations, we attempted to reconstruct the slowness surfaces for the L and TF modes (Figure 8.41). The slowness surface for the TS-mode is a normal sphere of radius $1/c_{TS}$. There are zones with a reduced curvature in the surfaces plotted. However, a considerable redistribution of the acoustic energy of an incident focused beam is atypical for collagen.

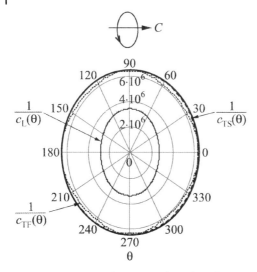

Figure 8.41 Acoustic slowness surfaces for collagen. Sections of the surface in the plane that goes through the axis of symmetry C. The surfaces proper arise as a result of the curve rotation around the C-axis.

The maximum deviation of the energy flux from the normal to the wave front amounts to about 13° (at $\theta \cong 49°$) for the quasi-longitudinal mode and to 6.5° (at $\theta \cong 30°$ and 70°) for the quasi-transverse mode. Thus, angle $\theta \cong 49°$ pertains to the direction of the basic mode for quasi-transverse waves: the phase and beam velocities are identical at this angle. This point in the diagram is the center of incipience of the acoustic energy concentration phenomenon; if the anisotropy increases (e.g., when the concentration of hydroxyapatite crystals in bone plates increases), the flat areas arise near the propagation angle.

Another important line of our studies is a theoretical analysis of the angular dependence of the reflection coefficient R at the collagen–liquid interface. It is the angular dependence of R which governs the peculiarities of output signal formation in the microscope and the mechanism of acoustic contrast.

The angular dependence of R is controlled by the sample surface orientation with respect to the fiber axis. Most frequently, the incident wave in a sample excites all three acoustic modes. In the case of a particular orientation, namely, when the liquid–collagen interface is perpendicular to the C-axis, only two modes (L and TF) participate in the formation of the transmitted and reflected waves at the interface. The angular dependence of R for neat collagen changes very little as the sample orientation varies because the anisotropy in collagen is not prominent. As the anisotropy increases (e.g., in solid tissues where collagen fibers have a certain orientation), discrepancies in R magnitudes evaluated at various sample orientations can be significant. We performed the appropriate calculations for a sample cross-section which comprises only transverse sections of collagen fibers. The refraction angles (more precisely, normal components of the wave vectors for waves being refracted) were calculated using the dispersion equations and then substituted into the formulas for amplitudes A_R, A_L, and A_{TF} of the reflected and refracted waves, in conformity to the boundary conditions. This allowed us to determine the value of the normalized amplitude of the reflected wave; in particular, the reflection coefficient R.

We also calculated [307, 426] the angular dependence of the so-called energy reflection and refraction coefficients V and W_L and W_{TF}, which are the ratios between the energy fluxes P_{refl}, P_L, and P_{TF} that spread together with the interface to the energy flux carried with the incident wave:

$$V = \frac{P_{refl}}{P_{inc}} = \left|\frac{A_R}{A_0}\right|^2 = |R|^2$$

$$W_L = \frac{P_L}{P_{inc}} = \frac{\rho}{\rho_0} \frac{c_{gz}^L}{c_0 \cos\theta} \left|\frac{A_L}{A_0}\right|^2$$

$$W_{TF} = \frac{P_{TF}}{P_{inc}} = \frac{\rho}{\rho_0} \frac{c_{gz}^{TF}}{c_0 \cos\theta} \left|\frac{A_{TF}}{A_0}\right|^2$$

where A_0 is the incident wave amplitude; A_R is the amplitude of the reflected wave in a liquid; A_L and A_{TF} are the amplitudes refracted in collagen; ρ_0 and c_0 are the density and speed of sound in the liquid; ρ is the collagen density; c_g^L and c_g^{TF} are the beam velocities of the acoustic modes in collagen; and c_{gz}^a ($a = $ L, TF) are their components perpendicular to the interface. The energy coefficients are calculated using the energy conservation law:

$$V + W_L + W_{TF} = 1$$

Finally, the angular dependence of the amplitude and energy coefficient was for three diverse immersion media applied in acoustic microscopy; namely, water, isopentane, and mercury. The acoustic properties of the liquids used as immersion fluids differ; therefore, the mechanisms of contrast formation for the same sample are dissimilar. Water is the standard and most popular immersion liquid used in acoustic microscopy. Isopentane has a low speed of sound, therefore, is best suited for excitation of surface Rayleigh waves. Mercury is an example of a heavy immersion liquid which emphasises the effect on image formation based on the inverse reflection of the energy carried by lateral waves which are incident on the sample surface. This can create a unique opportunity to form high-contrast acoustical images in the case of ultrasound beam reflection at the liquid mercury–collagen interface.

There are two critical angles in the collagen–water system; namely, the longitudinal ($\theta_L = \arcsin(c_0/c_L) = 30.6°$) and the transverse ($\theta_T = \arcsin(c_0/c_{TF}) = 73.1°$). The latter angle in this case is too large to fit into the lens aperture. Quasi-longitudinal waves are readily excited in collagen at angles ranging from $0°$ to $\theta_L \approx 30°$. Reflection in this angle range is also significant, as about 30% of the energy incident on the surface is reflected back into water and can be detected. At angles exceeding the first critical angle, $\theta > \theta_L$, a large fraction (more than 80%) of the incident energy is transformed into quasi-transverse waves. Thus, rays constituting a focused acoustic beam going at angles close to θ_L make a fundamental contribution to the formation of the output signal in an acoustic microscope.

Fast changes in the phase of the reflection coefficient near the critical angles provide evidence of a considerable contribution of some surface wave types to the output signal of the microscope. The range of variations in the phase φ of coefficient R in the water–collagen system is fairly small: $\Delta\varphi \approx 50°$ at $\theta = \theta_L$. Therefore, the contribution of lateral waves to the output signal can be assumed insignificant.

The Rayleigh angle $\theta_R \approx \theta_T$ is very large; therefore, Rayleigh waves affect the imaging process only slightly. Thus, when water is used as an immersion liquid, it is the direct reflection of the focused beam incident on the surface which makes the dominant contribution to the formation of the output signal in an acoustic microscope.

The values of the two critical angles are small when the speed of sound in immersion materials is low. In pentane ($c_0 = 1.016 \times 10^5$ cm s^{-1}, $\rho_0 = 1.0$ g cm^{-3}), they can both fit the aperture of wide-aperture lenses $\theta_L = \arcsin(c_0/c_L) = 21°$ and $\theta_T = \arcsin(c_0/c_{TF}) = 41°$. The overall reflection efficiency is higher when water serves as an immersion liquid: about 60% of the energy incident on the surface is reflected back into the liquid. The phase of the reflection coefficient changes significantly within a narrow range in the vicinity of the critical angle θ_T: $\Delta\varphi \approx 280°$. Owing to this, the surface Rayleigh waves dominate in acoustic imaging when wide-aperture lenses are used.

The critical angle in the mercury–collagen system is nearly identical to that in the water–collagen system. The surface Rayleigh waves contribute only slightly to the output signal formation. The Rayleigh angle goes beyond the aperture angle, and the probability of the excitation of Rayleigh waves is very low because of the high density of mercury. At the same time, excitation of the lateral waves comes into play: near the critical angle $\theta = \theta_L$, the phase value drops drastically. Thus, when mercury serves as an immersion liquid, the output signal results from interference of the direct reflected signal and lateral waves reflected back into the liquid.

8.8.2
Experimental Investigation into Acoustic Properties of an Isolated Collagen Thread

Unidirectional collagen fibers from a rat-tail tendon were studied using a transmission acoustic microscope operating at 450 MHz. The collagen fibers from the rat-tail tendon, assembled in bundles 50–150 μm in diameter, are aligned strictly with the long tendon axis. An individual bundle of fibers appears as a uniform dark strip in an acoustic image. The coefficient of ultrasound transmission through a cross-section of a collagen fiber bundle is very low ($\sim 8\%$). In fabricating a cross-section of homogeneous thickness, a fiber bundle was cut on a cryotome at 20 °C. Fresh, dried, 4–20 μm thick sections were used in measurements. To obtain acoustic images of living tissue, the samples were placed in a physiological solution that was used as an immersion medium. The major acoustic parameters were measured in the $A(z)$ regime.

This method is based on measuring the confocal focus position in a two-lens system. The sample was placed between the lenses. The output signal A is measured as a function of distance z between the lenses (see Figure 1.4). The $A(z)$ curves exhibit a single principal maximum corresponding precisely to the confocal position of the lenses. The curves decay on both sides of the maximum because of the conventional interference oscillations. When an object is placed between two lenses, the confocal configuration of the system changes because of the difference between the refraction coefficients. This effect is similar to the shift of the focus of a convergent light beam passing through a flat plate. The focus shift Δz is deduced from the $A(z)$ curves measured for two positions with and without the sample.

The measured difference between the peak values is used to assess ultrasound absorption in the sample. This ratio is approximate for actual wide-aperture signals. The appropriate theoretical substantiation of the method can be found in [125].

Using this method, we assessed the longitudinal velocity and absorption in a direction perpendicular to the collagen fiber axis in fresh, dried tendons. The speed of sound measured in fresh collagen fibers ($\rho = 1.25 \text{ g cm}^{-3}$) is $c_L^\perp = (1.8 \pm$

Table 8.10 Sonic velocities in oriented collagen fibers evaluated by various authors: $c_L^\parallel$ is the longitudinal velocity of an ultrasound wave spreading along the long fiber axis, and $c_L^\perp$ is the longitudinal velocity of an ultrasound wave spreading normal to the long fiber axis.

Speed of sound km s^{-1}	Source	Measurement technique	Experimental object
$c_L^\parallel = 2.11$	[425]	Scanning laser acoustic microscope, $f = 100$ MHz	Tendons from rat tail
$c_L^\perp = 1.73$	[412]	The same	The same
$c_L^\parallel = 2.64$, $c_L^\perp = 1.89$	[411]	Brillouin light scattering, hypersonic frequencies, $f = 10$ GHz	Fresh fibers from rat tail
$c_L^\parallel = 3.64$, $c_L^\perp = 2.94$	[411]	The same	Dried fibers from rat tail
$c_L^\parallel = 1.82$, $c_L^\perp = 1.69$	[428]	Ultrasound pulse method, $f = 10$ MHz	Collagen from femoral bone
$c_L^\perp = 1.8$	Our result	Transmission acoustic microscope, $f = 450$ MHz	Fresh fibers from rat tail
$c_L^\perp = 2.8$	Our result, [426]	Transmission acoustic microscope, $f = 450$ MHz	Dried fibers from rat tail

$0.08) \times 10^5$ cm s^{-1}, the absorption coefficient measured at 450 MHz is $\alpha = (1270 \pm 120)$ cm^{-1}.

The sound speed in a dried collagen sample ($\rho = 1.35$ g cm^{-3}) was $c_L^\perp = (2.8 \pm 0.2) \times 10^5$ cm s^{-1}. We performed additional studies to show that absorption in dry collagen was lower than in the fresh one: $\alpha = (890 \pm 90)$ cm^{-1}.

Our data are summarized together with the data reported by other authors in Table 8.10.

References

1. Sokolov, S. Ya. 1936, 31 Aug. Description of a device for detecting inhomogeneities in solid, liquid and gaseous media with the aid of ultrasound oscillations. Invention certificate 49426. USSR.
2. Sokoloff, S. 1937. Improvements in and relating to the detection of faults in solid, liquid or gaseous bodies. Patent No. 477,139 GB.
3. Sokoloff, S. 1937. Means for indicating flaws in materials. Patent No. 2.164.125 USA.
4. Sokolov, S. Ya. 1949. Ultrasound microscope. *Dokl. Akad. Nauk SSSR* 64:333–336.
5. Ludwig, G. D., and F. W. Struther. 1947. Ultrasound in medicine diagnostics. *Naval Medical Res. Inst. Project NM 004:001.* 4:1–23.
6. White, D. N., and R. E. Brown. 1977. *Ultrasound in medicine*. N.Y.: Plenum Press.
7. Baranskii, K. N. 1957. Physical grounds of ultrasound generation in semi-conducting films. *Dokl. Akad. Nauk SSSR* 114:517.
8. Dransfeld, K., and H. Botmel. 1958. Generating, receiving and attenuation of surface acoustic waves. *Phys. Rev. Lett.* 1:234–237.
9. Kessler, L. M. 1974. Review of progress and application in acoustic microscopy. *J. Acoust. Soc. Amer.* 55:909–918.
10. Lemons, R. A., and C. F. Quate. 1974. Acoustic microscope-scanning version. *Appl. Phys. Lett.* 24(2):163–165.
11. Krasilnikov, V. A., and L. A. Zarembo. 1966. *Introduction in nonlinear acoustics*. Moscow: Nauka Publ.
12. Auld, B. 1973. *Acoustic fields and waves in solids*. Vols. 1–2. N.Y.: J. Wiley.
13. Isakovich, M. A. 1973. *General acoustics*. Moscow: Nauka Publ.
14. Lependin, L. F. 1978. *Acoustics*. Moscow: Vysshaya Shkola.
15. Shutilov, V. A. 1980. *Fundamentals of ultrasound physics*. Leningrad: Izd. LGU.
16. Krasilnikov, V. A., and V. V. Krylov. 1984. *Introduction in physical acoustics*. Moscow: Nauka Publ.
17. Thurston, R. N. 1965. Wave propagation in fluids and normal solids. In: *Physical acoustics. Principles and methods.* Ed. W. P. Mason. N.Y.–London: Academic Press. Vol. 1. Part A. 1–110.
18. Sirotin, Yu. I., and M. P. Shaskolskaya. 1975. *Fundamentals of crystal physics*. Ch. VI. Moscow: Nauka Publ.
19. Nadeau, G. 1964. In: *Introduction in elasticity*. N.Y.: Holst, Rinehart & Winston. 24–54.
20. Landau, L. D., and E. M. Lifshits. 1987. *Theory of elasticity.* Ch. III. Moscow: Nauka Publ.
21. Baranskii, K. N. 1991. *Physical acoustics of crystals*. Moscow: Izd. MGU.
22. Starunov, V. S., E. V. Tiganov, and I. L. Fabelinskii. 1967. *Pisma v ZhTF* 5:317.
23. Baranskii, K. N., G. A. Sever, and T. S. Velichkina 1971. *Pisma v ZhTF* 13:52.
24. Taker, J., and V. Rampton. 1971. *Hyper-sound in solids*. N.Y.–London: Academic Press.

25 Mikhailov, I. G., V. A. Soloviev, and Yu. P. Syrnikov. 1964. *Basics of molecular acoustics*. Moscow: Khimiya Publ.

26 Klemens, P. G. 1965. Effect of thermal and phonon processes on ultrasound attenuation. In: *Physical acoustics. Principles and methods*. Ed. W. P. Mason. N.Y.–London: Academic Press. Vol. 3. Part B. 201–234.

27 Litovitz, T., and K. Devis. 1965. Thermal relaxation in liquids. In: *Physical acoustics. Principles and methods*. Ed. W. P. Mason. N.Y.–London: Academic Press. Vol. 2. Part A. 281–349.

28 Lamb, J. 1965. Thermal relaxation in liquids. In: *Physical acoustics. Principles and methods*. Ed. W. P. Mason. N.Y.–London: Academic Press. Vol. 2. Part A. 203–280.

29 Liebermann, L. N. 1965. Resonance absorbtion. In: *Physical acoustics. Principles and methods*. Ed. W. P. Mason. N.Y.–London: Academic Press. Vol. 4. Part A. 183–193.

30 Truell, R., C. Elbraum, and B. B. Chick. 1969. *Ultrasonic methods in solid state physics*. N.Y.–London: Academic Press.

31 Papadakis, E. P. 1965. Ultrasonic attenuation caused by scattering in polycrystalline media. In: *Physical acoustics*. Eds. W. P. Mason and D. Thurston. N.Y.: Academic Press. Vol. 4. 269–328.

32 Granato, A. V., and K. Lücke. 1965. The vibrating string model of dislocation damping. In: *Physical acoustics. Principles and methods*. Ed. W. P. Mason. N.Y.–London: Academic Press. Vol. 4. Part A. 226–276.

33 Mason, W. P., ed. 1964. *Physical acoustics. Principles and methods*. N.Y.–London: Academic Press. Vol. 3. Part A.

34 Gurevich, V. L. 1968. Theory of acoustic properties of piezoelectric semiconductors. *Fiz. Tekhn. Poluprovodnikov* 2:1557–1592.

35 McFee, J. H. 1965. Transmission and amplification of acoustic waves in piezoelectric semiconductors. In: *Physical acoustics. Principles and methods*. Ed. W. P. Mason. N.Y.–London: Academic Press. Vol. 4. Part A. 1–45.

36 Pustovoit, V. I. 1969. Interaction of electron fluxes with elastic waves. *Usp. Fiz. Nauk* 97(2):257–306.

37 Galperin, Yu. M., and V. L. Gurevich. 1978. *Acoustoelectronics of semiconductors and metals*. Moscow: Nauka Publ.

38 LeCraw, R. C., and R. L. Comstock. 1965. Magnetoelastic interactions in ferromagnetic insulators. In: *Physical acoustics. Principles and methods*. Ed. W. P. Mason. N.Y.–London: Academic Press. Vol. 3. Part B. 127–199.

39 Bolef, D. I. 1965. Interaction of acoustic waves with nuclear spins in solids. In: *Physical acoustics. Principles and methods*. Ed. W. P. Mason. N.Y.–London: Academic Press. Vol. 4. Part A. 113–182.

40 Gurevich, M. A., ed. 1977. *Magnetic quantum acoustics. Collection of papers*. Moscow: Nauka Publ.

41 Brekhovskikh, L. M. 1973. *Waves in stratified media*. Moscow: Nauka Publ. (Translation from Russian: Brekhovskikh, L. M. 1980. *Waves in layered media*. 2nd ed. N.Y.–London: Academic Press.)

42 Dransfeld, K., and E. Saltzmann. 1970. Excitation, detection, and attenuation of high-frequency elastic surface waves. In: *Physical acoustics*. Eds. W. P. Mason and R. N. Thurston. N.Y.: Academic Press. Vol. 7. 219–272.

43 Farnell, G. W. 1970. Properties of Elastic Surface Waves. In: *Physical acoustics*. Eds. W. P. Mason and R. N. Thurston. N.Y.: Academic Press. Vol. 6.

44 Farnell, W., and L. L. Adler. 1974. Elastic wave propagation in thin layers. In: *Physical acoustics*. Eds. W. P. Mason and R. N. Thurston. N.Y.: Academic Press. Vol. IX. 35–128.

45 Neubauer, W. G. 1975. Observation of acoustic radiation from plane and curved surfaces. In: *Physical acoustics*. Eds. W. P. Mason and R. N. Thurston. N.Y.: Academic Press. Vol. X. 61–126.

46 Uberall, H. 1975. Surface waves in acoustics. In: *Physical acoustics*. Eds. W. P. Mason and R. N. Thurston. N.Y.: Academic Press. Vol. X. 1–60.

47 Viktorov, I. A. 1981. *Acoustic surface waves in solids*. Moscow: Nauka Publ.

48. Brekhovskikh, L. M., and O. A. Godin. 1989. *Acoustics of laminated media.* Moscow: Nauka Publ. (Translation from Russian: Brekhovskikh, L. M., and O. A. Godin. 1990. *Acoustics of layered media: I. Plane and quasi-plane waves.* Berlin: Springer-Verlag.)
49. Biryukov, S. V., Yu. V. Gulyaev, V. V. Krylov, and V. P. Plesskii. 1991. *Surface acoustic waves in inhomogeneous media.* Moscow: Nauka Publ.
50. Volkenshtein, M. M., and V. M. Levin. 1986. Anomalous absorption of surface acoustic waves at the liquid–solid interface. *Pisma v ZhTF* 12(24):1438–1503.
51. Pettinen, A., and M. Luukkala. 1977. Absorption and dispersion in photoacoustics. *Ultrasonics* 15:205–210.
52. Morozov, A. I., and M. A. Kulakov. 1980. Single lens transmission scanning acoustic microscope. *Electronics Lett.* 16(15):596–597.
53. Bond, W. L., C. C. Cutler, R. A. Lemons, and C. F. Quate. 1975. Dark-field and stereo viewing with the acoustic microscope. *Appl. Phys. Lett.* 27(5):270–272.
54. Quate, C. F. 1979. The acoustic microscope. *Scientific American* 241(4):62–70.
55. Chubachi, N., and H. Okazaki. 1984. Theoretical analysis with line-focus beam in acoustic microscopy. *Electronics Lett.* 20(3):113–115.
56. Hadimioglu, B., and J. S. Foster. 1984. Advances in superfluid helium acoustic microscopy. *J. Appl. Phys.* 56(7):1976–1980.
57. Maev, R. G., and M. F. Pyshnyi. 1988. On the possibility of reconstructing 3D distributions of object acoustic properties studied by the acoustic microscope. *Akust. Zh.* 34(2):358–361.
58. Maev, R. G., and M. Hoppe, eds. 1985. *Microscope photometry and acoustic microscopy in science. Joint Russian–German Symposium (International) Proceedings.* Moscow: Nauka Publ.
59. Atalar, A. 1978. An angular-spectrum approach to contrast in reflection acoustic microscopy. *J. Appl. Phys.* 49(11):5130–5139.
60. Weglein, R. D., and R. G. Wilson. 1978. Characteristic material signatures by acoustic microscopy. *Electronics Lett.* 14(12):352–354.
61. Kushibiki, J., and N. Chubachi. 1985. Material characterization by line-focus-beam acoustic microscope. *IEEE Trans. Sonics Ultrason.* SU-32(2):189–212.
62. Levin, V. M., R. G. Maev, and T. A. Senjushkina. 1991. Scanning acoustic microscopy in biological medical research. In: *Physical characterization of biological cells.* Berlin: Verlag Gesundheit. 435–449.
63. Maev, R. G., V. M. Levin, and E. Y. Maeva. 1996. The application of scanning acoustic microscopy to dielectric polymer composite material study. *Conference on Electrical Insulation and Dielectric Phenomena.* San Francisco: IEEE Dielectrics and Electrical Insulation Society. USA.
64. Kolosov, O. V., V. M. Levin, R. G. Maev, and T. A. Senyushkina. 1987. Investigation into viscoelastic properties of biopolymers in a transmission acoustic microscope. *Transactions of the 5th USSR – W. Germany Symposium on New Methods and Instruments for Microscopy in Medicine and Biology.* Moscow. 76–81.
65. Enikolopov, N. S., O. V. Kolosov, E. Yu. Lagutenkova (Maeva), R. G. Maev, and D. D. Novikov. 1987. Studies of heterogeneity of polymer mixtures by means of scanning acoustic microscopy. *Dokl. Akad. Nauk SSSR* 292(6):1418–1422.
66. Maev, R. G. 1988. Scanning acoustic microscopy of polymeric materials and biological substances. Review. *Tutorial Archives of Acoustics* 13(1–2):13–43.
67. Kanevskii, I. N. 1977. *Focusing of acoustic and ultrasound waves.* Moscow: Nauka Publ.
68. Debye, P. 1909. Das Verhalten von Lichwellen im der Nahe eines Brennpunktes oder einen Brennlinie. *Ann. Phys. vierte Folge* 30:755–776.
69. O'Neyl, H. T. 1949. Theory of focusing radiators. *J. Acoust. Soc. Amer.* 21(5):516–526.

70. Lord Raleigh. 1945. *The theory of sound*. N.Y.: Dover Publs. 2 vols.
71. Lucas, B. G., and T. G. Muir. 1982. Field of a focusing source. *J. Acoust. Soc. Amer.* 72(4):1289–1295.
72. Ueda, M. 1985. Fast converging series expansion for velocity potential of circular concave piston source. *J. Acoust. Soc. Japan (E)* 6(1):35–39.
73. Prudnikov, A. P., E. A. Brychkov, and O. I. Marichev. 1983. *Integrals and series. Special functions.* Moscow: Nauka Publ.
74. Heiran, M. E. 1979. Measurements and visualization of acoustic wave fields. *TIIER* 67(4):10–24.
75. Gavrilov, L. R., V. N. Dmitriev, and L. V. Solontsova. 1986. Noninvasive method for studying acoustic fields of ultrasound focusing transducers. *Akust. Zh.* 32(5):670–675.
76. Gavrilov, L. R., V. N. Dmitriev, and L. V. Solontsova. 1988. Use of focused ultrasonic receivers for remote measurements in biological tissues. *J. Acoust. Soc. Amer.* 83(3):1167–1179.
77. Titov, S. A. 1988. Scanning acoustic microscope operating in the ultrasound field visualization regime. *Pisma v ZhTF* 14(1):22–24.
78. Goodman, J. 1968. *Introduction to Fourier optics*. N.Y.: McGraw–Hill Book Co.
79. Landau, L. D., and E. M. Lifshits. 1986. *Hydrodynamics*. Moscow: Nauka Publ.
80. King, L. V. 1934. On the acoustic radiation field of the piezoelectric transducers and the effect of viscosity on transmission. *Canadian J. Research* 11:125–155.
81. Maev, R. G. 1988. Scanning acoustic microscopy and its applications for material science. *6th Soviet – W. Germany Symposium Proceedings*. 35–51.
82. Maev, R. G., E. Yu. Maeva, and K. I. Maslov. 1997. Ultrasonic measurements of thickness of thin internal layers in highly absorptive layered polymer composite based on time domain inversion algorithm. *3rd International Workshop "Advances in Signal Processing for NDE of Materials."* USA. 3:33–37.
83. Kushibiki, I., A. Ohkubo, and N. Chubachi. 1982. Backscattering analysis from serial echoes in a heterotopic transplant model. In: *Acoust. Imaging*. N.Y.–London: Plenum Press. Vol. 12. 101–112.
84. Kushibiki, I., A. Ohkubo, and N. Chubachi. 1982. Theoretical study for $V(z)$ curves obtained by acoustic microscope with line-focus beam. *Electronics Lett.* 18:663–665.
85. Todd, K., and Y. Murata. 1977. Acoustic focusing device with interdigital transducer. *J. Acoust. Soc. Amer.* 62(4):1033–1036.
86. Guzhev, S. N., V. M. Levin, R. G. Maev, and I. M. Kotelinskii. 1984. On excitation of a Stoneley type surface acoustic wave at the interface of solid–liquid halfspaces with the aid of an interdigital transducer. *Zh. Tekhn. Fiz.* 54(7):1402–1404.
87. Joshi, S. G., and R. M. White. 1969. Excitation and detection of surface elastic waves in piezoelectric crystals. *J. Acoust. Soc. Amer.* 46(1):17–27.
88. Szillard, J. 1982. Ultrasonic Fresnel lenses. Part II. Cylindrical lenses. *Ultrasonics* 5:103–109.
89. Chen, W. H., H. J. Shaw, and L. T. Zitelli. 1978. A new grating acoustic scanner. *IEEE Ultrason. Symposium Proceedings*. 775–779.
90. Farnell, G. W., and C. K. Jen. 1981. Planar acoustic microscopy lens using Rayleigh waves to compressional convertion. *Electronics Lett.* 16:541–543.
91. Farnell, G. W., and C. K. Jen. 1981. Experiments with the planar acoustic microscope lens. *IEEE Ultrason. Symposium Proceedings*. 547–551.
92. Jen, C. K., and G. W. Farnell. 1982. Development of the planar microscope lens. *IEEE Ultrason. Symposium Proceedings*. 613–617.
93. Kikuchi, T., T. Shiokama, T. Nakamoto, T. Moriizumi, and T. Yasuda. 1982. Ultrasonic imaging system using interdigital transducer. *IEEE Ultrason. Symposium Proceedings*. 618–622.
94. Nakamoto, T., T. Nomura, T. Shiokama, T. Moriizumi, and T. Yasuda. 1982. Measurement of elastic

anisotropy using leaky SAW excited by IDT. *IECE Japan Tech. Group on Ultrasonics.* US82-53:19–24.

95 Nomura, T., T. Shiokama, T. Moriizumi, and T. Yasuda. 1983. Two-dimensional mapping of SAW propagation constants by using Frenel-phase-plate interdigital transducer. *IEEE Ultrason. Symposium Proceedings.* 621–626.

96 Nongailard, B., M. Ourck, J. M. Rouvaen, M. Houze, and E. Bridoux. 1984. A new focusing method for nondestructive evaluation by surface acoustic waves. *J. Appl. Phys.* 55(1):75–79.

97 Atalar, A., and H. Köyman. 1985. A new focusing technique for acoustic microscopy. *1st Joint Soviet – W. Germany Symposium (International) on Microscope Photometry and Acoustic Microscopy in Science Proceedings.* Moscow. 72–79.

98 Chen, W. H., F. C. Fu, and W. L. Lu. 1985. Scanning acoustic microscope utilizing SAW-BAW conversion. *IEEE Trans. Sonic. Ultrason.* SU-32(2):181–188.

99 Nomura, T., T. Shiokama, and T. Moriizumi. 1985. Measurement and mapping of elastic anisotropy of solids using a leaky SAW excited by an interdigital transducer. *IEEE Trans. Sonic. Ultrason.* SU-32(2):147–152.

100 Guzhev, S. N., R. G. Maev. 1990. Experimental investigation of the Stoneley SAW velocity at the solid–liquid interface. *Pisma v ZhTF* 16(17):77–81.

101 Holand, A., and P. Kleiborn. 1974. *TIIER* 62(5):41–65.

102 Warnes, L. A. 1982. The use of the autophased zones in an acoustic Fresnel lens for scanning sonar transmitter. *Ultrasonics* 20(4):184–188.

103 Wickramasinghe, H. K. 1983. Scanning acoustic microscopy. A review. *J. Microscopy* 129(1):63–73.

104 Briggs, A. 1985. *An introduction to scanning acoustic microscopy. Serial of Microscopy Handbooks.* Oxford: Alden Press. Vol. 12.

105 Wickramasinghe, H. K. 1978. Contrast in reflection acoustic microscopy. *Electronics Lett.* 14(10):305–306.

106 Quate, C. F., A. Atalar, and H. K. Wickramasinghe. 1979. Acoustic microscope with mechanical scanning – a review. *Proc. IEEE* 67(8):1092–1113.

107 Wickramasinghe, H. K. 1979. Contrast and imaging performance in the scanning acoustic microscope. *J. Appl. Phys.* 50:664–672.

108 Atalar, A. 1979. A physical model for acoustic signatures. *J. Appl. Phys.* 50(12):8237–8239.

109 Atalar, A., C. F. Quate, and H. K. Wickramasinghe. 1977. Phase imaging in reflection with the acoustic microscope. *Appl. Phys. Lett.* 31(12):791–793.

110 Hildebrand, J. A., K. Liang, and S. D. Bennet. 1983. Fourier transform approach to materials characterization with acoustic microscope. *J. Appl. Phys.* 54(12):7016–7019.

111 Atalar, A. 1985. Material dependent contrast response and penetration ability of the acoustic microscope. *Symposium (International) on Microscope Photometry and Acoustic Microscopy in Science Proceedings.* Moscow. 38–49.

112 Liang, K. K., G. S. Kino, and B. T. Khuri-Yakub. 1985. Material characterization by the inversion of $V(z)$. *IEEE Trans. Sonic. Ultrason.* SU-32(2):213–234.

113 Parmon, W., and H. L. Bertoni. 1979. Ray interpretation of the material signature in the acoustic microscope. *Electronics Lett.* 15(12):684–686.

114 Weglein, R. D. 1979. A model for predicting acoustic material signature. *Appl. Phys. Lett.* 34(3):179–181.

115 Weglein, R. D. 1980. Acoustic microscopy applied to SAW dispersion and film thickness measurement. *IEEE Trans. Sonic. Ultrason.* SU-27(2):82–96.

116 Bertoni, H. L. 1984. Ray-optical evaluation of $V(z)$ in the reflection acoustic microscope. *IEEE Trans. Sonic. Ultrason.* SU-31(1):105–116.

117 Wilson, R. G., and R. D. Weglein. 1984. Acoustic microscopy of materials and surface layers. *J. Appl. Phys.* 55(9):3261–3275.

118 Bertoni, H. L., and T. Tamir. 1973. Unified theory of Raleigh-angle phenomena for beams at liquid–solid interfaces. *J. Appl. Phys.* 2:157–172.

119 Pitts, L. E., T. J. Plona, and W. G. Mayer. 1976. Ultrasonic bounded beam reflection and transmission effects at liquid/solid plate/liquid interface. *J. Acoust. Soc. Amer.* 59(6):1324–1328.

120 Pitts, L. E., T. J. Plona, and W. G. Mayer. 1977. Theoretical similarities of Rayleigh and Lamb modes of vibration. *J. Acoust. Soc. Amer.* 60(2):374–377.

121 Pitts, L. E., T. J. Plona, and W. G. Mayer. 1977. Theory of nonspecular reflection effects for an ultrasonic beam incident on solid plate in a liquid. *IEEE Trans. Sonic. Ultrason.* SU-24(2):101–109.

122 Ng, K. W., T. D. Ngoc, J. E. Mc-Clure, and W. G. Mayer. 1981. Nonspecular transmission effects for ultrasonic beams incident on a solid plate in a liquid. *Acoustica* 48(3):168–173.

123 Kolosov, O. V., V. M. Levin, R. G. Maev, and T. A. Senjushkina. 1987. The use of acoustic microscopy for biological characterization. *J. Ultrasound Medicine Biology* 13(8):477–483.

124 Maev, R. G. 1988. Acoustic microscopy. *Vestnik AN SSSR* 2:74–84.

125 Levin, V. M., R. G. Maev, O. V. Kolosov, T. A. Senjushkina, and M. A. Bukhny. 1990. Theoretical fundamentals of quantitative acoustic microscopy. *Acta Phys. Slovaca* 40(3):171–184.

126 Fiorito, R., and H. Uberall. 1979. Resonance theory of acoustic reflection and transmission through a fluid layer. *J. Acoust. Soc. Amer.* 65(1):9–14.

127 Abramovitz, A. A., and I. M. Stigan, eds. 1964. *Handbook of mathematical functions with formulas, graphs and mathematical tables. Applied Math. Series.* Washington, DC: Govt. Print. Office. Vol. 55.

128 Liang, K. K., G. S. Kino, and B. T. Khuri-Yakub. 1985. Material characterization by the inversion of $V(z)$. *IEEE Trans. Sonics Ultrasonics* SU-32:213–224.

129 Briggs, G. A. D. 1992. In: *Acoustic microscopy.* Oxford: Clarendon Press. 105–152.

130 Atalar, A., H. Köymen, A. Bozkurt, and G. Yaralioglu. 1995. Lens geometries for quantitative acoustic microscopy. In: *Advances in acoustic microscopy.* Ed. A. Briggs. N.Y.: Plenum Press. Vol. 1. 117–151.

131 Titov, S., and R. Maev. 1997. Doppler continuous wave scanning acoustic microscope. *IEEE Ultrason. Symposium Proceedings.* 713–718.

132 Nadal, M.-H., P. Lebrun, and C. Gondard. 1998. Prediction of the impulse response of materials using a SAM technique in the MHz frequency range with a lensless cylindrical-focused transducer. *Ultrasonics* 36:505–512.

133 Lee, Y., and S. Cheng. 2001. Measuring Lamb dispersion curves of a bi-layered plate and its application on material characterization of coating. *IEEE Trans. Ultrason. Ferroelec. Freq. Contr.* 48:830–837.

134 Ono, Y., and J. Kushibiki. 2000. Experimental study of construction mechanism of $V(z)$ curves obtained by line-focus-beam acoustic microscopy. *IEEE Trans. Ultrason. Ferroelec. Freq. Contr.* 47:1042–1050.

135 Kushibiki, J., Y. Ono, and Y. Ohashi. 2000. Experimental consideration on water-couplant temperature for accurate velocity measurements by LFB ultrasonic material characterization system. *IEEE Ultrason. Symposium Proceedings.* 619–624.

136 Kushibiki, J., Y. Ono, Y. Ohashi, and M. Arakawa. 2002. Development of the line-focus-beam ultrasonic material characterization system. *IEEE Trans. Ultrason. Ferroelec. Freq. Contr.* 49:99–113.

137 Nakaso, N., K. Ohira, M. Yanaka, and Y. Tsukahara. 1994. Measurement of acoustic reflection coefficients by an ultrasonic microspectrometer. *IEEE Trans. Ultrason. Ferroelec. Freq. Contr.* 41:494–502.

138 Nakaso, N., Y. Tsukahara, and N. Chubachi. 1996. Evaluation of spatial resolution of spherical-planar-pair

lenses for elasticity measurement with microscopic resolution. *IEEE Trans. Ultrason. Ferroelec. Freq. Contr.* 43:422–427.

139 Vines, R. E., S. Tamura, and J. P. Wolfe. 1995. Surface acoustic wave focusing and induced Rayleigh waves. *Phys. Rev. Lett.* 74:2729–2732.

140 Every, A. G., A. A. Maznev, and G. A. D. Briggs. 1997. Surface response of a fluid loaded anisotropic solid to an impulsive point source: Application to scanning acoustic microscopy. *Phys. Rev. Lett.* 79(13):2478–2481.

141 Smith, R., D. A. Sinclair, and H. K. Wickramasinghe. 1980. Acoustic microscopy of elastic constants. *Ultrason. Symposium Proceedings.* 677–682.

142 Hauser, M. R., R. L. Weaver, and J. P. Wolfe. 1992. Internal diffraction of ultrasound in crystals: Phonon focusing at long wavelengths. *Phys. Rev. Lett.* 68(17):2604–2607.

143 Grill, W., K. Hillmann, K. U. Wurz, and J. Wesner. 1996. In: *Advances in acoustic microscopy.* Eds. A. Briggs and W. Arnold. N.Y.: Plenum Press. Vol. 2. 167–218.

144 Pluta, M., M. Schubert, J. Jahny, and W. Grill. 2000. Angular spectrum approach for the computation of group and phase velocity surface of acoustic waves in anisotropic materials. *Ultrasonics* 38:232–236.

145 Alleyne, D., and P. Cawley. 1991. A two-dimensional Fourier transform method for the measurement of propagating multimode signals. *J. Acoust. Soc. Amer.* 89:1159–1168.

146 Lobkis, O. I., and D. E. Chimenti. 1999. Three-dimensional transducer voltage in anisotropic materials characterization. *J. Acoust. Soc. Amer.* 106:36–45.

147 Hosten, B., M. Castaings, H. Tretout, and H. Voillaume. 2000. Identification of composite materials elastic moduli from Lamb wave velocities measured with single sided, contactless ultrasonic method. *Review of Progress in QNDE* 20:1023–1030.

148 Titov, S. A., R. G. Maev, and A. N. Bogachenkov. 2000. Measurement of the velocity and attenuation of leaky surface acoustic waves by an ultrasonic microscope with two focusing transducers. *Russ. Ultrason.* 30(6):269–274.

149 Titov, S. A., R. G. Maev, and A. N. Bogachenkov. 2003. Wide-aperture, line-focused ultrasonic material characterization system based on lateral scanning. *IEEE Trans. Ultrason. Ferroelec. Freq. Contr.* 50:1046–1056.

150 Yamanaka, K. 1983. Surface acoustic waves measurements using an impulsive converging beam. *J. Applied Physics* 54(8):4323–4329.

151 Xiang, D., N. N. Hsu, and G. V. Blessing. 1996. The design, construction and application of a large aperture lensless line-focus PVDF transducer. *Ultrasonics* 34(6):641–647.

152 Titov, S., R. Maev, and A. Bogatchenkov. 2001. Evaluation of the $V(x)$ temperature stability in time-resolved ultrasonic measurements. *IEEE Ultrason. Symposium Proceedings.* 497–500.

153 Liu, D., and C. Waag. 1997. Propagation and backpropagation for ultrasonic wavefront design. *IEEE Trans. Ultrason. Ferroelec. Freq. Contr.* 44:1–13.

154 Madisetti, V. K., and D. B. Williams, eds. 1998. *The digital signal processing handbook.* CRC Press. 1-1-1-28.

155 Birks, A. S., R. E. Green, Jr., and P. McIntire. 1991. Ultrasonic testing. In: *Nondestructive testing handbook.* Am. Soc. for Nondestructive Testing. Vol. 7. 836–841.

156 Brekhovskikh, L. M., and O. A. Godin. 1990. In: *Acoustics of layered media I.* Berlin: Springer. 98–112.

157 Chan, K. H., and H. L. Bertoni. 1991. Ray representation of longitudinal lateral waves in acoustic microscopy. *IEEE Trans. Ultrason. Ferroelec. Freq. Contr.* 38:27–34.

158 Tsukahara, Y., Y. Liu, C. Neron, C. K. Jen, and J. Kushibiki. 1994. Longitudinal critical angle singularities and their effect on $V(z)$ of the line-focus-beam acoustic microscope. *IEEE Trans. Ultrason. Ferroelec. Freq. Contr.* 41:458–465.

159 Del Grosso, V. A., and C. W. Mader. 1972. Speed of sound in pure water. *J. Acoust. Soc. Amer.* 52(5):1442–1446.

160 *CRC handbook of chemistry and physics.* 1993/1994. 74th ed. CRC Press. 12–165.

161 McSkimin, H. J. 1953. Measurement of elastic constants at low temperature by means of ultrasonic waves – data for silicon and germanium single crystals, and for fused silica. *J. Appl. Phys.* 24(8):988–997.

162 Tsukahara, Y., N. Nakaso, K. Ohira, and M. Yanaka. 1996. Interaction of acoustic waves with solid surfaces. In: *Advances in acoustic microscopy.* Eds. A. Briggs and W. Arnold. N.Y.: Plenum Press. Vol. 2. 103–165.

163 Bloembergen, N. 1996. *Nonlinear optics.* Singapore: World Scientific.

164 Bateman, T., M. P. Mason, and H. J. McSkimin. 1961. Third-order elastic moduli of germanium. *J. Appl. Phys.* 32:928–936.

165 Kino, G. S., D. M. Barnett, N. Grayeli, G. Herrmann, J. B. Hunter, D. B. Ilic, G. C. Johnson, R. B. King, M. P. Scott, J. C. Shyne, and C. R. Steel. 1980. Acoustic measurements of stress fields and microstructure. *J. Nondestructive Evaluation* 1:67–77.

166 Law, W. K., L. A. Frizzell, and F. Dunn. 1985. Determination of the nonlinearity parameter B/A of biological media. *Ultrasound Medicine Biology* 11:307–318.

167 Gong, X. F., Z. M. Zhu, T. Shi, and J. H. Huang. 1989. Determination of the acoustic nonlinearity parameter in biological media using FAIS and ITD methods. *J. Acoust. Soc. Amer.* 86(1):1–5.

168 Germain, L., R. Jacques, and J. D. N. Cheeke. 1989. Acoustic microscopy applied to nonlinear characterization of biological media. *J. Acoust. Soc. Amer.* 86(4):1560–1565.

169 Fassbender, S., M. Kroening, and A. Walter. 1996. Measurement of adhesion strength using nonlinear acoustics. In: *Nondestructive characterization of materials.* Part 2. Ed. A. L. Bartos. Zürich–Uetikon: Trans. Tech. Publications. Vol. VII. 783–790.

170 Buck, O., W. L. Morris, and J. M. Richardson. 1978. Acoustic harmonic generation at unbonded interfaces and fatigue cracks. *Appl. Phys. Lett.* 33(5):371–373.

171 Vella, P. J., T. C. Padmore, G. I. Stegeman, and V. M. Ristic. 1974. Nonlinear surface-wave interactions: Parametric mixing and harmonic generation. *J. Appl. Phys.* 45(5):1993–2006.

172 Shui, Y., and I. Solodov. 1988. Nonlinear properties of Rayleigh and Stoneley waves in solids. *J. Appl. Phys.* 64(11):6155–6165.

173 Solodov, I. 1998. Ultrasonics of nonlinear contacts: Propagation, reflection, and NDE applications. *Ultrasonics* 36:383-390.

174 Asainov, A. F., K. S. Len, and I. Y. Solodov. 1992. Experimental study of Stoneley waves at the interface of contacting solids. *Sov. Physics Acoustics* 38(3):293–295.

175 Severin, F. M., and I. Yu. Solodov. 1988. Polarization effects attending nonlinear reflection of acoustic SH-waves. *Vestnik Mosk. Univ. Series 3* 29(5):84–86.

176 Severin, F. M., I. Yu. Solodov, and Yu. N. Shkulanov. 1988. Experimental observation of nonlinear sound reflection from the solid interface. *Vestnik Mosk. Univ. Series 3* 29(4):94–96.

177 Rudenko, O. V., and S. Voo. 1953. Assessment of the height of unevenness on a rough surface with the use of acoustic measurements. *Vestnik Mosk. Gos. Univ.* 34(5):94–96.

178 Armyakov, D. V., A. F. Asainov, B. A. Korshak, and I. Yu. Solodov. 1998. Observations of cracks by the nonlinear acoustics method. *Defektoskopiya* 12:34.

179 Mao, Y. W., Y. Shui, W. Jiang, Z. Lu, and W. Wu. 1989. Second-harmonic generation of interface waves. *Appl. Phys. Lett.* 55(23):2394–2395.

180 Kompfner, R., and R. A. Lemons. 1976. Nonlinear acoustic microscopy. *Appl. Phys. Lett.* 28(6):295–297.

181 Wickramasinghe, H. K., and C. J. Yeack. 1977. *Appl. Phys. Lett.* 43:1006.

182 Hadimioglu, B., and C. F. Quate. 1983. Water acoustic microscopy at

suboptical wavelength. *Appl. Phys. Lett.* 43:1006–1007.

183 Rugar, D. 1984. Resolution beyond the diffraction limit in the acoustic microscope. A nonlinear effect. *J. Appl. Phys.* 56(5):1338–1340.

184 Germain, L., and J. D. N. Cheeke. 1988. Generation and detection of high-order harmonics in liquids using a scanning acoustic microscope. *J. Acoust. Soc. Amer.* 83(3):942–949.

185 Kozlowski, G. C., C. P. Silva, M. S. Muha, and A. A. Moulthrop. 1990. Acoustic microscopy and nonlinear effects in pressurized superfluid helium. *IEEE Ultrasonic Symposium*. Honolulu 2:907–912.

186 Karaki, K., T. Saito, K. Matsumoto, and Y. Okuda. 1990. Observation of new nonlinear phenomena in focused acoustic beam in pressurized superfluid helium-4. *Physica B: Physics of Condensed Matter* 165:131–132.

187 Karaki, K., T. Saito, K. Matsumoto, and Y. Okuda. 1991. Nonlinear resolution improvement and second-harmonic generation of a pressurized superfluid ^{4}He acoustic microscope. *Appl. Phys. Lett.* 59(8):908–910.

188 Muir, T. G., and E. L. Carstensen. 1980. Prediction of nonlinear acoustic effects in biomedical frequencies and intensities. *Ultrasound Medicine Biology* 6:345–357.

189 Carstensen, E. L., T. G. Muir, W. K. Law, and N. D. McKay. 1980. Demonstration of nonlinear acoustical effects at biomedical frequencies and intensities. *Ultrasound Medicine Biology* 6:359–368.

190 Starrit, H. C., F. A. Duck, A. J. Hawkins, and V. F. Humphrey. 1986. The development of harmonic distortion in pulsed finite-amplitude ultrasound passing through liver. *Phys. Med. Biol.* 31:1401–1409.

191 Ward, B., A. C. Baker, and V. F. Humphrey. 1997. Nonlinear propagation applied to the improvement of resolution in diagnostic medical ultrasound. *J. Acoust. Soc. Amer.* 101(1):143–154.

192 Uhlendor, V., and F. D. Scholle. 1996. Imaging of spatial distribution and flow of microbubbles using non-linear acoustic properties. *Acoustic Imaging* 22:233–238.

193 Ichida, N., T. Sato, and M. Linzer. 1983. Imaging the non-linear ultrasonic parameter of a medium. *Ultrasonic Imaging* 5:295–299.

194 Sato, T., A. Fukusima, N. Ichida, H. Ishikawa, H. Miwa, Y. Igarashi, T. Shimura, and K. Murakami. 1985. Nonlinear parameter tomography system using counterpropagating probe and pump waves. *Ultrasonic Imaging* 7(1):49–59.

195 Cain, C. A. 1986. Ultrasonic reflection mode imaging of the nonlinear parameter B/A: A theoretical basis. *J. Acoust. Soc. Amer.* 80(1):28–32.

196 Sato, T., E. Mori, K. Endo, Y. Yamakoshi, and M. Sase. 1992. A few effective signal processes for reflection-type imaging of nonlinear parameter N of soft tissues. *Acoustic Imaging* 19:363.

197 Nakagawa, Y., M. Nakagawa, M. Yoneyama, and M. Kikuchi. 1984. Nonlinear parameter imaging computed tomography by parametric acoustic array. *IEEE Ultrasonics Symposium Proceedings*. 673.

198 Sutin, A. M., and V. E. Nazarov. 1995. Nonlinear acoustic methods for crack diagnostics. *Radiofiz. Kvant. Elektronika* 38(3–4):109–120.

199 Tan, M. R. T., H. L. Ransom, Jr., C. C. Cutler, and M. Chodorow. 1985. Oblique, off-specular, linear, and nonlinear observations with a scanning micron wavelength acoustic microscope. *J. Appl. Phys.* 57(11):4931–4935.

200 Ransom, H. L., Jr., S. W. Meeks, and C. C. Cutler. 1986. Linear and nonlinear imaging of ferroelastic domains in neodymium pentaphosphate. *Ultrasonic Symposium Proceedings*. 731.

201 Severin, F. M., B. O'Neil, and R. G. Maev. 2000. Image of nonlinear structures of heterogeneous media using the second harmonic in scanning acoustic microscopy. In: *Review of progress in quantitative nondestructive evaluation*. Plenum Press. Vol. 19A. 881–889.

202. Maev, R. G. 2000. New development in materials characterization and vehicle quality control. *Physical Sciences and Advanced Vehicle Technologies. Symposium Proceedings.* Toronto–N.Y. 67–81.

203. Maev, R. G., V. M. Levin, R. M. Piliar, E. Yu. Maeva, and T. A. Senjushkina. 1995. Acoustic microscopy applications for observing microstructure of bones and bone-implant systems. *Acoustic Imaging* 22:323.

204. Maev, R. G., D. F. Watt, R. Pan, V. M. Levin, and K. I. Maslov. 1995. Development of high-resolution ultrasonic inspection methods for welding microdefectoscopy. *Acoustic Imaging* 22:779.

205. Zinin, P. V., V. M. Levin, O. I. Lobkis, and R. G. Maev. 1986. Forces of radiation pressure in the focal area of an acoustic microscope. *Akust. Zh.* 32(6):785–790.

206. Maev, R. G., and K. I. Maslov. 1989. Temperature effects in the focal area of an acoustic microscope. *Akust. Zh.* 35(1):84–90.

207. Yeach, C. G., M. Chodorov, and C. C. Cutler. 1980. Nonlinear acoustic off-axis imaging. *J. Appl. Phys.* 51(9):4631–4637.

208. Berezina, S. I. 1981. Investigation into physical characteristics and application of SAM. Candidate of Sci. Thesis in physics and math. Moscow.

209. Rudenko, O. V., S. I. Soluyan, and R. V. Khokhlov. 1973. Restricted quasi-planar beams of periodical perturbation in a nonlinear medium. *Akust. Zh.* 19(6):871–875.

210. Zabolotskaya, E. A., and R. V. Khokhlov. 1973. Quasi-planar waves in nonlinear acoustics of restricted beams. *Akust. Zh.* 15(1):40–47.

211. Miller, C. A., and L. E. Scriven. 1968. The oscillation of a fluid droplet immersed in another fluid. *J. Fluid Mech.* 32(3):417–435.

212. Marston, P. L., and R. E. Apfel. 1979. Acoustically forced shape oscillation of hydrocarbon drop levitated in water. *J. Colloid Interface Sci.* 68(2):280–286.

213. Marston, P. L. 1980. Shape oscillation and static deformation of drops and bubbles driven by modulated radiation stresses – theory. *J. Acoust. Soc. Amer.* 67(1):15–26.

214. Marston, P. L., and R. E. Apfel. 1980. Quadrupole resonance of drops driven by modulated acoustic radiation pressure – experimental properties. *J. Acoust. Soc. Amer.* 67(1):27–37.

215. Prosperetti, A. 1980. Normal mode analysis for the oscillations of a viscous liquid drop in an immersible liquid. *J. de Mechanique* 9(1):149–182.

216. Zinin, P. V., V. M. Levin, and R. G. Maev. 1981. Resonant interaction of ultrasound with biological microscopic objects. In: *Ultrasound in biology and medicine.* "UBIOMED V." Abstracts of papers presented at USSR Symposium with international attendance. Pushchino. 79.

217. Zinin, P. V., V. M. Levin, and R. G. Maev. 1983. Formation of shear waves at acoustic radiation scattering on microobjects. In: *Interaction of ultrasound with a biological medium.* Abstracts of papers presented at the USSR Conference. Erevan. 27.

218. Zinin, P. V., V. M. Levin, and R. G. Maev. 1987. Natural vibrations of biological microscopic objects. *Biofiz.* 32(1):186–192.

219. Berezina, S. I., V. E. Lyamov, and I. Yu. Solodov. 1977. Acoustic microscopy. *Vestnik MGU. Ser. Fizika, Astronomiya* 18(1):3–18.

220. Koledov, V. V., M. A. Kulakov, M. I. Levin, Ya. A. Monosov, A. I. Morozov, and V. A. Shakhunov. 1987. Visualization of acoustic radiation by a magnetorheological medium. *Akust. Zh.* 33(3):564–566.

221. Gavrilov, L. R., and E. M. Tsirulnikov. 1980. *Focused ultrasound in physiology and medicine.* Moscow: Nauka Publ.

222. Parker, K. J. 1985. Effects of heat conduction and sample size on ultrasonic absorption measurements. *J. Acoust. Soc. Amer.* 77(2):719–725.

223. Niborg, W. L. 1986. Sonically produced heat in a fluid with bulk viscosity and shear viscosity. *J. Acoust. Soc. Amer.* 80(4):1133–1139.

224. Love, L. A., and F. W. Kremkau. 1980. Intracellular temperature distribution

produced by ultrasound. *J. Acoust. Soc. Amer.* 67(3):1045–1050.

225 Kikoin, I. K., ed. 1976. *Tables of physical quantities. Handbook.* Moscow: Atomizdat.

226 Grishchenko, E. K. 1968. Miniature focusing emitter made of a single cadmium sulfide crystal operating at a 60-megahertz frequency. *Akust. Zh.* 14(3):385–389.

227 Eckart, C. 1948. Vortices and streams caused by sound waves. *Phys. Rev.* 73(1):68.

228 Budak, B. M., A. A. Samarskii, and A. N. Tikhonov. 1972. *Collection of mathematical physics problems.* Moscow: Nauka Publ.

229 Gradshtein, I. S., and I. M. Ryzhik. 1971. *Tables of integrals, sums, series, and products.* Moscow: Nauka Publ.

230 Yakovkin, I. B., and D. V. Petrov. 1979. *Light diffraction on surface acoustic waves.* Novosibirsk: Nauka Publ.

231 Bereiter-Hahn, J. 1985. Acoustic microscope investigation of living cells. *1st Joint Soviet – W. Germany Symposium (International) on Microscope Photometry and Acoustic Microscopy in Science Proceedings.* Moscow. 124–130.

232 Romanenko, E. V. 1967. Ultrasound sensors and the methods for their calibration. In: *Sources of powerful ultrasound.* Moscow: Nauka Publ. Part 2. 327.

233 Beyer, R. T. 1978. Radiation pressure – the history of a mislabeled tensor. *J. Acoust. Soc. Amer.* 63:1025–1030.

234 Hagezawa, T., M. Ochi, and K. Matsuzawa. 1981. Acoustic radiation force on a solid elastic sphere in a spherical wave field. *J. Acoust. Soc. Amer.* 69:937–943.

235 Beissner, K. 1984. Acoustic radiation pressure in the near field. *J. Sound Vibr.* 93(4):537–548.

236 Alekseev, V. N. 1983. About the radiation pressure acting on a sphere. *Akust. Zh.* 29(2):129–136.

237 Sutin, A. M. 1979. Influence of nonlinear effect on the properties of acoustic focusing systems. *Akust. Zh.* 24(2):593–601.

238 Flammer, K. 1962. *Tables of spheroidal wave functions.* Moscow: Computer Center USSR Acad. Sci.

239 Morse, P. M., and H. Feshbach. 1953. *Methods of theoretical physics.* N.Y.: McGraw–Hill.

240 Kolosov, O. V., L. F. Matsiev, E. Yu. Lagutenkova (Maeva), R. G. Maev, T. A. Senyushkina, and M. F. Pyshnyi. 1986, 2 June. Method for investigation of object structures in a transmission acoustic microscope. IPCS USSR Acad. Sci. Patent SU 1409915 A 1 (4073800).

241 Awatany, J. 1955. Studies on acoustic radiation pressure. Radiation pressure on a circular disk. *J. Acoust. Soc. Amer.* 27:282–286.

242 Zinin, P. V., O. I. Lobkis, and R. G. Maev. 1986. Sound scattering on a disc in the focal area of an acoustic microscope. *Akust. Zh.* 32(5):685–688.

243 Borgnis, F. E. 1953. On the force due to acoustic wave use in the measurement of acoustic intensity. *J. Acoust. Soc. Amer.* 25:546–548.

244 Miller, D. L. 1986. Effects of high amplitude 1 MHz standing ultrasonic field on algae hidroction. *IEEE Trans. Ultrason. Ferroelec. Freq. Contr.* 33(2):165–170.

245 Westervelt, P. J. 1951. Propagation of power ultrasound in condensed matter. *J. Acoust. Soc. Amer.* 23(4):312–315.

246 Goldberg, Z. A. 1968. Sound pressure. In: *Powerful ultrasound fields.* Ed. L. D. Rozenberg. Moscow: Nauka Publ. 51–86.

247 Danilov, S. D. 1985. The average force acting on a small sphere in a field of an acoustic wave spreading in viscous liquid. *Akust. Zh.* 31(1):45–49.

248 Danilov, S. D. 1986. The average force acting on a small body in an axially symmetric sound field in a real medium. *Mekh. Zhidk. Gaza* 5:161–169.

249 Rotenberg, M., and R. Bivins. 1959. *The 3j and 6j symbols.* N.Y.: Technology Press.

250 Landau, L. D., and E. M. Lifshits. 1974. *Quantum mechanics. Nonrel-*

ativistic theory. 3rd issue. Moscow: Nauka Publ.

251 Mikhailov, I. G., and L. I. Tarutina. 1950. Absorption of acoustic waves in gelatine solutions. *Dokl. Akad. Nauk SSSR* 74(1):41–43.

252 Mikhailov, I. G. 1962. Absorption of ultrasound waves in concentrated solutions of some polymers. *Akust. Zh.* 8:478.

253 Hoppe, M., and J. Bereiter-Hahn. 1985. Applications of scanning acoustic microscopy – survey and new aspects. *IEEE Trans. Ultrasonics* SU-32(2):289–301.

254 Hoppe, M. 1985. Applications of high resolution acoustic microscopy in material sciences. *1st Joint Soviet – W. Germany Symposium (International) on Microscope Photometry and Acoustic Microscopy in Science Proceedings.* Moscow. 94–101.

255 Horie, Y., Y. Terashi, and S. Mase. 1989. Ultrasonic studies and phonon modes of (RE)Ba$_2$YCu$_3$O$_7$. *J. Phys. Soc. Japan* 58:279–290.

256 Bukhny, M. A., L. A. Chernozatonskii, and R. G. Maev. 1990. Acoustic imaging of high-temperature superconducting materials. *J. Microscopy* 160(3):299–313.

257 Kim, T. J., B. Liuthi, M. Schwarz, U. Kuhnberger, B. Wolf, G. Hampel, and D. Nike. 1988. Valence fluctuation aspects of high temperature superconductors. *J. Magn. Mat.* 76–77:604–606.

258 Gupta, H. C. 1988. Zone center phonons in high TcBa$_2$YCu$_3$O$_7$. *Solid State Commun.* 65:495–496.

259 Weglein, R. D. 1982. Rayleigh wave absorption via acoustic microscopy. *Electronics Lett.* 48:20–21.

260 Dransfeld, K., and I. Salzman. 1970. Excitation, detection and attenuation HF elastic surface waves. In: *Physical acoustics. Principles and methods.* Eds. V. P. Mason and R. N. Thurston. London: Academic Press. Vol. 7. Ch. 4.

261 Vlasko-Vlasov, V. K., M. V. Indenbom, and Yu. A. Osipian. 1982. Polarized optical contrast in superhigh conductors YBaCuO. *Pisma v ZhETF* 47:312–315.

262 Ma, Q. Y., T. J. Licata, X. Wu, *et al.* 1988. High Tc superconducting thin films by rapid thermal annealing of Cu/BaO/Y$_2$O$_3$ layered structures. *Appl. Phys. Lett.* 53:2229–2231.

263 Chin-Au-Chang, Y., C. C. Tsuei, T. R. McGuire, D. C. Yee, J. P. Boresh, H. R. Lilienthal, and C. E. Farrel. 1988. YdFCuO superconducting films on SiO$_2$ substrate formed from Cu/BaO/Y$_2$O$_3$/Ag layers structures. *Appl. Phys. Lett.* 53:916–918.

264 Maslov, K. I., R. G. Maev, and V. M. Levin. 1993. New methods and technical principles of low frequency scanning acoustic microscopy. *Acoustic Imaging* 20:756–760.

265 Lipatov, Yu. S. 1977. In: *Physical chemistry of filled polymers.* Moscow: Khimiya Publ. 196–199.

266 Nilsen, L. 1978. *Mechanical properties of polymers and polymer composition.* Moscow: Khimiya Publ.

267 Kuleznev, V. N. 1980. *Polymer blends.* Moscow: Khimiya Publ.

268 Paul, D., and S. Newman. 1978. *Polymer blends.* N.Y.: Academic Press.

269 Bespalov, Yu. A., and N. G. Konovalenko. 1981. *Multicomponent polymeric system.* Leningrad: Khimiya Publ.

270 Bartenev, G. M., and Yu. V. Zelenev. 1976. *Course of polymer physics.* Leningrad: Nauka Publ.

271 Marikhin, I. A., and N. P. Myasnikova. 1977. *Supramolecular structure of polymers.* Leningrad: Nauka Publ.

272 Ferry, J. D. 1980. *Viscoelastic properties of polymers.* N.Y.–London: Wiley & Sons.

273 Brostow, W., and R. Corneliussen. 1986. *Failure in plastics.* Munich–N.Y.: Hansen Publ. in cooperation with Macmillan Publ.

274 Philipoff, W. 1965. Relaxations in polymer solutions, liquids and gels. In: *Physical acoustics. Principles and methods.* Ed. W. P. Mason. N.Y.–London: Academic Press. Vol. 2. Part B. 1–90.

275 Hopkins, I., and K. Kirkjenal. 1970. In: *Relaxation spectra and relaxation processes in solid polymers and glasses.* Moscow: Mir. 110–192.

276 Perepechko, I. I. 1973. *Acoustic methods for investigation of polymers.* Moscow: Khimiya Publ.

277 Bashyam, M. 1990. Thickness-compensation technique for ultrasonic evaluation of composite materials. *Nondestructive Materials Evaluation* 9:1360–1364.

278 Briggs, G. A. D. 1995. *Advances in acoustic microscopy.* N.Y.–London: Plenum Press.

279 Maev, R. G., M. P. Pyshnyi, E. Yu. Maeva, K. I. Maslov, and V. M. Levin. 1997. Imaging of deep internal layers in layered polymer system using ultra-short pulse acoustic microscope. In: *Review of progress in quantitative nondestructive evaluation.* N.Y.: Plenum Press. Vol. 16B. 1845–1851.

280 Maev, R. G., and I. Yu. Solodov. 1999. Nonlinear acoustics: A new tool for composite materials characterization. *6th Conference (International) on Composites Engineering Proceedings.* Orlando, USA. 525–527.

281 Perepechko, I. I. 1967. Application of ultra-acoustics in investigation of materials. *Akust. Zh.* 13(1):143.

282 Perepechko, I. I. 1967. Noninvasive control of polymers and polymer articles. *Ultrazvuk. Tekhn.* 5(1):1.

283 Balodis, A. A., and V. A. Latishenko. 1966. Application of ultra-acoustic methods in physicochemical investigations. *Mekh. Polimer* 6:923.

284 McSkimin, H. J. 1965. Ultrasonic methods for measuring the mechanical properties of liquids and solids. In: *Physical acoustics. Principles and methods.* Ed. W. P. Mason. N.Y.: Academic Press. Vol. 1. Part A. 272–334.

285 Bergman, L. 1954. Der Ultraschal und seine Anwendung in Wissenschaft und Technik. Zurich.

286 Nozdrev, V. F. 1958. *Application of ultra-acoustics in molecular physics.* Moscow: Fizmatizdat.

287 Brazhnikov, N. I. 1965. *Ultrasound methods for controlling the composition and properties of materials.* Moscow: Energiya Publ.

288 Brazhnikov, N. I. 1968. *Ultrasound phasometry.* Moscow: Energiya Publ.

289 Hung, B. N., and A. Goldstein. 1983. Acoustic parameters of commercial plastics. *Trans. IEEE* SU-30(4):249–253.

290 Selfridge, A. R. 1985. Approximate material properties in isotropic materials. *Trans. IEEE* SV-32(3):381–393.

291 Kanai, H., T. Kimura, and N. Chubachi. 1995. Accurate determination of transit time of ultrasound in thin layers. *Electronics Lett.* 31(13):1109–1112.

292 Williams, J., and J. Lambs. 1958. *J. Acoust. Soc. Amer.* 30(4):308.

293 Ivanov, V. E., L. G. Merkulov, and V. A. Shchukin. 1965. Ultrasound apparatus for studying anisotropic materials. *Ultrazvuk. Tekhn.* 3(2).

294 McSkimin, H. J. 1961. *J. Acoust. Soc. Amer.* 33(1):12.

295 McSkimin, H. J., and R. P. Chambers. 1964. *IEEE Trans. Sonic. Ultrason.* 11(2):74.

296 Ermakova, G. A., and M. A. Gorbunova. 1969. *Uch. Zapiski MOPI* 24:80.

297 Meyer, R. 1967. *Kunststoffe* 57(4):257.

298 Yanovskii, V. G., and E. A. Dzyura. 1969. *Zvodsk. Lab.* 35(1):107.

299 Atalar, A., and M. Hoppe. 1979. High-performance acoustic microscope. *Ann. Rev. Mater. Sci.* 9:255–281.

300 Somekh, M. G., G. A. D. Briggs, and C. Ilett. 1984. The effect of elastic anisotropy on contrast in the acoustic microscope. *Philosoph. Magazine A* 49(2):179–204.

301 Enikolopian, N. S., O. V. Kolosov, E. Y. Lagutenkova (Maeva), R. G. Maev, and D. D. Novikov. 1985. *1st Joint Soviet – W. Germany Symposium (International) on Microscope Photometry and Acoustic Microscopy in Science.* Moscow. 106–110.

302 Lemons, R. A., and C. F. Quate. 1975. Acoustic microscopy. Biomedical application. *Science* 188:905–914.

303 Hildebrand, J. A., D. Rugar, R. N. Johnston, and C. F. Quate. 1981. Acoustic microscopy of living cells. *Biophysics* 78(3):1656–1660.

304 Piruzyan, L. A., O. V. Kolosov, V. M. Levin, R. G. Maev, and T. A. Senyushkina. 1985. Acoustic mi-

croscopy of organic and biological materials. *Dokl. Akad. Nauk SSSR* 280(6):1115–1117.

305 Maev, R. G. 1985. Principles and future of acoustic microscopy. *1st Joint Soviet – W. Germany Symposium (International) on Microscope Photometry and Acoustic Microscopy in Science Proceedings.* Moscow. 1–12.

306 Kolosov, O. V., E. Yu. Lagutenkova, V. M. Levin, R. G. Maev, and T. A. Senyushkina. 1989. Acoustic microscopy of biological and organic materials. *USSR Seminar on Acoustoelectric and Photoacoustic Methods for Investigation of Materials.* Kiev. 25–27.

307 Maev, R. G., and V. M. Levin. 1997. Principles of local sound velocity and attenuation measurements using transmission acoustic microscope. *IEEE Trans. Ultrason. Ferroelec. Freq. Contr.* 44(6):1224–1232.

308 Kuleznev, V. N., A. V. Grachev, and Yu. P. Miroshnikov. 1976. The effect of visco-elastic components of polymer blends on the size of the disperse phase. *Kolloid. Zh.* 38(2):265–270.

309 Razinskaya, I. N. 1981. Phase structure and properties of thermoplast mixtures. *Plasmassy* 2:32–35.

310 Leps, G., J. Sachse, J. Bohse, and R. Stephan. 1983. Quantitative Morphologieanaluse zur Aufstellung von Morfologie-Eigenschaftskorrelationen on Schlagrahem PVC. *Prakt. Metallogr.* 20(6):297–305.

311 Tovmasyan, Yu. M., V. A. Topolkaraev, V. G. Oshmyan, A. A. Berlin, E. F. Oleinik, and N. S. Enikolopov. 1985. Fluctuations in the filler distribution and their role in formation of certain composite properties. *Dokl. Akad. Nauk SSSR* 283(3):681–684.

312 Tovmasyan, Yu. M., V. A. Topolkaraev, A. A. Berlin, I. L. Zhuravlev, and N. S. Enikolopyan. 1986. Structural filler arrangement in thermoplasts. Method of description and modeling. *Vys. Molek. Soed.* 28(A)(2):321–328.

313 Maev, R. G. 1988. Scanning acoustic microscopy of polymeric and biological substances. *Tutorial Archives of Acoustics* 13(1–2):13–43.

314 Barentsen, W. M., and S. D. Heikens. 1973. Mechanical properties of polystyrene blended with low density polyethylene. *Polymer* 14(11):579–583.

315 Baramboim, N. K., and V. F. Rakityanskaya. 1974. Formation of the structure of solid-phase polymer dispersions. *Kolloid. Zh.* 36(1):129–132.

316 Miroshnikov, Yu. P., and H. L. Williams. 1982. Disperse structure and mechanical properties of extruded polypropylene-polystyrene blends. *Vysok. Molek. Soed.* 24(A)(8):1606–1614.

317 Manevich, L. I., V. G. Oshmyan, Yu. M. Tovmasyan, and V. A. Topolkaraev. 1983. Mathematical modeling of elastic-plastic deformation of a filled composite material. *Dokl. Akad. Nauk SSSR* 270:806–809.

318 Wu, S. 1985. Phase structure and adhesion in polymer blends. *Polymer* 26(11):1855–1863.

319 Glagolev, A. A. 1941. *Geometric methods for quantitative analysis of aggregates under microscope.* Lvov: Gosgeolizdat.

320 Razinskaya, I. N., L. I. Batueva, and B. P. Shtarkman. 1974. Quantitative methods for assessment of phase morphology parameters of polymer blends. *Kolloid. Zh.* 36(2):291–297.

321 Hanneman, S. E., and V. K. Kinra. 1992. A new technique for ultrasonic nondestructive evaluation of adhesive joints: Part 1. Theory. *SEM J. Experimental Mechanics* 32:323–331.

322 Kinra, V. K., P. T. Jaminet, C. Zhu, and V. R. Iyer. 1994. Simultaneous measurement of the acoustical properties of a thin-layered medium: The inverse problem. *J. Acoust. Soc. Amer.* 95:3059–3074.

323 Mazeika, L., R. Kazys, E. Jasiuniene, and R. Sliteris. 2003. Ultrasonic imaging technique for visualization in hot metals. *World Congress of Ultrasound Proceedings.* Paris. 163.

324 Lemons, R. A., and C. F. Quate. 1974. Integrated circuits as viewed with an acoustic microscope. *Appl. Phys. Lett.* 25:251–253.

325 Okawai, H., M. Tanaka, F. Dunn, N. Chubachi, and K. Honda. 1988. Quantitative display of acoustic proper-

ties of the biological tissue elements. *Acoustical Imaging* 17:193–201.

326 Van der Steen, A. F. W., J. M. Thijssen, G. P. J. Ebben, and P. C. M. deWilde. 1992. Effects of tissue-processing techniques in acoustic (1–2 MHz) and light microscopy. *Histochem. J.* 97:195–199.

327 Sasaki, H., Y. Saijo, M. Tanaka, and S. Nitta. 2002. Influence of fat components and tissue preparation on the high frequency acoustic properties. In: *Acoustical Imaging Symposium.* N.Y.–London–Moscow: Kluwer Academic/Plenum Publ. Vol. 26. 161–166.

328 Bamber, J. C., C. R. Hill, J. A. King, and F. Dunn. 1979. Ultrasonic propagation through fixed and unfixed tissues. *Ultrasound Medicine Biology* 5:159–165.

329 Kremkau, F. W., R. W. Barnes, and C. P. McGraw. 1981. Ultrasonic attenuation and propagation speed in normal human brain. *J. Acoust. Soc. Amer.* 70:29–38.

330 Van der Steen, A. F. W., M. H. M. Cuypers, J. M. Thijssen, and P. C. M. deWilde. 1991. Influence of histochemical preparation on acoustic parameters of liver tissue, a 5-MHz study. *Ultrasound Medicine Biology* 17:879–891.

331 Sasaki, H., Y. Saijo, M. Tanaka, S. Nitta, Y. Terasawa, T. Yambe, and Y. Taguma. 1997. Acoustic properties of dialysed kidney by scanning acoustic microscopy. *Nephrol. Dial. Transplant.* 12:2151–2154.

332 Morozov, A. I., and M. A. Kulakov. 1980. Single lens transmission scanning acoustic microscope. *Electronics Lett.* 15:596–598.

333 Korpel, A., L. M. Kessler, and P. R. Palermo. 1971. Acoustic microscope operating at 100 MHz. *Nature* 232:110–111.

334 Sinclair, D. A., and I. R. Smith. 1982. Tissue characterization techniques using acoustic microscopy. In: *Acoustic imaging*. Eds. E. A. Asdh and C. R. Hill. N.Y.: Plenum Press. Vol. 12. 505–516.

335 Sinclair, D. A., I. R. Smith, and S. D. Bennett. 1984. Elastic constants measurement with a digital acoustic microscope. *IEEE Trans. Sonic. Ultrason.* SU-31(4).

336 Tervola, K. 1985. Tissue characterization techniques using a 100 MHz scanning laser acoustic microscope. *Annual Proc. of University of Oulu.* 38–42.

337 Okawai, H., M. Tanaka, N. Chubachi, and J. Kushibiki. 1987. Noncontact simultaneous measurement of thickness and acoustic properties of a biological tissue using focused wave in a scanning acoustic microscope. *Jap. J. Appl. Phys.* 26(26-1):52–54.

338 O'Brien, W. D., J. W. Erdman, and T. B. Hebner. 1988. Ultrasonic propagation properties (at 100 MHz) in excessively fatty rat liver. *J. Acoust. Soc. Amer.* 83:1159–1166.

339 Carson, P. L., E. H. Chiang, J. M. Rubin, C. R. Meyer, H. F. Andersen, and T. I. Marks. 1991. Pre- to postnatal reduction in ultrasound attenuation coefficient of the liver. *Invest. Radiol.* 26:8–12.

340 Fields, S., and F. Dunn. 1973. Correlation of echographic visualisibility of tissue with biological composition and physiological state. *J. Acoust. Soc. Amer.* 54:809–812.

341 O'Brien, W. D. 1977. The relationships between collagen and ultrasound attenuation and velocities in tissue. *Ultrasonics (International) Proceedings*. Gruildford, GB: IPS Science and Technology Press Ltd. 194–205.

342 Ham, A. W., and D. H. Cormack. 1979. *Histology*. 8th ed. N.Y.: Harper & Row.

343 O'Brien, W. D., J. Olerud, J. M. Reid, and K. K. Shung. 1981. Quantitative acoustical assessment of wound maturation with acoustic microscopy. *J. Acoust. Soc. Amer.* 69(2):575–579.

344 Olerud, J. E., W. D. O'Brien, M. A. Riederer-Henderson, D. L. Steiger, J. R. Debel, and G. F. Odland. 1990. Correlation of tissue constituents with the acoustic properties of skin and wound. *Ultrasound Medicine Biology* 16:55–64.

345 Itoh, K., G. Gosung, E. Jeno, K. Kasashara, and L. Zhao. 1983. Studies

346 Saijo, Y., M. Tanaka, H. Okawai, and F. Dunn. 1991. The ultrasonic properties of gastric cancer tissues obtained with a scanning acoustic microscope system. *Ultrasound Medicine Biology* 17:709–714.

345 (continued) of the relationship between acoustic patterns produced by liver carcinoma in ultrasonography and in scanning acoustic microscopy. *Asian Medical J.* 26(9):585–597.

347 Yuhas, D. E., and L. W. Kessler. 1979. Acoustic microscopic analysis of myocardium. Ultrasonic tissue characterization. Part 2. Ed. M. Linzer. In: *National Bureau of Stand. Special publ.* Vol. 525. 73–79.

348 Saijo, Y., M. Tanaka, H. Okawai, H. Sasaki, S. Nitta, and F. Dunn. 1997. Ultrasonic tissue characterization of infracted myocardium by scanning acoustic microscopy. *Ultrasound Medicine Biology* 23:77–85.

349 Saijo, Y., H. Sasaki, H. Okawai, S. Nitta, and M. Tanaka. 1998. Acoustic properties of atherosclerosis of human aorta obtained with high-frequency ultrasound. *Ultrasound Medicine Biology* 24:1061–1064.

350 Saijo, Y., A. Iguchi, K. Tabayashi, K. Kobayashi, H. Sasaki, A. Tanaka, N. Hozumi, and S. Nitta. 2002. Detecting cardiac allograft rejection by acoustic microscopy. In: *Acoustical Imaging Symposium Proceedings*. N.Y.–London–Moscow: Kluwer Academic/Plenum Publ. Vol. 26. 105–110.

351 Mottley, J. G., and J. G. Miller. 1990. Anisotropy of the ultrasonic attenuation in soft tissues measured *in vitro*. *J. Acoust. Soc. Amer.* 88:1203–1210.

352 Marmor, M. F., H. K. Wickramasinghe, and R. A. Lemons. 1977. Acoustic microscopy of the human retina and pigment epithelium. *Invest. Opth. Vis. Sci.* 16(7):660–666.

353 Rugar, D., J. Heiserman, S. Minden, and C. F. Quate. 1980. Acoustic microscopy of human metaphase chromosomes. *J. Microscopy* 120:193–199.

354 Hildebrand, J. A., D. Rugar, and C. F. Quate. 1982. Biological acoustic microscopy living cells at 37 °C and fixed cells in cryogenic liquids. *40th Annual Electron Microscopy Soc. Amer. Proceedings*. Washington, DC. 174–177.

355 Hildebrand, J. A., and D. Rugar. 1984. Measurement of cellular elastic properties by acoustic microscopy. *J. Microscopy* 134(3):245–260.

356 Cachon, J., M. Cachon, and J. Brunton. 1981. An ultrastructural study of the effects of very high frequency ultrasounds on a microtubular system. *Biol. Cell* 40(1):69–72.

357 Lindner, A., S. Winkelhaus, and M. Hauser. 1992. In: *Acoustical imaging*. Eds. H. Ermert and H.-P. Harjes. N.Y.: Plenum Press. Vol. 19. 523–528.

358 Hildebrand, J. A. 1985. Observation of cell-substrate attachment with the acoustic microscope. *IEEE Trans. Sonics Ultrason.* SU-32(2):332–340.

359 Bereiter-Hahn, J. 1987. Scanning acoustic microscopy visualizes cytomechanical responses to citochalasin D. *J. Microscopy* 146:29–39.

360 Litniewski, J., and J. Bereiter-Hahn. 1990. Measurements of cells in culture by scanning acoustic microscopy. *J. Microscopy* 158:95–107.

361 Lüers, H. K., J. Bereiter-Hahn, and J. Litniewski. 1991. SAM investigations: The structural basis of cell surface stiffness of cultured cells. *Acoustical Imaging* 19:511–516.

362 Lüers, H., K. Hillmann, J. Litniewski, and J. Bereiter-Hahn. 1992. Acoustic microscopy of cultured cells: Distribution of forces and cytoskeletal elements. *Cell Biophysics* 18:279–293.

363 Bereiter-Hahn, J., and H. Lüers. 1994. The role of elasticity in the motile behaviour of cells. In: *Mechanics of actively locomoting cells*. Ed. N. Akhas. *NATO ASI Ser. H: Cell Biol.* Vol. 84. 181–230.

364 Karl, I., and J. Bereiter-Hahn. 1998. Cell contraction caused by microtubule disruption is accompanied by shape changes and an increased elasticity measured by scanning acoustic microscopy. *Cell Biochem. Biophys.* 29:225–241.

365 Bereiter-Hahn, J., F. Berghofer, T. Kundu, C. Penzkofer, and K. Hillmann. 1992. Evaluation of mechanical properties of cells by scanning

acoustic microscopy using $V(z)$-characteristics. In: *Acousto-optics and acoustic microscopy*. Eds. J. Gracewski and T. Kundu. Vol. 140. 71–80.

366 Briggs, G. A. D., J. Wang, and R. Gundle. 1993. Quantitative acoustic microscopy of individual living human cells. *J. Microscopy* 172:3–12.

367 Bereiter-Hahn, J., I. Karl, H. Lüers, and M. Voth. 1995. Mechanical basis of cell shape: Investigation with the scanning acoustic microscope. *Biochem. Cell Biol.* 73:337–348.

368 Bereiter-Hahn, J., C. Blase, T. Kundu, and O. Wagner. 2002. Cells as seen with the acoustic microscope. In: *Acoustical imaging*. N.Y.–London–Moscow: Kluwer Academic/Plenum Press. Vol. 26. 83–90.

369 Sherar, M. D., M. B. Noss, and F. S. Foster. 1987. Ultrasound backscatter microscopy images of the internal structure of living tumour spheroids. *Nature* 330:493–495.

370 Berube, L. R., K. Harasievicz, F. S. Foster, E. Dobrowsky, M. D. Sherar, and A. M. Rauth. 1992. Use of a high-frequency ultrasound microscope to image the action of 2-nitroimidazoles in multicellular spheroids. *Brit. J. Cancer.* 65:633–640.

371 Gertner, M. R., B. C. Wilson, and M. D. Sherar. 1998. High-frequency ultrasound properties of multicellular spheroid during heating. *Ultrasound Medicine Biology* 24(3):461–468.

372 Saijo, Y., H. Sasaki, N. Kataoka, M. Sato, S. Nitta, and M. Tanaka. 1998. Morphological and acoustical changes of endothelium by fluid shear stress. *Proc. IEEE Ultrasonics.* 1333–1336.

373 Czarnota, G. J., M. C. Kolios, H. Vaziri, S. Benchimol, F. P. Ottensmeyer, M. D. Sherar, and J. W. Hunt. 1996. Ultrasonic biomicroscopy of viable, dead and apoptotic cells. *Ultrasound Medicine Biology* 23(6):961–965.

374 Heizerman, J. E., and C. F. Quate. 1986. The scanning acoustic microscope. Frontiers in Physical Acoustics. XCIII Corso. 343–394.

375 Jipson, V. B., and C. F. Quate. 1978. Acoustic microscopy at optical wavelengths. *Appl. Phys. Lett.* 32(12):789–791.

376 Attal, J., and C. F. Quate. 1976. Investigation of some low ultrasonic absorption liquids. *J. Acoust. Soc. Amer.* 1:69–73.

377 Jipson, J. B. 1979. Acoustic microscopy of interior planes. *Appl. Phys. Lett.* 35(5):385–387.

378 Lemons, R. F., and C. F. Quate. 1979. Acoustic microscopy. In: *Physical acoustics*. Eds. W. P. Mason and R. N. Thurston. N.Y.: Academic Press. Vol. XIV. Ch. 1. 1–92.

379 Attal, J. 1980. Acoustic microscopy: Imaging microelectronic circuits with liquid metals. In: *Scanned image microscopy*. Ed. E. A. Ash. N.Y.: Academic Press. 97–118.

380 Heizerman, J., D. Rugar, and C. F. Quate. 1980. Cryogenic acoustic microscopy. *J. Acoust. Soc. Amer.* 67(5):1629–1637.

381 Levin, V. M., and L. A. Chernozatonskii. 1985. Theoretical approach to acoustic microscopy of crystals and other solids. *Symposium (International) on Microscopy in Science Proceedings*. Moscow. 54–61.

382 Roskin, G. I. 1951. *Microscopic technique*. Moscow: Izd. Sovetskaya Nauka.

383 Fossum, J. O., and J. D. Cheeke. 1986. Acoustic microscopy applied to measurements of sound absorption in liquid propane. *J. Acoust. Soc. Amer.* 79(4):983–985.

384 Beck, K., and J. Bereiter-Hahn. 1981. Evaluation of reflection interference contrast microscopy images of living cells. *Micros. Acta* 84:153.

385 Bereiter-Hahn, J., and H. Lüers. 1990. Shape changes and force distribution in locomoting cells. Investigation with reflected light and acoustic microscopy. *Eur. J. Cell Biol.* 53(31):85.

386 Lüers H., K. Hillmann, J. Litniewski, and J. Bereiter-Hahn. 1992. Acoustic microscopy of cultured cell: Distribution of forces and cytoskeletal elements. *Cell Biophys.* 18:279.

387 Bereiter-Hahn, J., J. Litniewski, K. Hillmann, A. Krapohl, and L. Zylberberg. 1992. What can scanning

acoustic microscopy tell about animal cells and tissues? *Acoust. Imaging* 17:27.

388 Bereiter-Hahn, J., R. Strohmeier, and K. Beck. 1983. Determination of the thickness profile of cells with the reflection contrast microscope. *Scient. Techn. Inf.* 8:125.

389 Strohmeier, R., and J. Bereiter-Hahn. 1987. Hydrostatic pressure in epidermal cells is dependent on Ca-mediated contraction. *J. Cell Sci.* 88:631.

390 Daft, C. M. W., and G. A. D. Briggs. 1989. The elastic microstructure of various tissues. *J. Acoust. Soc. Amer.* 85:416.

391 Daft, C. M. W., G. A. D. Briggs, and W. D. O'Brien. 1989. Frequency dependence of tissue attenuation by acoustic microscopy. *J. Acoust. Soc. Amer.* 85:2194.

392 Akashi, N., J. Kushibiki, and N. Chubachi. 1989. Quantitative characterization of biological tissues by acoustic microscopy-effect of multiple reflection and viscosity. Technical Report of the IECE 43.80 Ev. 43. 767.

393 Kundu, T., J. Bereiter-Hahn, and K. Hillmann. 1989. Calculating acoustical properties of cells: Influence of surface topography and liquid layer between cell and substrate. *J. Acoust. Soc. Amer.* 85:2086.

394 Wang, J., R. Gundle, and G. A. D. Briggs. 1990. The measurement of acoustic properties of living human cells. *Trans. Roy. Microsc. Soc.* Vol. 1. Ed. H. Y. Elder. 91.

395 Tittmann, B. R., C. Miyasaka, A. Mastro, and R. R. Mercer. 2007. Study of cellular adhesion with scanning acoustic microscopy. *IEEE Trans. Ultrason. Ferroelec. Freq. Contr.* 54(8):1502–1513.

396 Miyasaka, C., R. Halter, B. R. Tittmann, and N. Nicholas. 2001. High-power acoustic insult to cells as studied by acoustic microscopy. In: *Ultrasonic Symposium Proceedings.* Ed. D. E. Yates. Vol. 2. 1273–1276.

397 Kundu, T., J. Bereiter-Hahn, and K. Hillmann. 1991. Measuring elastic properties of cells by evaluation of scanning acoustic microscopy $V(Z)$ values using simplex algorithm. *Biophys. J.* 56:1194.

398 Sasaki, Y., T. Endo, T. Yamagishi, and M. Sakai. 1992. Thickness measurement of thin-film layer on an anisotropic substrate by phase-sensitive acoustic microscope. *IEEE Trans. Ultrason.* 39(5):638.

399 Endo, T., Y. Sasaki, T. Yamagishi, and M. Sakai. 1992. Determination of sound velocities by high frequency complex $V(z)$ measurements in acoustic microscopy. *Jpn. J. Appl. Phys.* Suppl. 31(1):160.

400 Endo, T., C. Abe, M. Sakai, and M. Ohno. 1993. Measurement of dispersion relation in acoustic velocities for the thin film on an anisotropic substrate using acoustic microscope. *IEEE Ultrason. Inter. '93 Proceedings.* 45.

401 Nishida, M., and T. Endo. 1993. Measurement of elastic moduli in local area by scanning acoustic microscope. *Experimental Theoretical Mechanics '93 Proceedings.* 290.

402 Tittmann, B. R., and C. Miyasaka. 2003. Imaging and quantitative data acquisition of biological cells and soft tissues with scanning acoustic microscopy. In: *Science, technology and education of microscopy: An overview.* Badajoz, Spain: Fomatex. 325–344.

403 Tittmann, B. R., and C. Miyasaka. 2002. Thermal and acoustical insult to cells as studied by in-vivo acoustic microscopy. *ASME PVP.* 450/NDE-Vol 22 NDE Engineering: Applications. 43–48.

404 Miyasaka, C., J. Du, and B. R. Tittmann. 2002. Characterization of thin biological tissue with scanning acoustic microscopy. *ASME PVP.* 450/NDE-Vol 22 NDE Engineering: Applications. 27–32.

405 Foster, F. S., C. J. Pavlin, K. A. Harasiewicz, D. A. Christopher, and D. H. Turnbull. 2000. Advances in ultrasound biomicroscopy. *Ultrasound Medicine Biology* 26(1):1–27.

406 Goldman, D. E., and T. F. Hueter. 1956. Tabular data of the velocity and absorption of high frequency sound

in mammalian tissue. *J. Acoust. Soc. Amer.* 28:35–37.

407 Chivers, R. C. 1981. Review paper: Tissue characterization. *Ultrasound Medicine Biology* 7:1–20.

408 Johnston, R. L., S. A. Goss, V. Maynard, J. K. Brady, L. A. Frizzell, W. D. O'Brien, and F. Dunn. 1979. *Elements of tissue characterization. Ultrasonic tissue characterization II.* Washington, DC: U.S. Government Printing Office Bureau.

409 Piruzian, L. A., T. A. Senjushkina, O. V. Kolosov, V. M. Levin, and R. G. Maev. 1985. Acoustic microscopy of biological tissues. *1st Joint Soviet – W. Germany Symposium (International) on Microscope Photometry and Acoustic Microscopy in Science Proceedings.* Moscow. 137–146.

410 Kolosov, O. V., V. M. Levin, R. G. Maev, and T. A. Senjushkina. 1987. The use of acoustic microscopy for biological tissue characterization. *Ultrasound Medicine Biology* 13(8):477–483.

411 Čusak, S., and A. Miller. 1979. Determination of the elastic constants of collagen by Brillouin light scattering. *J. Mol. Biol.* 135:39–51.

412 Goss, S. A., and W. D. O'Brien. 1979. Direct ultrasonic velocity measurements of mammalian collagen threads. *J. Acoust. Soc. Amer.* 65:507–511.

413 Saugozsis, Yu. Zh., and R. Yu. Volkolakova. 1983. The role of mechanostructural peculiarities of the fibrous eye shell in changes of its shape. *Sovremennye Problemy Biomekhaniki* 1:180–202.

414 Mennier, J., L. Katz, P. Christel, and L. Sedel. 1988. A reflection scanning acoustic microscope for bone and bone–biomaterials interface studies. *J. Orthopaedic Research* 6(5):770.

415 Pilliar, R., H. Cameron, J. Welsh, et al. 1981. Radiographic and morphologic studies of load-bearing porous-surfaced structural implants. *J. Clinical Orthopedics* 6(5):770.

416 Pilliar, R. 1987. Porous-surfaced metallic implants for orthopaedic application. *J. Biomed. Material Research: Applied Biomaterials* 21(A1):1.

417 Baum, G., I. Greenwood, S. Slawski, and R. Smirnow. 1963. Observation of internal structures of teeth by ultrasonography. *Science* 139:495–496.

418 Kossoff, G., and C. J. Sharp. 1966. Examination of the contents of the pulp cavity in teeth. *Ultrasonics* 4:77–83.

419 Barber, F. E., S. Lees, and R. R. Lobene. 1969. Ultrasonic pulse-echo measurement in teeth. *Archs. Oral Biol.* 14:745–760.

420 Reich, F. R., B. B. Brenden, and N. S. Porter. 1967. Ultrasonic imaging of teeth. Report of Batelle Memorial Institute. Pacific Northwest Laboratory. Richland, Washington.

421 Peck, S. D., and G. A. D. Briggs. 1986. A scanning acoustic microscopy study of the small caries lesion in human enamel. *Caries Res.* 20:356–360.

422 Peck, S. D., J. M. Rowe, and G. A. D. Briggs. 1989. Studies of sound and carious enamel with the quantitative acoustic microscope. *J. Dent. Res.* 68(2):107–112.

423 Harkness, R. D. 1966. *Collagen.* Oxford: Sci. Progress. 54:257–274.

424 Miller, A. 1976. Molecular packing in collage fibrils. In: *Biochemistry of collagen.* Eds. P. Ramachpandran and B. Reddi. N.Y.: Plenum Press. 85–136.

425 Lees, S., J. D. Heely, and J. M. Ahern 1982. Axial phase velocity in rat tail tendon fibers at 100 MHz by ultrasonic microscopy. *Proc. IEEE Ultrasonics Symp.* 638–639.

426 Kolosov, O. V., V. M. Levin, R. G. Maev, and T. A. Senyushkina. 1986. Acoustic microscopy of collagen tissues. *Med. Biomekh.* 1:200–205.

427 Northrop, G. A., and J. P. Wolf. 1984. *Phys. Rev. Lett.* 52:2156.

428 Lees, S., J. Heely, and P. Cleary 1981. Some properties of the organic matrix of a bovine bone. *Calc. Tissue Int.* 33:83–86.

Additional Reading

Adler, L., X. Jia, and G. Quentin. 1995. *Proc. SPIE* 26:228.

Agner, T., and J. Serup. 1989. Quantification of the DMSO-response – a test for assessment of sensitive skin. *Clin. Exp. Dermatol.* 14:214–217.

Agner, T., and J. Serup. 1989. Skin reactions to irritants assessed by non-invasive bioengineering methods. *Contact Dermatitis* 20:352–359.

Agner, T., and J. Serup. 1989. Seasonal variation of skin resistance to irritants. *Br. J. Dermatol.* 121:323–328.

Agner, T., and J. Serup. 1990. Individual and instrumental variations in irritant patch-test reactions – clinical evaluation and quantification by bioengineering methods. *Clin. Exp. Dermatol.* 15:29–33.

Alexander, H., and D. L. Miller. 1979. Determining skin thickness with pulsed ultrasound. *J. Invest. Dermatol.* 72:17–19.

Alfano, R. R., W. Lam, H. J. Zarrabi, M. A. Alfano, J. Cordero, D. B. Tata, and C. E. Swenberg. 1984. Human teeth with and without caries studied by laser scattering, fluorescence and absorption spectroscopy. *IEEE J. Quantum Electronics* 20:1512–1515.

Altmeyer, P., K. Hoffman, M. Stucker, S. Goetz, and S. El-Hammal. 1992. General phenomena of ultrasound in dermatology. In: *Ultrasound in dermatology*. Berlin: Springer-Verlag. 55–79.

Aristizabal, O., D. A. Christofer, F. S. Foster, and D. H. Turnbull. 1997. Measuring blood flow in the mouse embryo. *IEEE Ultrasonics Symposium Proceedings* 2:1489–1492.

Aslanides, I. M., P. E. Libre, R. H. Silverman, D. Z. Reinstein, D. R. Lazzaro, M. J. Rondeau, G. K. Harmon, and D. J. Coleman. 1995. High frequency ultrasound imaging in papillary block glaucoma. *Brit. J. Ophthalmol.* 79:972–976.

Atalar, A. A. 1985. Penetration depth of the scanning acoustic microscope. *IEEE Trans. Sonic. Ultrason.* SU-32(2):164–167.

Baker, A. C. 1991. *Phys. Med. Biol.* 36:1457.

Bereiter-Hahn, J. 1987. Comparison of the appearance of cultured cells observed using scanning acoustic microscopy with that obtained by interference and fluorescence microscopy. In: *Scanning imaging*. Techn. SPIE Vol. 809. 162–165.

Bereiter-Hahn, J. 1995. Probing biological cells and tissues with acoustic microscopy. In: *Advances in acoustic microscopy*. N.Y.–London: Plenum Press. Vol. 1. 79–115.

Bereiter-Hahn, J., C. H. Fox, and B. Thorell. 1979. Quantitative reflection contrast microscopy of living cells. *J. Cell Biol.* 82:767–779.

Berson, M., L. Vaillant, F. Patat, and L. Pourcelot. 1992. High-resolution real-time ultrasonic scanner. *Ultrasound Med. Biol.* 18:471–478.

Breazeale, M. A., L. Adler, and G. W. Scott. 1977. Interaction of ultrasonic waves incident at the Rayleigh angle onto a liquid–solid interface. *J. Appl. Phys.* 48:530–537.

Breazeale, M. A., and J. Philip. 1980. In: *Physical acoustics*. Ed. W. P. Mason. N.Y.: Academic Press. Vol. 17. 1–60.

Briggs, G. A. D. 1984. Scanning electron microscopy and scanning acoustic microscopy: A favorable comparison. *Scanning Electron Microscopy* 3:1041–1052.

Briggs, G. A. D. 1990. Ultrasound – its chemical, physical and biological effects. In: *Interdisciplinary Sci. Rev.* 15:190–191.

Briggs, G. A. D., and M. Hoppe. 1991. Acoustic microscopy. In: *Images of materials*. Eds. D. B. Williams, A. R. Pelton, and R. Gronsky. Oxford University Press.

Broby-Johansen, U., T. Karlsmark, L. J. Peterson, and J. Serup. 1990. Ranking of the antipsoriatic effect of various topical corticosteroids applied under a hydrocolloid dressing: skin thickness, blood flow and colour measurements compared to clinical assessments. *Clin. Exp. Dermatol.* 15:343–348.

Bukhny, M. A., O. V. Kolosov, V. M. Levin, and R. G. Maev. 1990. Formation of $A(z)$-dependence in transmission acoustic microscope. JASA.

Cantrell, J. H., and G. J. Zhang. 1998. *Appl. Phys.* 84:5469.

Cervenka, P., and P. J. Alais. 1990. *J. Acoust. Soc. Amer.* 88:473.

Chernozatonsky, L. A., V. M. Levin, R. G. Maev, T. A. Senjushkina, and M. A. Bukhny. 1988. Analysis of microstructure of superconductive materials with the use of scanning acoustic microscope. *IEEE Ultrasonic Symposium Proceedings.* Chicago 2. 1017–1019.

Chertov, A., and R. Gr. Maev. 2005. Onedimensional model of acoustic wave propagation in the multilayered structure of the spot weld. *IEEE Transaction on Ultrasonics, Ferroelectrics, and Frequency Control* 52(10):1783–1790.

Chertov, A., R. Gr. Maev, and F. Severin. 2007. Acoustic microscopy of internal structure of resistance spot welds. *IEEE Transaction on Ultrasonics, Ferroelectrics, and Frequency Control* 54(8):1521–1529.

Cook, B. D. 1961. *IASA* 33:832.

Dace, G. E., O. Buck, and R. B. Thompson. 1992. In: *Vibro-acoustic characterization of materials and structures.* N.Y.: ASME, NCA. Vol. 14. 221.

Darling, A. I. 1956. Studies of the early lesion of enamel caries with transmitted light, polarised light and radiography. *British Dent. J.* 101:289–297, 329–341.

Denisov, A. A., C. M. Shakarji, B. B. Lawford, R. Gr. Maev, and J. M. Paille. 2004. Spot weld analysis with 2D ultrasonic array. *J. of Research of the National Institute of Standards and Technology* 2:109.

Denisov, A. F., E. Yu. Bakulin, N. M. Livanova, A. A. Popov, R. Gr. Maev, and L. Denisova. 2006. Investigation of the microstructure and physical-mechanical properties of mixtures of butadiene–nitrile rubber with PVC with different contents by methods of acoustical microscopy. *Chemical Physics* 7:243.

Denisova, L., R. Gr. Maev, E. Khramcova, O. Dadasheva, A. F. Denisov, and E. Snetkova. 2006. Application of SAM for investigation of embryonic growth of Japanese quail. *J. of Technology for Living Systems* 3(1):56–63.

Denisova, L., R. Gr. Maev, I. Yu. Poyurovskaya, A. F. Denisov, T. V. Grineva, E. Yu. Maeva, and E. Yu. Bakulin. 2003. Comparative investigation of the glassionomer cements "Dentis" microstructure and mechanical properties. *J. of Mechanical Composites Materials* 9(1):24–33.

Denisova, L., R. Gr. Maev, I. Ya. Poyurovskaya, T. Grineva, A. F. Denisov, E. Yu. Maeva, and E. Bakulin. 2004. The use of acoustic microscopy to study the mechanical properties of glass–Ionomer cement. *Dental Materials* 20:358–363.

Denisova, L., N. Nasirova, R. Gr. Maev, A. S. Grigorian, L. A. Grigorianz, A. A. Denisov, T. V. Grineva, E. Yu. Maeva, and F. M. Severin. 2003. Investigation of the microstructure of dental sealing materials using methods of acoustical microscopy. *J. of Stomatology* 81(1):26–31.

De Rigal, J., C. Escoffier, B. Querleux, P. Agache, and J. L. Leveque. 1989. Assessment of aging of the human skin *in vivo.* Ultrasound imaging. *J. Invest. Dermatol.* 93:621–625.

De Rigal, J., and J. L. Leveque. 1985. *In vivo* measurement of the stratum corneum elasticity. *Bioeng. Skin* 1:13–23.

Diridollou, S., M. Berson, V. Varbe, D. Black, B. Karlsson, F. Auriol, J. M. Gregorire, C. Yvon, L. Vaillant, Y. Gall, and F. Patat. An *in vivo* method for measuring the mechanical properties of the skin using ultrasound. *Ultrasound Med. Biol.* 24(2):215–224.

Du, G., and M. A. Breazeale. 1986. *J. Acoust. Soc. Amer.* 80:212.

Du, G., and M. A. Breazeale. 1987. *J. Acoust. Soc. Amer.* 81:51.

Edwards, C., and P. A. Payne. 1984. Ultrasound velocities in skin components. International society for bioengineering and the skin: Ultrasound in dermatology. *Symposium Proceedings.* Liege. 187–189.

Eggleton, R. C., and F. S. Vinson. 1977. Heart model supported in organ culture and analysed by acoustic microscopy. *Acoust. Holography* 7:31–35.

El-Hammal, S., K. Hoffman, T. Auer, M. Korten, P. Altmeyer, A. Hoss, and H. Ermert. A 50 MHz high-resolution ultrasound imaging system for dermatology. In: *Ultrasound in dermatology.* Berlin: Springer-Verlag. 297–322.

Enicolopian, N. S., and E. Y. Maeva. 1990. Application of scanning acoustic microscopy for research and technology of new polymer composite materials. *Symposium (International) "MashTec 90.".* Dresden, Germany. 2:333.

Escoffier, C., B. Querleux, J. De Rigal, and J. L. Leveque. 1986. *In vitro* study of the velocity of ultrasound in the skin. *Bioeng. Skin* 2:87–94.

Finlay, A. Y., H. Moseley, and T. C. Duggan. 1987. Ultrasound transmission time: An *in vivo* guide to nail thickness. *Br. J. Dermatol.* 117:765–770.

Fornage, B. D., and J. L. Deshayes. 1986. Ultrasound of normal skin. *J. Clin. Ultrasound* 14:619–622.

Foster, F. S., C. J. Pavlin, G. R. Lockwood, L. K. Ryan, K. A. Harasiewicz, L. Berube, and A. M. Rauth. 1993. Principles and application of ultrasound backscatter microscopy. *IEEE Transaction UFFC* 40:608–616.

Foster, F. S., C. J. Pavlin, B. Starkoski, and K. A. Harasiewicz. 1990. Ultrasound backscatter microscopy of the eye *in vivo*. *IEEE Ultrasonics Symposium Proceedings.* 1481–1484.

Foster, J. S., and D. Rugar. 1983. High resolution acoustic microscopy in superfluid helium. *Appl. Phys. Lett.* 42:869–871.

Foster, J. S., and D. Rugar. 1985. Low-temperature acoustic microscopy. *IEEE Trans. Sonic. Ultrasonics* SU-32(2):139–151.

Foster, F. S., M. Y. Zhang, Y. Q. Zhou, G. Liu, J. Mehi, E. Cherin, K. A. Harasiewicz, B. G. Starkoski, L. Zan, D. A. Knapik, and S. L. Adamson. 2002. A new ultrasound instrument for *in vitro* imaging of mice. *Ultrasound Med. Biol.* 28(9):1165–1172.

Fournier, C., S. L. Bridal, G. Berger, and P. Laugier. 2001. Reproducibility of skin characterization with backscattered spectra (12–25 MHz) in healthy subjects. *Ultrasound Med. Biol.* 27(5):603–610.

Gilmore, R. A., R. P. Pollack, and J. L. Katz. 1970. Elastic properties of bovine dentine and enamel. *Arch. Oral Biol.* 15:787–796.

Gniadecka, M., J. Serup, and J. Sondergaard. 1994. Age-related diurnal changes of dermal oedema: Evaluation by high-frequency ultrasound. *Br. J. Dermatol.* 131:849–855.

Gupta, A. K., D. H. Turnbull, and F. S. Foster. 1996. High frequency 40 MHz ultrasound. A possible non-invasive method for the assessment of the boundary of basal cell carcinomas. *Dermatol. Surg.* 22:131–136.

Harland, C. C., S. G. Kale, P. Jackson, P. S. Mortimer, and J. C. Bamber. 2000. Differentiation of common benign pigmented skin lesions from melanoma by high-resolution ultrasound. *Br. J. Dermatol.* 143:1–10.

Heiserman, J. E., and C. F. Quate. 1986. The scanning acoustic microscopy. In: *Frontiers in physical acoustics.* Vol. XCIII Corso. 343–394.

Henneke, E. G. 1972. Reflection–refraction of a stress wave at a plane boundary between anisotropic media. *J. Acoust. Soc. Amer.* 51(1–2):210–217.

Hirai, T., and M. Fumiiri. 1995. Ultrasonic observation of the nail matrix. *Dermatol. Surg.* 21:158–161.

Hoffman, K., El Hammal S., K. Winkler, J. Jung, K. Pistorius, P. Almeyer. 1992. Skin tumours in high-frequency ultrasound. In:

Ultrasound in dermatology. Berlin: Springer-Verlag. 181–202.

Hollis, R., and R. Hammer. 1980. Defect detection for microelectronics by acoustic microscopy. In: *Scanned image microscopy.* Ed. E. Ash. Academic Press. 155–164.

Hollis, R. L., and R. Hammer. 1984. Subsurface imaging of glass fibres in a polycarbonate by acoustic microscopy. *J. Mater. Sci. (GB)* 19(6):1897–1903.

Hurley, D. C., W. T. Yost, E. S. Boltz, and C. M. Fortunko. In: *Review of progress in QNDE.* Eds. D. O. Thompson and D. E. Chimenti. N.Y.: Plenum Press. Vol. 16. 1383.

Ichida, N., T. Sato, H. Miwa, and K. Murakami. 1984. *IEEE Trans. Sonic. Ultrasonics* 31:635.

Jemec, G. B. E., and J. Serup. 1989. Ultrasound structure of the human nailplate. *Arch. Dermatol.* 125:643–646.

Joa, X., A. Boumiz, and G. Quentin. 1993. *Appl. Phys. Lett.* 63:2192.

Johnson, P. A., A. Migliori, and T. J. Shankland. 1991. *J. Acoust. Soc. Amer.* 89:598.

Kolosov, O. V. 1985. Transmission raster acoustic microscope with quantitative characterization facilities. *Symposium (International) on Microscope Photometry and Acoustic Microscopy in Science Proceedings.* Moscow. 26–31.

Kushibiki, J., K. L. Ha, H. Kato, N. Chubachi, and F. Dunn. 1987. Application of acoustic microscopy to dental materials characterization. *IEEE Ultrasound Symposium.* N.Y. 837–842.

Kushibiki, J., A. Ohakubo, and N. Chubachi. 1981. Anisotropy detection in sapphire by acoustic microscopy using line-focus beam. *Electronics Lett.* 17:534–536.

Laugier, P., E. Laplace, J. L. Lefaix, and G. Berger. 1998. *In vivo* results with a new device for ultrasound monitoring of pig skin cryosurgery: The echographic cryoprobe. *J. Invest. Dermatol.* 111:101–106.

Laugier, P., J. L. Lefaix, and G. Berger. 1998. A new echographic cryoprobe for *in vivo* ultrasonic monitoring of skin cryosurgery. *Proc. IEEE Ultrasonics Symposium.* 1337–1340.

Lebertre, M., F. Ossant, J. Bouyer, L. Vaillant, S. Diridollou, and F. Patat. 2001. Ultrasound skin characterization: An *in vivo* study of intra and inter individual variations. *IEEE Ultrasonics Symposium.* 1241–1244.

Lee, H. T., M. Wang, R. Gr. Maev, E. Yu. Maeva. 2003. A study on using scanning acoustic microscopy and neural network techniques to evaluate the quality of resistance spot welding. *Intern. J. of Advanced Manufacturing Technology* 22:727–732.

Lees, S. 1968. Specific acoustic impedance of enamel and dentine. *Arch. Oral. Biol.* 13:1491–1500.

Lees, S., and F. E. Barber. 1968. Looking into teeth with ultrasound. *Science* 161:477–478.

Lees, S., F. E. Barber, and R. R. Lobene. 1970. Dental enamel: Detection of surface changes by ultrasound. *Science* 169:1314–1316.

Lees, S., F. B. Gerbard, and F. G. Oppenheim. 1973. Ultrasonic measurement of dental enamel demineralization. *Ultrasonics.* 269–273.

Lees, S., and F. R. Rollins. 1972. Anisotropy in hard dental tissues. *J. Biomechanics* 5:557–566.

Lemons, R. A., and C. F. Quate. 1974. Integrated circuits as viewed with an acoustic microscope. *Appl. Phys. Lett.* 25:251–253.

Len, K. S., and I. Yu. Solodov. 1994. Acad. Sci. DPR Korea. 3:29.

Levin, V. M., K. I. Maslov, T. A. Senjushkina, I. G. Grigorieva, and I. Baranchikova. 1991. In: *Acoustic imaging.* Eds. H. Ermert and H.-P. Harjes. N.Y.: Plenum Press. 651.

Liu, A., A. L. Joyner, and D. H. Turnbull. 1998. Alteration of limb and brain patterning in early mouse embryos by ultrasound guided injection of Shh-expressing cells. *Med. Dev.* 75:107–115.

Lizzi, F. L., C. X. Deng, and S. K. Alam. 2000. Ocular tumor treatments with focused ultrasound: Effects of beam geometry, tissue morphology, and adjacent tissues. *IEEE Ultrasonics Symposium Proceedings.* 1299–1301.

Lizzi, F. L., E. J. Feleppa, A. Kalisz, R. H. Silverman, and D. J. Coleman. 2000. High-resolution 3-dimensional visualization and morphological assays of the in vivo ciliary body. *Proc. IEEE Ultrasonics Symposium.* 1421–1423.

Lucassen, G. W., W. L. N. Van der Sluys, J. J. Van Herk, A. M. Nuijs, P. E. Wierenga, A. O. Barel, and R. Lambrecht. 1997. The effectiveness of massage treatment on cellulite as monitored by ultrasound imaging. *Skin Res. Technol.* 3:154–160.

Lussi, A. 1991. Validity of diagnostic and treatment decisions of fissure caries. *Caries Res.* 25:296–303.

Maev, R. Gr. 2005. *Acoustic microscopy: Manuscript.* Moscow: TORUS PRESS.

Maev, R. Gr. 2006. Development of novel principles and methods of high resolution acoustical imaging for materials characterization: Review. *Physics in Canada* 62(2):91–98.

Maev, R. Gr., guest ed. 2007. Special issue on high resolution ultrasonic imaging in industry, materials and biomaterials applications. *IEEE Transaction on Ultrasonics, Ferroelectrics and Frequency Control* 54(8).

Maev, R. Gr., L. Denisova, E. Yu. Maeva, A. Denisov, A. Krasnov, V. Popov, A. Popova, and E. Bakulin. 2002. Investigation of the microstructure of medical polyamide and hydroxiapatite polymer composites using acoustic microscopy methods. *News in Dentistry* 101(1):84–90.

Maev, R. Gr., R. E. Green, Jr., and A. M. Siddiolo. 2006. Review of advanced acoustical imaging techniques for nondestructive evaluation of art objects. *Research in Nondestructive Evaluation* 17(4):191–204.

Maev, R. G., O. V. Kolosov, and O. I. Lobkis. 1990. Investigation of the confocal system of the transmission acoustic microscope. *Trans. Royal Microscopial Soc. MICRO 90* 1:107–110.

Maev, R. Gr., H. Shao, and E. Yu. Maeva. 1998. Measurement using ultrasonic pulse-echo method of a curved multilayered polymer system. *J. of Material Characterization* 1(2–3).

Maeva, E. Yu., I. Bruno, B. Zielinsky, M. Docker, F. Severin, and R. Gr. Maev. 2004. The use of pulse-echo acoustic microscopy to invasively determine sex of living larval sea lamprey, petromyzon marinus. *J. of Fish Biology* 65:148–156.

Maeva, E. Yu., I. Severina, S. Bondarenko, G. Chapman, B. O'Neill, F. Severin, and R. Gr. Maev. 2004. Acoustical methods for investigation of adhesively bonded structures: A review. *Canadian J. of Physics* 82:36–58.

Marks, R. 1990. Methods for assessment of cutaneous ageing. *Int. J. Cosmet. Sci.* 12:153–163.

Maslov, K. I. 1992. Acoustic scanning microscope for investigation of surface defects. *Acoustic Imaging* 19:645–649.

Melngailis, J., A. A. Maradudin, A. Seeger. 1963. *Phys. Rev.* 131:172.

Milner, S. M., O. M. Memar, G. Gherardini, J. D. C. Bennet, and L. G. Phillips. 1997. The histological interpretation of high-frequency cutaneous ultrasound imaging. *J. Dermatol. Surg.* 23:43–45.

Moreau, A. J. 1995. *J. Acoust. Soc. Amer.* 98:2745.

Morozov, G. V., R. Gr. Maev, and G. W. F. Drake. 2001. Reflection of plane electromagnetic waves from two-layered periodic structures with fluctuations in layer thickness. *Physical Review D* 64:3456–3467.

Ng, S. Y., P. A. Payne, and M. W. J. Ferguson. 1993. Ultrasonic imaging of experimentally induced tooth decay. *Acoustic Sensing and Imaging, Conference Proceedings*. IEEE No. 369. 82–86.

Nikoonahad, M. 1984. Recent advances in high resolution acoustic microscopy. *Contemp. Phys.* 25(2):129–158.

Nongaillard, B., J. M. Rouvaen, E. Bridoux, R. Terquet, and C. Brunnel. 1979. Visualization of thick specimens using a reflection acoustic microscope. *J. Appl. Phys.* 50(3):1245–1249.

O'Brien, W. D., and L. W. Kessler. 1975. Examination of mouse embryological development with an acoustic microscope. *Am. Zool.* 15:807–814.

O'Neill, B., and R. Gr. Maev. 1998. Integral approximation method for calculating ultrasonic beam propagation in anisotropic materials. *Physical Review B* 58:5479–5485.

O'Neill, B., and R. Gr. Maev. 2006. Acousto-elastic measurement of the fatigue damage in waspaloy. *J. of Research in Nondestructive Evaluation* 17(3):121–135.

O'Neill, B., and R. Gr. Maev. 2006. Application of a nonlinear boundary condition model to adhesion interphase damage and failure. *J. of the Acoustical Society of America* 120(6):3509–3517.

Olsson, M., K. Campbell, and D. H. Turnbull. 1977. Specification of mouse telencephalic and mid-hindbrain progenitors following heterotopic ultrasound guided embryonic transplantation. *Neuron*. 19:761–822.

Pan, L., L. Zan, and F. S. Foster. 1997. In vivo frequency ultrasound assessment of skin elasticity. *IEEE Ultrasonics Symposium Proceedings*. 1088–1091.

Pavlin, C. J., M. Easterbrook, J. J. Hurwitz, K. A. Harasiewicz, and F. S. Foster. 1993. Ultrasound biomicroscopy in the assessment of anterior scleral disease. *Amer. J. Ophtalmol.* 116:628–635.

Pavlin, C. J., and F. S. Foster. 1994. High frequency ultrasound biomicroscopy. *Ophtalmol. Clin. North. Amer.* 7:509–522.

Pavlin, C. J., and F. S. Foster. 1995. *Ultrasound biomicroscopy of the eye*. N.Y.: Springer-Verlag.

Pavlin, C. J., R. Harasiewicz, and F. S. Foster. 1991. Clinical application of ultrasound biomicroscopy. *Ophthalmology* 98:287–295.

Pavlin, C. J., R. Harasiewicz, and F. S. Foster. 1992. Ultrasound biomicroscopy of anterior segment structures in normal and glaucomatous eyes. *Amer. J. Ophtalmol.* 113:381–389.

Pavlin, C. J., McWhae, and F. S. Foster. 1992. Ultrasound biomicroscopy of anterior segment tumours. *Ophthalmology* 99:1220–1228.

Peck, S. D., and G. A. D. Briggs. 1987. The caries lesion under the scanning acoustic microscope. *Adv. Dent. Res.* 1:50–63.

Potash, S., C. Tello, J. Liebmann, and R. Ritch. 1994. Ultrasound biomicroscopy in pigment dispersion syndrome. *Ophthalmology* 101:322–329.

Pulgiese, P. T. 1989. Use of ultrasound in evaluation of skin care products. *Cosmet. Toil.* 104:61–75.

Ramachandran, G. N., and A. H. Reddi, eds. 1976. *Biochemistry of collagen*. N.Y.

Rippon, M. G., K. Springett, R. Walmsley, K. Patrick, and S. Millson. 1998. Ultrasound assessment of skin and wound tissue: Comparison with histology. *Skin Res. Technol.* 4:147–154.

Rokhlin, S. I., T. K. Bolland, and L. Alder. 1986. Reflection and refraction of elastic waves on a plane interface between two generally anisotropic media. *J. Acoust. Soc. Amer.* 79(4):906–918.

Sadler, J., and R. Gr. Maev. 2007. Experimental and theoretical basis of lamb waves and their applications in material sciences: Review. *Canadian J. of Physics* 85(7):707–731.

Sadler, J., B. O'Neill, and R. Gr. Maev. 2005. Ultrasonic wave propagation across a thin nonlinear anisotropic layer between two halfspaces. *J. of the Acoustical Society of America* 118(1):51–59.

Saied, A., E. Bossy, A. Watrin, B. Pellaumail, D. Loeuille, P. Laugier, P. Netter, and G. Berger. 2000. Quantitative assessment of arthritic cartilage using high-frequency ultrasound. *IEEE Ultrasonics Symposium Proceedings*. 1375–1378.

Saied, A., E. Cherin, H. Gaucher, P. Laugier, P. Gillet, J. Floquet, P. Netter, and G. Berger. 1997. Assessment of articular cartilage and subchondral bone subtle and progressive changes in experimental osteoarthritis using 50 MHz echography *in vitro*. *J. Bone Mineral Res.* 12:1378–1386.

Saied, A., B. Dehecq, M. Savoldelli, B. Briat, J. M. Legeais, and G. Berger. 1997. Evaluation of keratoprosthesis biointegration *in situ* with quantitative ultrasound backscatter microscopy. *IEEE Ultrasonics Symposium Proceedings*. 1093–1096.

Saied, A., H. Gaucher, C. Guingamp, P. Laugier, B. Terlain, P. Gillet, P. Netter, and G. Berger. 1995. Detection of early bone and cartilage remodeling in a rat model of osteoarthritis by high resolution echography. *Inflammatory Res.* 44:255–257.

Saied, A., L. Jachino, M. Boudinet, P. Giat, P. Laugier, J. M. Legeais, and G. Berger. 1995. *In vitro* high frequency ultrasound characterization of colonization of microporous polymer used to support artificial cornea. *IEEE Ultrasonics Symposium Proceedings*. 1299–1303.

Saied, A., P. Laugier, N. Bovo, G. Berger, D. Chevrier, B. Terlain, P. Gillet, and P. Netter. 1994. Experimental osteoarthritic cartilage. *In vitro* visualization of lesions and three-dimensional surface reconstruction using 50-MHz ultrasound microscope. *IEEE Ultrasonic Symposium Proceedings*. 1475–1478.

Sanghvi, N. I., A. M. Snoddy, S. L. Myers, K. D. Brandt, C. R. Reilly, and T. D. Franklin. 1990. Characterization of normal and osteoarthritic cartilage using 25 MHz ultrasound. *IEEE Ultrasonics Symposium Proceedings* 3:1413–1416.

Schoch, A. 1952. Der Schalldurchgang durch platten. *Acoustica* 2(1):1–17.

Schoch, A. 1952. Seitliche versetzung eines total reflectierten strahls bei ultrashall wellen. *Acoustica* 2(1):18–19.

Seidenary, S., A. Pagnoni, A. Di Nardo, and A. Giannetti. 1994. Echographic evaluation with image analysis of normal skin: Variations according to age and sex. *Skin Pharmacol.* 7:201–209.

Serup, J. 1984. Decreased skin thickness of pigmented spots appearing in localized scleroderma. Measurement of skin thickness by 15 MHz pulsed ultrasound. *Arch. Dermatol. Res.* 276:135–137.

Serup, J. 1984. Diameter, thickness, area and volume of skinprick histamine weals. *Allergy* 39:359–364.

Serup, J. 1984. Localized scleroderma: Thickness of sclerotic plaques as measured by 15 MHz pulsed ultrasound. *Acta Dermatol. Venereol. (Stockholm)* 64:214–219.

Serup, J. 1984. Noninvasive quantification of psoriasis plaques. Measurement of skin thickness with 15 MHz pulsed ultrasound. *Clin. Exp. Dermatol.* 9:502–508.

Serup, J. 1984. Quantification of acrosclerosis: measurement of skin thickness and skin phalanx distance in females with 15 MHz pulsed ultrasound. *Acta Dermatol. Venereol.* 64(1):33–40.

Serup, J. 1986. Localized scleroderma. Clinical, physiological, biochemical and ultrastructural studies with particular reference to quantification of scleroderma. *Acta Dermatol. Venereol. (Stockholm)* 65(122):1–61.

Serup, J. 1992. Ten years experience with high frequency ultrasound examination of the skin: Development and refinement of technique and equipment. In: *Ultrasound in dermatology*. Berlin: Springer-Verlag. 41–54.

Serup, J., P. Holm, I. M. Stender, and J. Pichard. 1987. Skin atrophy and telaniectasia after topical corticosteroids as measured noninvasively with high-frequency ultrasound, evaporimetry and laser Doppler flowmetry. Methodological aspects including evaluation of regional differences. *Bioengin. Skin.* 3:43–58.

Serup, J., and B. Staberg. 1987. Ultrasound for assessment of allergic and irritant patch test reactions. *Contact Dermatitis* 17:80–84.

Serup, J., B. Staberg, and P. Klemp. 1984. Quantification of cutaneous oedema in patch test reactions by measurement of skin thickness with high-frequency pulsed ultrasound. *Contact Dermatitis* 10:88–93.

Severin, F., and R. Gr. Maev. 2005. Adaptive filtration of ultrasonic signals in the evaluation of adhesive bonds in automotive bodies. *Insight, J. of the British Institute of Non-Destructive Testing* 47(1):8–11.

Shaum, C. T., C. L. Anthony, and G. F. Ian. 1993. *J. Acoust. Soc. Amer.* 93:148.

Sherar, M. D., B. G. Starkoski, W. B. Taylor, and F. S. A. Foster. 1989. A 100 MHz B-scan ultrasound backscatter microscope. *Ultrason. Imaging* 11:95–104.

Silverman, R. H., D. Z. Reinstein, T. Raevsky, and D. J. Coleman. 1997. Improved system for sonographic imaging and biometry of the cornea. *J. Ultrasound Med. Biol.* 16:117–124.

Silverman, R. H., M. J. Rondeau, F. L. Lizzi, and D. J. Coleman. 1995. Three-dimensional high-frequency ultrasonic parameter imaging of anterior segment pathology. *Ophthalmology* 102:837–843.

Slavkin, H. C., and R. L. Kirschtein. 2000. Oral health in America. A report of Surgeon General. U.S. Department of Health and Human Services. National Institute of Dental and Craniofacial Research. 302.

Slobin, J. A., D. L. Stocum, and W. D. O'Brien. 1986. Amphibian limb regeneration curves generated by the scanning laser acoustic microscope. *J. Histochem. Cytochem.* 34(1):53–56.

Smith, I. R., and H. K. Wickramasinghe. 1982. Differential phase contrast in the acoustic microscope. *Electronics Lett.* 18:92–94.

Solodov, I. Yu., A. F. Asainov, and K. S. Len. 1993. *Ultrasonics* 31:91.

Somekh, H. L., H. L. Bertoni, G. A. D. Briggs, and N. A. Burton. 1985. A two-dimensional imaging theory of surface discontinuities with the scanning acoustic microscope. *Proc. Royal Soc. London A* 401:29–51.

Sondergaard, J., J. Serup, and G. Tikjob. 1985. Ultrasound A- and B-scanning in clinical and experimental dermatology. *Acta Dermatol. Venereol. (Stockholm)* 65(120):76–82.

Stiller, M. J., J. Driller, J. L. Shupak, C. G. Gropper, M. C. Rorke, and F. L. Lizzi. 1993. Three-dimensional imaging for diagnostic ultrasound in dermatology. *J. Amer. Acad. Dermatol.* 29:171–175.

Sulewski, P. E., and D. J. Bishop. 1983. Acoustic contrast of dislocation line in an isotropic medium. *J. Appl. Phys.* 54(10):5715–5717.

Sulewski, P. E., R. C. Dynes, S. Mahajan, and D. J. Bishop. 1983. Study of defects in optoelectronic materials using a scanning acoustic microscope. *J. Appl. Phys.* 54(10):5711–5714.

Tan, C. Y., E. Roberts, B. Statham, and R. Marks. 1981. *Brit. J. Dermatol.* 105:25–26.

Tervola, K. M. U., S. G. Foster, and W. D. O'Brien. 1985. Ultrasonic attenuation measurement techniques at 100 MHz with the scanning laser acoustic microscope. *Proc. IEEE Trans. Sonic. Ultrason.* SU-32(2):259.

Tikjob, G., V. Kassis, and J. Sondergaard. 1984. Ultrasonic B-scanning of the human skin. An introduction of a new ultrasonic skin-scanner. *Acta Dermatol. Venereol. (Stockholm)* 64:67–90.

Titov, S., R. Maev, and A. Bogatchenkov. 2001. Suppression of edge waves by apodization of wide band cylindrical transducers. *Acoust. Imag. Symposium* 26:437–442.

Titov, S. R. Gr. Maev, and A. Bogatchenkov. 2003. Wide-aperture, line-focused ultrasonic material characterization system based on lateral scanning. *IEEE Transaction on Ultrasonics, Ferroelectrics, and Frequency Control* 50(8):1046–1056.

Titov, S., R. Gr. Maev, and A. Bogachenkov. 2004. Use of ultrasonic transducer array for measuring the velocity and attenuation of leaky acoustic waves. *Technical Physics Letters* 30(10):883–885.

Titov, S., R. Gr. Maev, and A. Bogachenkov. 2004. Application of array transducer for ultrasonic wave velocity measurements and measuring of attenuation of acoustical surface leaky waves. *Sov. Physics J. of Experimental and Theoretical Physics (JETP) Letters* 30(20):89–94.

Titov, S. A., R. Gr. Maev, and A. N. Bogachenkov. 2006. Measurements of velocity and attenuation of leaky waves using an ultrasonic array. *J. of Ultrasonics* 44(2):182–187.

Trop, G., C. Pavlin, A. Bau, C. Baumal, and F. S. Foster. 1994. Malignant glaucoma: Clinical and ultrasound biomicroscopic characterization. *Ophthalmology* 101:1030–1035.

Tsai, C. S., S. K. Wang, and C. C. Lee. 1977. Visualization of solid material joints using a transmission type scanning microscope. *Appl. Phys. Lett.* 31(5):317–320.

Tucker, P. A., and R. G. Wilson. 1980. Acoustic microscopy of polymers. *J. Polym. Sci.* 18:97–103.

Turnbull, D. H. 1999. In utero ultrasound backscatter microscopy of early stage mouse embryos. *Comput. Med. Imaging Graphics* 23:25–31.

Turnbull, D. H., T. S. Bloomfield, F. S. Foster, and A. L. Joyner. 1995. Ultrasound backscatter microscope analysis of early mouse embryonic brain development. *Proc. National Acad. Sci.* 92:2239–2243.

Turnbull, D. H., J. A. Ramsay, G. S. Shivji, T. S. Bloomfield, L. From, D. N. Sauder, and F. S. Foster. 1996. Ultrasound backscatter microscope analysis of mouse melanoma progression. *Ultrasound Med. Biol.* 22(7):845–853.

Turnbull, D. H., B. G. Starkoski, K. A. Harasievicz, J. L. Semple, L. From, A. K. Gupta, D. N. Sauder, and F. S. Foster. 1995. A 40–100 MHz B-scan ultrasound backscatter microscope for skin imaging. *Ultrasound Med. Biol.* 21(1):79–88.

Van Dorp, C. S. F., R. A. M. Exterkate, and J. M. Ten Cate. 1988. The effect of dental probing on subsequent enamel demineralization. *J. Dent. Child.* 55:343–347.

Wachtmann, J. B., Jr., W. E. Tefft, D. G. Lam, Jr., and R. P. Stinchfield. 1960. Elastic constants of synthetic single crystal corundum at room temperature. *J. Res. Nat. Bur. Stand.* 64A:213–228.

Weaver, J. M. R., M. G. Somekh, A. D. Briggs, S. D. Peck, and C. Ilet. 1985. Applications of the scanning reflection acoustic microscope to the study of materials science. *IEEE Trans. Sonic. Ultrason.* SU-32(2):302–312.

Weglein, R. D. 1983. Integrated circuit inspection via acoustic microscopy. *IEEE Trans. Sonic. Ultrason.* SU-30(1):40–42.

Weglein, R. D. 1985. Acoustic micrometrology. *IEEE Trans. Sonic. Ultrason.* SU-32(2):225–234.

Wichard, R., J. Schlegel, R. Haak, J. F. Roulet, and R. M. Schmitt. 1996. Dental diagnosis by high frequency ultrasound. *Acoustical Imaging* 22:329–334.

Wickramasinghe, H. K., and M. Hall. 1976. Phase imaging with the scanning acoustic microscope. *Electronics Lett.* 12(24):637–638.

Wilson, R. G., R. D. Weglein, and D. M. Bonnel. 1977. Scanning acoustic microscopy for integrated circuit diagnostics. In: *Semicond. Silicon*. Eds. H. R. Huff and E. Sirtl. New Jersey: Electrochem. Soc. Princeton. 77(2):431–440.

Wollina, U., M. Berger, and K. Karte. 2001. Calculation of nail plate and nail matrix parameters by 20 MHz ultrasound in healthy volunteers and patients with skin disease. *Skin Res. Technol.* 7:60–64.

Yamanaka, K., and Y. Emonoto. 1982. Observation of surface cracks with scanning acoustic microscope. *J. Appl. Phys.* 53(2):846–850.

Ye, S. G., K. A. Harasievicz, C. J. Pavlin, and F. S. Foster. 1995. Ultrasound characterization of ocular tissue in the frequency range from 50 MHz to 100 MHz. *IEEE Ferroelectric Frequency Control Symposium Proceedings.* 42:8–14.

Index

a

absorption 9–12, 18, 21, 23, 47, 48, 50, 51, 54, 56, 83, 90, 94–97, 100, 107, 108, 133, 134, 138, 141, 142, 144, 145, 147, 149, 159, 175, 177, 187, 188, 190, 192, 194, 212, 224, 240
acoustic field 21, 29
acoustic parameter 122, 188, 212, 213, 239
adhesion 18, 86, 136, 183
angular spectrum 66
anisotropic 16, 19, 65, 66
anisotropy 39, 94, 189, 224, 232, 234, 236
A-Scan 51
attenuation 15, 18, 19, 23, 55, 59, 67, 86, 119, 120, 122–125, 127, 134, 148, 149, 151, 155–158, 191, 192, 194, 205, 211–213, 215, 216
$A(z)$ 17, 18, 56, 58, 120, 122, 150, 152, 153, 155–157, 239

b

bandwidth 74
biological cell 108, 109, 115, 194, 197
biological sample 55, 96, 100, 115, 116
biological tissue 108, 187, 197, 205, 206, 212, 231
blood vessel 188
bond 91
bone 212, 219, 236
bone tissue 212, 219
boundary conditions 24–26, 28, 41, 104, 113, 207, 237
B-Scan 51, 120

c

cardiac muscle 218
cell 108, 109, 111, 118, 122, 123, 187, 188–190, 197, 211
collagen 56, 119, 189, 215, 216, 219, 232, 237–240

composite 119, 126, 137, 147, 159, 160
contrast 15, 16, 53, 55, 87, 91, 119, 131, 135, 147, 149, 155–157, 159, 187–189, 192, 195, 197, 212, 216, 221, 222
crack 46, 91, 133, 134
critical angle 67, 72, 76, 192, 237, 238
C-Scan 120, 138, 139, 231
cylindrical lens 19, 45, 224

d

deformation 10, 108, 110, 113, 118, 139
dental tissue 222
diffraction 23, 55, 83, 103, 106, 107

f

field structure 21
films 18, 19, 37, 54–56, 119, 120, 122, 124, 136, 147, 154, 156, 157, 167, 169, 174
focused transducer 21, 24, 65, 69
focused ultrasound 108
frequency 9–11, 14, 19, 22, 23, 29, 33, 37, 41, 44, 45, 48, 49, 58, 66, 70, 71, 74, 75, 84, 86–88, 90–93, 96, 97, 108, 109, 111, 112, 114–119, 121, 122, 124–126, 138, 141–143, 145, 148, 156–158, 174–177, 179, 181, 182, 187–192, 194, 197, 198, 205, 210, 212, 213, 219, 220

g

gates 134, 179, 196
grain 19, 126, 127, 160, 161, 163, 170, 172
grain size 164, 170, 172
grain structure 161
group velocity 66

h

harmonic 13, 81, 83–85, 87, 89, 93, 103, 104, 113–115, 145
harmonic immersion 88
heart 190

i

immersion   13, 14, 16–18, 21–23, 39, 46, 55, 58, 59, 61, 68, 91–98, 100, 108, 119–121, 125, 128, 129, 131, 132, 137, 145, 153, 174, 176, 179, 183, 186, 189, 191, 211, 224, 227, 231, 237–239

immersion liquid   13, 14, 16–18, 21–23, 39, 46, 55, 56, 58, 59, 61, 68, 88, 91–101, 108, 119–121, 125, 128–132, 134, 137, 148, 153, 174, 176, 179, 184, 186, 191, 192, 224, 237, 238

immersion material   14

immersion medium   189, 191, 211, 237, 239

impedance   13, 15, 18, 77, 145, 148, 149, 175, 188, 193, 202, 222, 224, 229

implant   219, 222

incident angle   207

intensity   95, 96, 101, 104, 138, 217, 226, 227

interface   12, 39, 43, 46–49, 51, 53, 54, 66, 68, 74, 75, 81, 108, 112, 131, 153, 182–184, 198, 212, 224, 233, 236, 237

isotropic   9, 46, 57, 65, 132, 206, 222

l

Lamb wave   18, 53, 65, 66, 75, 76

leaky surface acoustic wave (LSAW)   65–68, 70, 72, 73, 75, 78

leaky surface wave   16, 53

leaky wave   16, 18, 46, 49, 71, 74–78, 128, 135

liver   187, 217

longitudinal wave   9, 18, 47, 49, 84, 128, 134, 148, 149, 207, 208, 233

Love wave   12

m

matching layer   196

microstructure   81, 119, 124, 127, 141, 145, 165, 187, 189, 212, 213, 227, 231

modulus   10, 54, 55, 58, 110, 112, 143, 232

n

nonlinear   14, 81, 86, 112

nonlinearity   89, 210

p

parametric   87, 90, 91

phonon   11

piezoelectric   11, 12, 15, 17, 19, 37, 39, 48, 84, 89, 145, 176, 196

piezoelectric effect   11

piezoelectric transducer   15, 17, 37, 89, 145, 176, 196

planar transducer   26, 34, 36, 66

plane wave   29, 53, 57, 101–103, 105, 106, 196

point source   22, 25, 28, 66, 69

polarization   9, 233, 234

polymer   54–56, 141, 144, 147, 157, 158, 165, 167, 168, 171, 174, 178, 182, 184, 187, 221

polymer blends   144, 146, 147, 157, 158, 165, 167, 168, 171

r

ray model   66, 121

Rayleigh angle   238

Rayleigh velocity   50

Rayleigh wave   12, 14, 16, 19, 46, 49, 50, 74, 75, 77, 78, 127, 128, 131–135, 205, 237, 238

resonance   54, 58, 81, 84, 86, 145

resonance frequency   111, 115, 116, 175

s

scanning acoustic microscope (SAM)   9, 21, 46, 53, 54, 88, 93–96, 124, 127, 131, 134, 137, 141, 187, 189, 194, 197, 200, 204, 205, 224

sclera   189, 215

Sezawa wave   53

shear wave   9, 12, 18, 47, 49, 50, 58, 77, 78, 85, 128–131, 148, 207, 208, 233, 234, 236

skin   122, 189, 200, 213

slowness surface   235, 236

soft tissue   194, 211–213

spherical lens   21, 137, 179

spherical transducer   25, 29, 30, 34, 36, 101, 102, 105

Stoneley–Scholte wave   12, 46, 81, 85

subsurface   130, 131, 133, 197

surface acoustic wave (SAW)   12, 14, 39, 46, 81, 85, 205, 209, 210

surface wave   16, 85

t

temperature effects   94

tone burst   65

tooth   212, 222, 223, 226, 230, 231

transducer   13, 15–17, 21, 24–31, 33–37, 39, 43–45, 49, 51, 57, 66, 68–72, 74, 82, 83, 87, 89, 91, 101, 102, 104, 105, 107, 108, 115, 128, 145, 159, 176, 179, 183, 191, 196, 223

transmission acoustic microscopy   13, 19, 53, 54, 87, 119, 122, 216

transmitted wave 151
transverse wave 9, 12, 18, 47, 49, 50, 58, 77, 78, 85, 128–131, 148, 207, 208, 233, 234, 236

v

viscosity 10, 11, 50, 53, 110, 142, 157, 190, 191

visualization 29, 51, 54, 86, 87, 89–92, 135, 165, 189, 197
$V(x)$ 66, 68, 73, 78
$V(x,t)$ 66, 70–73, 76–78
$V(z)$ 16, 65, 66, 68, 79, 126, 129–133, 136, 197, 205, 208, 210
$V(z,t)$ 65, 66, 70, 71